"In this important work Dembski applies to evolutionary theory the conceptual apparatus of the theory of intelligent design developed in his acclaimed book *The Design Inference*. He gives a penetrating critical analysis of the current attempt to underpin the neo-Darwinian synthesis by means of mathematics. Using recent information-theoretic "no free lunch" theorems, he shows in particular that evolutionary algorithms are by their very nature incapable of generating the complex specified information which lies at the heart of living systems. His results have such profound implications, not only for origin of life research and macroevolutionary theory, but also for the materialistic or naturalistic assumptions that often underlie them, that this book is essential reading for all interested in the leading edge of current thinking on the origin of information."

—John C. Lennox, Green College and the
Mathematical Institute, Oxford, England

"This is a book for the twenty-first century. In the information age, life science is finally leaving behind the nineteenth century mindset that constrained it for most of the twentieth century, and coming to grips with the information technology that underlies the biochemistry of life. Dr. Dembski's formidable intellect has presented a rigorous and persuasive case for the law of conservation of information, the implications of which are revolutionary in their significance for the study of life sciences in the new millennium. This is a book that can be read at two levels: on the one hand, Dembski's examples and analogies explain the theories very well to the lay reader; at another level, Dembski's mathematical proofs argue his case convincingly to the professional and the skeptic."

—Andrew Ruys, Queen Elizabeth II Fellow (bioceramic
engineering), University of Sydney, Australia

"What we know as the 'laws of nature' represent boundary conditions which define how nature works. Great advancements in science depend on, and have been made by, understanding these laws. In *No Free Lunch*, Dembski convincingly demonstrates another of these boundary conditions; namely, that the creation of information requires intelligence. Currently, this 'law of information' is routinely ignored in science by acceptance of the unproven idea that all of life, including the awesome information content in the genome of any living cell, has been created by the purely naturalistic process of random mutation and selection. As with the discovery of all the 'laws of nature,' Dembski's 'law of information' will lead to a greater understanding of nature and to better science. This book, as well as Dembski's other books (*Intelligent Design* and *The Design Inference*), is a must read for both the lay person as well as the professional scientist."

—Russell W. Carlson, professor of biochemistry and molecular biology,
adjunct professor of microbiology, and technical director,
Complex Carbohydrate Research Center, University of Georgia

"Eventually, every thoughtful person wonders about the Big Questions—*Who are we? Where did we come from? Where are we going?* Modern evolutionary theory offers powerful answers to these questions, but that same power has encouraged wholesale dismissal of alternative hypotheses. *No Free Lunch* is one alternative that is worthy of serious scientific consideration. William Dembski provides a coherent and rigorous way of approaching a key issue underlying the Big Questions: *How do we know if an observable is due to pure chance or is intelligently designed?* Sections of *No Free Lunch* will be accessible to general readers, but at its core the book is a scholarly manifesto, packed with intriguing ideas. I give this book my strongest recommendation."

—Dean Radin, senior scientist, Institute of Noetic Sciences,
author of *The Conscious Universe*

"This is a very interesting book that I have looked forward to reading on the bus each day. Since Dembski believes the universe is complex, whereas I think it is simple, I disagree with many of his arguments, but I do find his ideas intriguing to consider and debate. Anyone who is interested in these issues and in the Design movement that Dembski spearheads should find *No Free Lunch* a stimulating and provocative book."

—Don N. Page, professor of physics, University of Alberta

"In this important book, the question of how to infer intelligent design with an objective and scientific criterion is answered in a way that is accessible to a broad audience. Perhaps the most striking and insightful extension of Dembski's earlier work on this topic is the proposed 'fourth law of thermodynamics' that deals directly with the conservation of complex specified information."

—Fred Skiff, professor of physics, University of Iowa

"Using impeccable information- and computation-theoretic arguments, Professor Dembski demonstrates that complex specified information cannot originate solely through the actions of natural laws, chance, or any combination of the two. This book does not question the many successes enjoyed by evolutionary algorithms in numerous fields of science and engineering. Rather, it illustrates the role of intelligent agency in the problem formulations that enables their convergence to useful solutions and calls for a renewed recognition of design as a primary, not a derivative, aspect of our world."

—Terry Rickard, machine intelligence expert, senior vice president,
ORINCON Corporation International

"Biology has been ill-served by the mindless insistence that blind natural mechanisms account for the totality of biological complexity and diversity. In this book mathematician/philosopher William A. Dembski develops a novel information-theoretic framework that powerfully illuminates this problem, yet without shortchanging Darwinian and self-organizational theories."

—Philip S. Skell, Evan Pugh Professor of Chemistry (Emeritus)
Pennsylvania State University, member of the National Academy of Sciences

# No Free Lunch

# No Free Lunch

*Why Specified Complexity Cannot Be
Purchased without Intelligence*

William A. Dembski

**ROWMAN & LITTLEFIELD PUBLISHERS, INC.**
Lanham • Boulder • New York • Oxford

ROWMAN & LITTLEFIELD PUBLISHERS, INC.

Published in the United States of America
by Rowman & Littlefield Publishers, Inc.
4720 Boston Way, Lanham, Maryland 20706
www.rowmanlittlefield.com

12 Hid's Copse Road
Cumnor Hill, Oxford OX2 9JJ, England

British Library Cataloguing in Publication Information Available

**Library of Congress Cataloging-in-Publication Data**

Dembski, William A., 1960–
    No free lunch : why specified complexity cannot be purchased without intelligence / William A. Dembski.
        p. cm.
    Includes bibliographical references.
    ISBN 0-7425-1297-5 (hard : alk. paper)
        1. Evolution (Biology)—Philosophy. 2. Biological systems. 3. Causation. I. Title.
QH360.5 .D46 2002
576.8'01—dc21                                                                                       2001031791

Printed in the United States of America

∞ ™ The paper used in this publication meets the minimum requirements of American National Standard for Information Sciences—Permanence of Paper for Printed Library Materials, ANSI/NISO Z39.48-1992.

From marks of intelligence and wisdom in effects, a wise and intelligent cause may be inferred.

—Thomas Reid

To my beloved children,
Chloe Marie, John Daniel, and William Michael

Psalm 127

~

# Contents

# Illustrations

~

# Preface

How a designer gets from thought to thing is, at least in broad strokes, straightforward: (1) A designer conceives a purpose. (2) To accomplish that purpose, the designer forms a plan. (3) To execute the plan, the designer specifies building materials and assembly instructions. (4) Finally, the designer or some surrogate applies the assembly instructions to the building materials. What emerges is a designed object, and the designer is successful to the degree that the object fulfills the designer's purpose. In the case of human designers, this four-part design process is uncontroversial. Baking a cake, driving a car, embezzling funds, and building a supercomputer each presuppose it. Not only do we repeatedly engage in this four-part design process, but we have witnessed other people engage in it countless times. Given a sufficiently detailed causal history, we are able to track this process from start to finish.

But suppose a detailed causal history is lacking and we are not able to track the design process. Suppose instead that all we have is an object, and we must decide whether it emerged from such a design process. In that case, how do we decide whether the object is in fact designed? If the object in question is sufficiently like other objects that we know were designed, then there may be no difficulty inferring design. For instance, if we find a scrap of paper with writing on it, we infer a human author even if we know nothing about the paper's causal history. We are all familiar with humans writing on scraps of paper, and there is no reason to suppose that this scrap of paper requires a different type of causal story.

Nevertheless, when it comes to living things, the biological community holds that a very different type of causal story is required. To be sure, the biological community admits that biological systems *appear* to be designed. For instance, Richard Dawkins writes, "Biology is the study of complicated things that give the appearance of having been designed for a purpose."[1] Likewise, Francis Crick writes, "Biologists must constantly keep in mind that what they see was not designed, but rather evolved."[2] Or consider the title of Renato Dulbecco's biology text: *The Design of Life.*[3] The term "design" is everywhere in the biological literature. Even so, its use is carefully regulated. According to the biological community, the appearance of design in biology is misleading. This is not to deny that biology is filled with marvelous contrivances. Biologists readily admit as much. Yet as far as the biological community is concerned, living things are not the result of the four-part design process described above.

But how does the biological community know that living things are only apparently and not actually designed? According to Francisco Ayala, Charles Darwin provided the answer: "The functional design of organisms and their features would therefore seem to argue for the existence of a designer. It was Darwin's greatest accomplishment to show that the directive organization of living beings can be explained as the result of a natural process, natural selection, without any need to resort to a Creator or other external agent. The origin and adaptation of organisms in their profusion and wondrous variations were thus brought into the realm of science."[4] Is it really the case, however, that the directive organization of living beings can be explained without recourse to a designer? And would employing a designer in biological explanations necessarily take us out of the realm of science? The purpose of this book is to answer these two questions.

The title of this book, *No Free Lunch*, refers to a collection of mathematical theorems proved in the past five years about evolutionary algorithms. The upshot of these theorems is that evolutionary algorithms, far from being universal problem solvers, are in fact quite limited problem solvers that depend crucially on additional information not inherent in the algorithms before they are able to solve any interesting problems. This additional information needs to be carefully specified and fine-tuned, and such specification and fine-tuning is always thoroughly teleological. Consequently, evolutionary algorithms are incapable of providing a computational justification for the Darwinian mechanism of natural selection and random variation as the primary creative force in biology. The subtitle, *Why Specified Complexity Cannot Be Purchased without Intelligence*, refers to that form of information, known as *specified complexity* or *complex specified information*, that is increasingly coming to be regarded as a reliable empirical marker of purpose, intelligence, and design.

What is specified complexity? An object, event, or structure exhibits specified complexity if it is both complex (i.e., one of many live possibilities) and specified (i.e., displays an independently given pattern). A long sequence of randomly strewn Scrabble pieces is complex without being specified. A short sequence spelling the word "the" is specified without being complex. A sequence corresponding to a Shakespearean sonnet is both complex and specified. In *The Design Inference: Eliminating Chance through Small Probabilities*,[5] I argued that specified complexity is a reliable empirical marker of intelligence. Nevertheless, critics of my argument have claimed that evolutionary algorithms, and the Darwinian mechanism in particular, can deliver specified complexity apart from intelligence.[6] I anticipated this criticism in *The Design Inference* but did not address it there in detail. Filling in the details is the task of the present volume.

*The Design Inference* laid the groundwork. This book demonstrates the inadequacy of the Darwinian mechanism to generate specified complexity. Darwinists themselves have made possible such a refutation. By assimilating the Darwinian mechanism to evolutionary algorithms, they have invited a mathematical assessment of the power of the Darwinian mechanism to generate life's diversity. Such an assessment, begun with the No Free Lunch theorems of David Wolpert and William Macready (see section 4.6), will in this book be taken to its logical conclusion. The conclusion is that Darwinian mechanisms of any kind, whether in nature or in silico, are in principle incapable of generating specified complexity. Coupled with the growing evidence in cosmology and biology that nature is chock-full of specified complexity (cf. the fine-tuning of cosmological constants and the irreducible complexity of biochemical systems), this conclusion implies that naturalistic explanations are incomplete and that design constitutes a legitimate and fundamental mode of scientific explanation.

In arguing that naturalistic explanations are incomplete or, equivalently, that natural causes cannot account for all the features of the natural world, I am placing natural causes in contradistinction to intelligent causes. The scientific community has itself drawn this distinction in its use of these twin categories of causation. Thus, in the quote earlier by Francisco Ayala, "Darwin's greatest accomplishment [was] to show that the directive organization of living beings can be explained as the result of a natural process, natural selection, without any need to resort to a Creator or other external agent."[7] Natural causes, as the scientific community understands them, are causes that operate according to deterministic and nondeterministic laws and that can be characterized in terms of chance, necessity, or their combination (cf. Jacques Monod's *Chance and Necessity*).[8] To be sure, if one is more liberal about what one means by natural causes and includes among natural causes telic processes that are not reducible to chance and necessity (as the ancient Stoics did by endowing nature with immanent teleology), then my claim that natural causes are incomplete dissolves. But that is not how the scientific community by and large understands natural causes.

The distinction between natural and intelligent causes now raises an interesting question when it comes to embodied intelligences like ourselves, who are at once physical systems and intelligent agents: Are embodied intelligences natural causes? Even if the actions of an embodied intelligence proceed solely by natural causes, being determined entirely by the constitution and dynamics of the physical system that embodies it, that does not mean the origin of that system can be explained by reference solely to natural causes. Such systems could exhibit derived intentionality in which the underlying source of intentionality is irreducible to natural causes (cf. a digital computer). I will argue that intelligent agency, even when conditioned by a physical system that embodies it, cannot be reduced to natural causes without remainder. Moreover, I will argue that specified complexity is precisely the remainder that remains unaccounted for. Indeed, I will argue that the defining feature of intelligent causes is their ability to create novel information and, in particular, specified complexity.

Design has had a turbulent intellectual history. The chief difficulty with design to date has consisted in discovering a conceptually powerful formulation of it that will fruitfully advance science. While I fully grant that the history of design arguments warrants misgivings, they do not apply to the present project. The theory of design I envision is not an atavistic return to the design arguments of William Paley and the Bridgewater Treatises. William Paley was in no position to formulate the conceptual framework for design that I will be developing in this book. This new framework depends on advances in probability theory, computer science, the concept of information, molecular biology, and the philosophy of science—to name but a few. Within this framework design promises to become an effective conceptual tool for investigating and understanding the world.

Increased philosophical and scientific sophistication, however, is not alone in separating my approach to design from Paley's. Paley's approach was closely linked to his prior religious and metaphysical commitments. Mine is not. Paley's designer was nothing short of the triune God of Christianity, a transcendent, personal, moral being with all the perfections commonly attributed to this God. On the other hand, the designer that emerges from a theory of intelligent design is an intelligence capable of originating the complexity and specificity that we find throughout the cosmos and especially in biological systems. Persons with theological commitments can co-opt this designer and identify this designer with the object of their worship. But this move is strictly optional as far as the actual science of intelligent design is concerned.

The crucial question for science is whether design helps us understand the world, and especially the biological world, better than we do now when we systematically eschew teleological notions from our scientific theorizing. Thus, a scientist may view design and its appeal to a designer as simply a fruitful device for understanding the world, not attaching any significance to questions such as whether a theory of design is in some ultimate sense true or whether the designer actually exists. Philosophers of science would call this a *constructive empiricist* approach to design. Scientists in the business of manufacturing theoretical entities like quarks, strings, and cold dark matter could therefore view the designer as just one more theoretical entity to be added to the list. I follow here Ludwig Wittgenstein, who wrote, "What a Copernicus or a Darwin really achieved was not the discovery of a true theory but of a fertile new point of view."[9] If design cannot be made into a fertile new point of view that inspires exciting new areas of scientific investigation, then it deserves to wither and die. Yet before that happens, it deserves a fair chance to succeed.

One of my main motivations in writing this book is to free science from arbitrary constraints that, in my view, stifle inquiry, undermine education, turn scientists into a secular priesthood, and in the end prevent intelligent design from receiving a fair hearing. The subtitle of Richard Dawkins's *The Blind Watchmaker* reads *Why the Evidence of Evolution Reveals a Universe without Design*. Dawkins may be right that design is absent from the universe. But science needs to address not only the evidence that reveals the universe to be without design but also the evidence that reveals the universe to be with design. Evidence is a two-edged sword: claims capable of being refuted by evidence are also capable of being supported by evidence. Even if design ends up being rejected as an unfruitful explanatory tool for science, such a negative outcome for design needs to result from the evidence for and against design being fairly considered. Darwin himself would have agreed: "A fair result can be obtained only by fully stating and balancing the facts and arguments on both sides of each question."[10] Consequently, any rejection of design must not result from imposing arbitrary constraints on science that rule out design prior to any consideration of evidence.

Two main constraints have historically been used to keep design outside the natural sciences: methodological naturalism and dysteleology. According to methodological naturalism, in explaining any natural phenomenon, the natural sciences are properly permitted to invoke only natural causes to the exclusion of intelligent causes. On the other hand, dysteleology refers to inferior design—typically design that is either evil or incompetent. Dysteleology rules out design from the natural sciences on account of the inferior design that nature is said to exhibit. In this book, I will address methodological naturalism. Methodological naturalism is a regulative principle that purports to keep science on the straight and narrow by limiting science to natural causes. I intend to show that it does nothing of the sort but instead constitutes a straitjacket that actively impedes the progress of science.

On the other hand, I will not have anything to say about dysteleology. Dysteleology might present a problem if all design in nature were wicked or incompetent and continually flouted our moral and aesthetic yardsticks. But that is not the case. To be sure, there are microbes that seem designed to do a number on the mammalian nervous system and biological structures that look cobbled together by a long trial-and-error evolutionary process. But there are also biological examples of nano-engineering that surpass anything human engineers have concocted or entertain hopes of concocting. Dysteleology is primarily a theological problem.[11] To exclude design from biology simply because not all examples of biological design live up to our expectations of what a designer should or should not have done is an evasion. The problem of design in biology is real and pervasive, and needs to be addressed head on and not sidestepped because our presuppositions about design happen to rule out imperfect design. Nature is a mixed bag. It is not William Paley's happy world of everything in delicate harmony and balance. It is not the widely caricatured Darwinian world of nature red in tooth and claw. Nature contains evil design, jerry-built design, and exquisite design. Science needs to come to terms with design as such and not dismiss it in the name of dysteleology.

A possible terminological confusion over the phrase "intelligent design" needs to be cleared up. The confusion centers on what the adjective "intelligent" is doing in the phrase "intelligent design." "Intelligent" can mean nothing more than being the result of an intelligent agent, even one who acts stupidly. On the other hand, it can mean that an intelligent agent acted with consummate skill and mastery. Critics of intelligent design often understand the "intelligent" in intelligent design in the latter sense and thus presume that intelligent design must entail optimal design. The intelligent design community, on the other hand, understands the "intelligent" in intelligent design simply to refer to intelligent agency (irrespective of skill, mastery, or cleverness) and thus separates intelligent design from optimality of design. But why then place the adjective intelligent in front of the noun design? Does not design already include the idea of intelligent agency, so that juxtaposing the two becomes redundant? Redundancy is avoided because intelligent design needs also to be distinguished from apparent design. Because design in biology so often connotes apparent design, putting intelligent in front of design ensures that the design we are talking about is not merely apparent but also actual. Whether that intelligence acts cleverly or stupidly, wisely or unwisely, optimally or suboptimally are separate questions.

Who will want to read *No Free Lunch*? The audience includes anyone interested in seriously exploring the scope and validity of Darwinism as well as in learning how the emerging theory of intelligent design promises to supersede it. Napoleon III remarked that one never destroys a thing until one has replaced it. Similarly, Thomas Kuhn, in the language of paradigms and paradigm shifts, claimed that for a paradigm to shift, there has to be a new paradigm in place ready to be shifted into. Throughout my work, I have not been content merely to critique existing theory but have instead striven to provide a positive more-encompassing framework within which to reconceptualize phenomena inadequately explained by existing theory. Much of *No Free Lunch* will be accessible to an educated lay audience. Many of the ideas have been presented in published articles and public lectures. I have seen how the ideas in this book have played themselves out under fire. The chapters are therefore tailored to questions people are actually asking. The virtue of this book is filling in the details. And the devil is in the details.

I have tried to keep technical discussions to a minimum. I am no fan of notation-heavy prose and avoid it whenever possible. A book of this sort, however, poses a peculiar challenge. Forms of thinking that turn biological complexity into a free lunch pervade science and are deeply entrenched. It does no good therefore to speak in generalities or point to certain obvious tensions (e.g., how can intelligence arise out of an inherently unintelligent Darwinian process? or how can we have any confidence in the reliability of our cognitive faculties if we are the result of a brute natural process for which survival and reproduction is everything and truth-seeking is incidental?). Make the book too obvious, and no one will pay it any mind. Make it too technical, and no one will read it. My strategy in writing this book, therefore, has been to include just enough technical discussion so that experts can fill in the details as well as sufficient elaboration of the technical discussion so that nonexperts feel the force of the design inference. Whether I have been successful is for others to judge.

*No Free Lunch* has the following logical structure. Chapter 1 presents a nontechnical summary of my work on inferring design and makes the connection between my previous work and Darwinism explicit. Chapter 2 rebuts critics who argue that specified complexity is not a well-defined concept and cannot form the basis for a compelling design inference. In particular, I offer there a simplified account of specification. Chapter 3 translates the design-inferential framework of chapters 1 and 2 into a more powerful information-theoretic framework. Chapter 4 shows how this information-theoretic approach to design withstands and then overturns the challenge of evolutionary algorithms. In particular, I show that evolutionary algorithms cannot generate specified complexity. Chapter 5 then shows how the theoretical apparatus developed in the previous chapters can be applied to actual biological systems. Finally, chapter 6 examines what intelligent design means for science.

What follows is a chapter by chapter summary of the book.

**Chapter 1: The Third Mode of Explanation.** How is design empirically detectable and thus distinguishable from the two generally accepted modes of scientific explanation, chance and necessity? To detect design, two features must be present: complexity and specification. Complexity guarantees that the object in question is not so simple that it can readily be attributed to chance. Specification guarantees that the object exhibits the right sort of pattern associated with intelligent causes. *Specified complexity* thus becomes a criterion for detecting design empirically. Having proposed a theoretical apparatus for detecting design, I next consider the challenge that Darwin posed to design historically and indicate why his challenge is viewed among many scientists as counting decisively against design. Essentially, Darwin opposed to design the joint action of chance and necessity and therewith promised to explain the complex ordered structures in biology that prior to him were attributed to design.

**Chapter 2: Another Way to Detect Design?** Many in the scientific and philosophical community have staked their hopes on explaining specified complexity by means of evolutionary algorithms. Yet even without evolutionary algorithms to explain specified complexity, few are prepared to embrace design. One approach, now increasingly championed by the philosopher of science Elliott Sober, is to attack specified complexity head-on and claim that it is a spurious concept, incapable of rendering design testable in the case of natural objects, and that a precise probabilistic and complexity-theoretic analysis of specified complexity vitiates the concept entirely. In critiquing my approach to detecting design, Sober has tied himself to a likelihood framework for probability that is itself highly problematic. This chapter demonstrates that specified complexity is a well-defined concept and that it readily withstands the criticisms raised by Sober and his colleagues.

**Chapter 3: Specified Complexity as Information.** Intelligent design can be formulated as a theory of information. Within such a theory, specified complexity becomes a form of information that reliably signals design. As a form of information specified complexity also becomes a proper object for scientific investigation. This chapter takes the ideas of chapters 1 and 2 and translates them into an information-theoretic framework. This reframing of intelligent design within information theory powerfully extends the design-inferential framework developed in chapter 1 and makes it possible accurately to assess the power (or lack thereof) of the Darwinian mechanism. The upshot of this chapter is a conservation law governing the origin and flow of information. From this law it follows that specified complexity is not reducible to natural causes and that the origin of specified complexity is best sought in intelligent causes. Intelligent design thereby becomes a theory for detecting and measuring information, explaining its origin, and tracing its flow.

**Chapter 4: Evolutionary Algorithms.** This chapter is the climax of the book. Here I examine evolutionary algorithms, which constitute the mathematical underpinnings of Darwinism. I show that evolutionary algorithms are in principle incapable of generating specified complexity. Whereas this result follows immediately from the conservation of information law in chapter 3, this law involves a high level of abstraction, so that simply applying the law does not make clear just how limited evolutionary algorithms really are. In this chapter I therefore examine the nuts and bolts of the evolutionary algorithms: phase spaces, fitness landscapes, and optimization algorithms. An elementary combinatorial analysis shows that evolutionary algorithms can no more generate specified complexity than can five letters fill ten mailboxes.

**Chapter 5: The Emergence of Irreducibly Complex Systems.** Specified complexity as a reliable empirical marker of intelligence is all fine and well, but if there are no complex specified systems in nature, what then? The previous chapters establish that specified complexity reliably signals design, not that specified complexity is actualized in any concrete physical system. This chapter examines how we determine whether a physical system exhibits specified complexity. The key to this determination, at least in biology, is Michael Behe's notion of irreducible complexity. Irreducibly complex biological systems exhibit specified complexity. Irreducible complexity is therefore a special case of specified complexity. Because specified complexity is a probabilistic notion, determining whether a physical system exhibits specified complexity requires being able to calculate probabilities. One of the objections to intelligent design becoming a viable scientific research program is that one cannot calculate the probabilities needed to confirm specified complexity for actual systems in nature. This chapter shows that even though precise calculations may not always be possible, setting bounds for the relevant probabilities is possible, and that this is adequate for establishing specified complexity in practice.

**Chapter 6: Design as a Scientific Research Program.** Having shown that specified complexity is a reliable empirical marker of intelligence and having overturned the main scientific objections raised against it, I conclude this book by examining what science will look like once design is readmitted to full scientific status. The worry is that attributing design to natural systems will stultify science in the sense that once a scientist concedes that some natural system is designed, all the scientist's work is over. But this is not the case. Design raises a host of novel and interesting research questions that it does not make sense to ask within a strictly Darwinian or naturalistic framework. One such question is teasing apart the effects of natural and intelligent causation. For instance, a rusted old Cadillac is clearly designed but also shows the effects of natural causes (i.e., weathering). Intelligent design is capable of accommodating the legitimate insights of Darwinian theory. In particular, intelligent design admits a place for the Darwinian mechanism of natural selection and random variation. But as a framework for doing science, intelligent design offers additional tools for investigating nature that render it conceptually more powerful than Darwinism.

Ideally, this book should be read from start to finish. Nevertheless, because this is not always possible, let me offer the following suggestions for reading the book. Chapter 1 is the most accessible chapter in the book and is prerequisite for everything that follows. This material needs to be under the reader's belt. Sections 1.1 to 1.7 present a nontechnical summary of my previous work on inferring design, and readers familiar with it can skip these sections without loss. On the other hand, sections 1.8 to 1.10 are new and make explicit the connection between my previous work and Darwinism. Readers definitely need to read these sections. Chapter 2 is primarily directed at critics. This is the most technical chapter, and readers persuaded by my previous work may want to skip it on their initial reading. Chapters 3 and 4 translate the design-theoretic framework of chapters 1 and 2 into an information-theoretic framework. Chapter 3 presents the general theory whereas chapter 4 looks specifically at evolutionary algorithms. For nontechnical readers, I recommend a light perusal of chapter 3 and then a careful examination of chapter 4. Chapter 5 brings theory in contact with biological reality. This is where most of the current controversy lies, and readers will not want to miss this chapter. Chapter 6, on the other hand, looks at the broader implications of intelligent design for science and can be read at leisure.

There are nontechnical readers who can comfortably wade past technical mathematical discussions without being intimidated; and then there are math phobics whose eyes glaze over and brains shut down at the sight of technical mathematical discussions. This book can also be read with profit by math phobics. I suggest reading sections 1.1–1.10, 5.1–5.7, 5.9, and 6.1–6.10 in order. The only thing one needs to know about mathematics to read these sections is that powers of ten count the number of zeroes following a one. Thus $10^3$ is 1,000 (a thousand has three zeroes after the initial one), $10^6$ is 1,000,000 (a million has six zeroes after the initial one), etc. Reading these sections will provide a good overview of the current debate regarding intelligent design, particularly as it relates to Michael Behe's work on irreducibly complex molecular machines. Math phobics who then want to see why evolutionary algorithms cannot do the design work that Darwinists regularly attribute to these algorithms can read sections 4.1–4.2 and 4.7–4.9.

One final caution: Even though much in this book will look familiar to readers acquainted with my previous work, this familiarity can be deceiving. I have already noted that sections 1.1 to 1.7 present a nontechnical summary of my work on inferring design and that readers familiar with it can skip these sections without loss. But other sections, though apparently covering old ground, in fact differ markedly from previous work. For instance, two of my running examples in *The Design Inference* were the Caputo case (an instance of apparent ballot-line fraud) and algorithmic information theory. The case studies in sections 2.3 and 2.4 re-examine these examples in light of criticisms brought against them. Except for chapter 1, arguments and topics revisited are in almost every instance reworked or beefed up.

My debts to friends, foes, colleagues, and institutions are many. Let me begin with the Templeton Foundation. In the fall of 1999 I received one of seven book awards from the Templeton Foundation to write a book titled *Being as Communion: The Science and Metaphysics of Information*. After making the proposal and receiving the award, it became clear to me that the science of information (and specifically the science of complex specified information) required a book of its own. Indeed, before one can take seriously the metaphysics of information one must take seriously the science of information (perhaps this is why editions of Aristotle's work always list his *Physics* before his *Metaphysics*). I therefore decided to divide this project in two, handling the science of information in the present volume and the metaphysics of information in a subsequent volume, to be titled *Being as Communion: The Metaphysics of Information.*

In addition to generously supporting me in the writing of this and the follow-up volume, the Templeton Foundation has also sponsored various conferences and symposia at which I have participated, notably a conference titled "The Nature of Nature" at Baylor University in April of 2000 and a symposium titled "Complexity, Information, and Design: A Critical Appraisal," which took place in Santa Fe in October of 1999 and was organized by Paul Davies. The Santa Fe symposium was enormously helpful in taking me to the next stage in my thinking about design inferences. Indeed, the talk I presented there and the feedback I received were the direct impetus for the present volume. Conversations at this symposium with Charles Bennett, Gregory Chaitin, Paul Davies, Niels Gregersen, Stuart Kauffman, Harold Morowitz, and Ian Stewart are etched in my mind and have left their imprint all over this volume. It was a privilege to interact with them.

I also wish to thank the staff and administration at the Templeton Foundation for their competence and warmth. I am especially grateful to Charles Harper for taking an interest in my work and making it possible for me to attend the Santa Fe symposium. Finally, I want to thank Sir John Templeton for some lovely conversations over dinner at the Santa Fe symposium concerning his life and aspirations. Although well into his eighties, he attended virtually all the working sessions of the symposium. These were very full days, and he was tracking the presentations and interactions intently. Should I reach his age, I hope to have the same stamina and clarity of mind.

Next I want to commend the Discovery Institute, and especially its Center for the Renewal of Science and Culture of which I am a fellow. Bruce Chapman, the president of Discovery, Stephen Meyer, the director of the center, and John West, the associate director of the center, have been a constant encouragement to me. They, along with the fellows of the center, have been among my best conversation partners in stimulating my thinking about intelligent design. Stephen Meyer and Paul Nelson stand out. I have now been collaborating with them for almost a decade on writing projects, academic conferences, and media events, all relating to intelligent design. Along with Steve and Paul, I also want to single out Michael Behe, David Berlinski, Phillip Johnson, Jay Richards, and Jonathan Wells. In addition, I want to thank the staff at the Discovery Institute for all their help with practical matters, especially Doug Bilderback, Mark Edwards, and Steve Jost.

One event that the Discovery Institute sponsored deserves special mention here. Intelligent design has numerous detractors, and among the criticisms of intelligent design as an intellectual movement has been the charge that proponents of intelligent design are not sufficiently self-critical, motivated more by a political agenda than by pure love of inquiry. A symposium on design reasoning sponsored by the Discovery Institute and organized by Timothy McGrew, at least to my mind, puts to rest this charge. On May 22 and 23, 2001, at Calvin College in Grand Rapids, Michigan, eight design theorists, many sharply critical of each other's work and mine in particular, met to hash out the logic of design reasoning. I am grateful for this severe scrutiny. Besides me, the symposium participants included Robin Collins, Rob Koons, Lydia McGrew, Timothy McGrew, Steve Meyer, Paul Nelson, and Del Ratzsch. Except for Del, who moderated our sessions, each of us took turns presenting a paper and responding to another paper. Rob Koons was my respondent, responding to an earlier draft of chapter 2 of this book. I especially want to thank him for his careful reading of this chapter and for finding a number of mistakes that fortunately I was able to fix in time for the publication of this book. Rob is one of the most insightful philosophers that I know and has been an enormous stimulus for my own work. Also, I want to commend Timothy McGrew for pulling off this meeting and for his ongoing work to edit a volume of proceedings from this symposium (it will make for an interesting volume). Finally, I want to thank Jay Richards for doing the spadework to organize this symposium. Unfortunately, he was not able to attend because of the death of his son Josiah.

Billy Grassie as director of the science and religion website www.metanexus.net, John Wilson as editor of *Books & Culture*, and Fr. Richard John Neuhaus as editor of *First Things* have each provided a forum for me to present my ideas about intelligent design. I thank them for opening their doors to me. These forums have been especially helpful for testing and clarifying my ideas, and the present volume would have been considerably poorer without the interaction they afforded. Billy Grassie's science and religion website has been especially helpful in this regard. Having at times had to wait two to three years for a peer-reviewed paper to appear in print once it was accepted for publication, I find it refreshing to post articles through www.metanexus.net and see them appear and receive feedback virtually instantaneously.

My host institution Baylor University has provided me uninterrupted time to devote to my research on intelligent design. As an associate research professor with no teaching duties, I am in a unique position to explore what it will take to bring intelligent design into the academic mainstream. I thank Robert Sloan, president of Baylor University, for this opportunity. President Sloan, at great personal and professional expense, hired me on a five-year trial basis to see whether intelligent design can, as it were, "produce the goods." The results are not yet in and much work remains to be done, but Baylor and President Sloan are to be commended for giving this work a chance to succeed. In this respect I also want to thank Provost Donald Schmeltekopf, Michael Beaty, David Lyle Jeffrey, and Bruce Gordon. Bruce and I go back a long way. He is an outstanding conversation partner and careful reader of my work. His imprint can be felt throughout this book.

Other institutions and individuals with whom I have had direct contact and who have contributed significantly to this volume include: Paul Allen, Dean Anderson, Larry Arnhart, Art Battson, John Bracht, James Bradley, Walter Bradley, J. Budziszewski, Jon Buell, Anna Mae Bush, Eli Chiprout, Isaac Choi, Calvin College, John Angus Campbell, Center for Theology and the Natural Sciences, William Lane Craig, Ted Davis, Richard Dawkins, Michael Denton, Wesley Elsberry, Fieldstead and Company, David Fogel, Foundation for Thought and Ethics, John Gilmore, Guillermo Gonzalez, Steve Griffith, Roland Hirsch, Muzaffar Iqbal, Steve Jones, Gert Korthof, Robert Larmer, Neil Manson, John H. McDonald, Angus Menuge, Todd Moody, Gregory Peterson, Phylogenists, John Mark Reynolds, Terry Rickard, Douglas Rudy, Michael Ruse, Jeff Schloss, Kerry Schutt, Eugenie Scott, Michael Shermer, Fred Skiff, Elliott Sober, John L. Stahlke, Karl Stephan, Charlie Thaxton, Frank Tipler, Royal Truman, Regina Uhl, Howard Van Till, Deryck Velasquez, Richard Wein, John Wiester, and Ben Wiker.

Finally, I wish to commend my family for always standing behind me in my work on intelligent design. Here I want especially to thank my beloved wife, Jana, who encourages me without indulging me and who loves me without dissimulation. I dedicate this book to our three children, Chloe Marie, John Daniel, and William Michael.

# Notes

1. Richard Dawkins, *The Blind Watchmaker* (New York: Norton, 1987), 1.

2. Francis Crick, *What Mad Pursuit: A Personal View of Scientific Discovery* (New York: Basic Books, 1988), 138.

3. Renato Dulbecco, *The Design of Life* (New Haven, Conn.: Yale University Press, 1990).

4. Francisco J. Ayala, "Darwin's Revolution," in *Creative Evolution?!*, eds. J. H. Campbell and J. W. Schopf (Boston: Jones and Bartlett, 1994), 4. The subsection from which this quote is taken is titled "Darwin's Discovery: Design without Designer."

5. William A. Dembski, *The Design Inference: Eliminating Chance through Small Probabilities* (Cambridge: Cambridge University Press, 1998).

6. See, for instance, Taner Edis, "Darwin in Mind: 'Intelligent Design' Meets Artificial Intelligence," *Skeptical Inquirer* 25(2) (March/April 2001): 35–39. Edis writes:

"[Dembski] has recently gathered his arguments in a book that claims to put ID [i.e., intelligent design] on a solid footing. . . . Though dead wrong in his overall conclusions, he makes interesting mistakes, and his errors highlight how powerful an idea Dawinian evolution is, in biology and beyond" (36). Thus my work is supposed to "highlight what is correct" about Darwinian evolution (39). Edis concludes: "Dembski's criteria [for detecting design] do reveal a special kind of order. . . . The irony is, what these criteria actually detect is that there were Darwinian processes at work" (39). To this Wesley Elsberry adds, "Dembski argues that [his criteria] reliably [diagnose] the action of an intelligent agent, yet [they fail] to exclude natural selection. . . . Somehow, I doubt that natural selection is what Dembski had in mind for the agent of biological design." Quoted from http://inia.cls.org/~welsberr/zgists/wre/papers/dembski7.html (last accessed 6 June 2001). Indeed, natural selection is not the agent I had in mind for biological design, nor have my criteria for detecting design stumbled on yet another confirmation of Darwinism—they do not constitute an unexpected vindication of natural selection. When properly understood and applied, these criteria demonstrate the inherent limitations of the Darwinian mechanism.

7. Ayala, "Darwin's Revolution," in *Creative Evolution?!*, 4.

8. Jacques Monod, *Chance and Necessity* (New York: Vintage, 1972).

9. Ludwig Wittgenstein, *Culture and Value*, ed. G. H. von Wright, trans. P. Winch (Chicago: University of Chicago Press, 1980), 18e.

10. Charles Darwin, *On the Origin of Species*, facsimile 1st ed. (1859; reprinted Cambridge, Mass.: Harvard University Press, 1964), 2.

11. See Cornelius Hunter, *Darwin's God: Evolution and the Problem of Evil* (Grand Rapids, Mich.: Brazos Press, 2001); and Paul Nelson, "The Role of Theology in Current Evolutionary Reasoning," *Biology and Philosophy* 11 (1996): 493–517.

# CHAPTER ONE

~

# The Third Mode of Explanation

## 1.1   Necessity, Chance, and Design

In ordinary life we find it important to distinguish between three modes of explanation: necessity, chance, and design. Did she fall or was she pushed? And if she fell, was her fall accidental or unavoidable? To say she was pushed is to impute design. To say her fall was accidental or unavoidable is to impute respectively chance or necessity. More generally, given an event, object, or structure, we want to know: Did it have to happen? Did it happen by accident? Did an intelligent agent cause it to happen? In other words, did it happen by necessity, chance, or design?

Such everyday distinctions between necessity, chance, and design are informal and thus inadequate for constructing a scientific theory of design. It is therefore fair to ask whether there is a principled way to distinguish these modes of explanation. Philosophers and scientists have disagreed not only about how to distinguish these modes of explanation but also about their very legitimacy. In antiquity the Epicureans gave pride of place to chance. The Stoics, on the other hand, emphasized necessity and design but rejected chance. In the Middle Ages Moses Maimonides contended with the Islamic interpreters of Aristotle who viewed the heavens as, in Maimonides words, "the necessary result of natural laws."[1] Where the Islamic philosophers saw necessity, Maimonides saw design.

In arguing for design in his *Guide for the Perplexed*, Maimonides looked to the irregular distribution of stars in the heavens. For him that irregularity demonstrated contingency (i.e., the occurrence of an event that did not have

1

to happen and therefore was not necessary). But was that contingency the result of chance or design? Neither Maimonides nor the Islamic interpreters of Aristotle were sympathetic to Epicurus and his views on chance. For them chance could never be fundamental but was at best a placeholder for ignorance. Thus, for Maimonides and his Islamic colleagues the question was whether a principled distinction could be drawn between necessity and design. The Islamic philosophers, intent on keeping Aristotle pure of theology, said no. Maimonides, arguing from observed contingency in nature, said yes. His argument focused on the distribution of stars in the night sky:

> What determined that the one small part [of the night sky] should have ten stars, and the other portion should be without any star? . . . The answer to [this] and similar questions is very difficult and almost impossible, if we assume that all emanates from God as the necessary result of certain permanent laws, as Aristotle holds. But if we assume that all this is the result of design, there is nothing strange or improbable; the only question to be asked is this: What is the cause of this design? The answer to this question is that all this has been made for a certain purpose, though we do not know it; there is nothing that is done in vain, or by chance. . . . How, then, can any reasonable person imagine that the position, magnitude, and number of the stars, or the various courses of their spheres, are purposeless, or the result of chance? There is no doubt that every one of these things is . . . in accordance with a certain design; and it is extremely improbable that these things should be the necessary result of natural laws, and not that of design.[2]

Modern science has also struggled with how to distinguish between necessity, chance, and design. Newtonian mechanics, construed as a set of deterministic physical laws, seemed only to permit necessity. Nevertheless, in the General Scholium to his *Principia*, Newton claimed that the stability of the planetary system depended not only on the regular action of the universal law of gravitation, but also on the precise initial positioning of the planets and comets in relation to the sun. As he explained:

> Though these bodies may, indeed, persevere in their orbits by the mere laws of gravity, yet they could by no means have at first derived the regular position of the orbits themselves from those laws. . . . [Thus] this most beautiful system of the sun, planets, and comets, could only proceed from the counsel and dominion of an intelligent and powerful being.[3]

Like Maimonides, Newton saw both necessity and design as legitimate explanations, but gave short shrift to chance.

Newton published his *Principia* in the seventeenth century. By the nineteenth century, the scientific community no longer accepted Newton's approach to scientific explanation. To be sure, necessity was still in and chance was still out; nonetheless, design had lost much of its appeal. When asked by Napoleon where God fit into his equations of celestial mechanics, Laplace famously replied, "Sire, I have no need of that hypothesis." In place of a designing intelligence that precisely positioned the heavenly bodies, Laplace proposed his nebular hypothesis, which accounted for the origin of the solar system strictly through natural gravitational forces.[4]

Since Laplace's day, science has largely dispensed with design. Certainly Darwin played a critical role here by eliminating design from biology. Yet at the same time science was dispensing with design, it was also dispensing with Laplace's vision of a deterministic universe (recall Laplace's famous demon who could predict the future and retrodict the past with perfect precision provided that present positions and momenta of particles were fully known).[5] With the rise of statistical mechanics and then quantum mechanics, the role of chance in physics came to be regarded as ineliminable. Especially convincing here has been the failure of the Bell inequality.[6] Consequently, a deterministic, necessitarian universe has given way to a stochastic universe in which chance and necessity are both regarded as fundamental modes of scientific explanation, neither being reducible to the other. To sum up, contemporary science allows a principled distinction between necessity and chance, but repudiates design.

## 1.2   Rehabilitating Design

But was science right to repudiate design? In this book I will argue that design is a legitimate and fundamental mode of scientific explanation, on a par with chance and necessity. In arguing this claim, however, I want to avoid prejudging the implications of design for science. In particular, it is not my aim to force a religious doctrine of creation upon science. Design, as I develop it, cuts both ways and might just as well be used to empty such religious doctrines of empirical content by clarifying the superfluity of design in nature (if indeed it is superfluous). My aim is not to find design in any one place or to gain ideological mileage, but to open up possibilities for finding design as well as for shutting it down.

My aim, then, is to rehabilitate design as a legitimate mode of scientific explanation without prejudice about its applicability to actual physical systems. Given that aim, it will help to review why design was removed from modern science. Design, in the form of Aristotle's formal and final causes, had after all once occupied a perfectly legitimate role within natural philosophy, or what we now call science. With the rise of modern science, however, these causes fell into disrepute.

We can see how this happened by considering Francis Bacon. Bacon, a contemporary of Galileo and Kepler, though himself not a scientist, was a terrific propagandist for science. Bacon concerned himself much about the proper conduct of science, providing detailed canons for experimental observation, recording of data, and inferences from data. What interests us here, however, is what he did with Aristotle's four causes. For Aristotle, to understand any phenomenon properly, one had to understand its four causes, namely its material, efficient, formal, and final cause.[7]

To illustrate Aristotle's four causes consider a statue—say Michelangelo's David. The material cause is what it is made of—marble. The efficient cause is the immediate activity that produced the statue—Michelangelo's actual chipping away at a marble slab with hammer and chisel. The formal cause is its structure—it is a representation of David and not some random chunk of marble. And finally, the final cause is its purpose—presumably, to beautify some Florentine palace.

Two points about Aristotle's causes are relevant to this discussion. First, Aristotle gave full weight to all four causes. In particular, Aristotle would have regarded any inquiry that omitted one of his causes as deficient. Second, Bacon adamantly opposed including formal and final causes within science (see his *Advancement of Learning*).[8] For Bacon, formal and final causes belonged to metaphysics and not to science. Science, according to Bacon, needed to limit itself to material and efficient causes, thereby freeing science from the sterility that inevitably results when science and metaphysics are conflated. This was Bacon's line, and he argued it forcefully.

We see Bacon's line championed in our own day by atheists and theists alike. In *Chance and Necessity*, biologist and Nobel laureate Jacques Monod argued that chance and necessity alone suffice to account for every aspect of the universe. Now, whatever else we might want to say about chance and necessity, they provide at best a reductive account of Aristotle's formal causes and leave no room whatever for Aristotle's final causes. Indeed, Monod explicitly denies any place for purpose within science: "The cornerstone of the scientific method is the postulate that nature is objective. In other words, the *systematic* denial that 'true' knowledge can be got at by interpreting phenomena in terms of final causes—that is to say, of 'purpose'."[9]

Monod was an outspoken atheist. Nevertheless, as outspoken a theist as Stanley Jaki will agree with Monod about this aspect of science. Jaki is as theologically conservative a historian of science and Catholic priest as one is likely to find. Yet in his published work he explicitly states that purpose is a purely metaphysical notion and cannot legitimately be included within science. He writes: "I want no part whatever with the position . . . in which science is surreptitiously taken for a means of elucidating the utterly metaphysical question of purpose."[10] Jaki's exclusion of purpose, and more generally of design, from science has practical implications. For instance, it leads him to regard as misguided Michael Behe's project of inferring biological design from irreducibly complex biochemical systems.

Now I do not want to give the impression that I am advocating a return to Aristotle's theory of causation. There are problems with Aristotle's theory, and it needed to be replaced. My concern, however, is with what replaced it. By limiting scientific inquiry to material and efficient causes, which are of course perfectly compatible with chance and necessity, Bacon championed a view of science that could only end up excluding design.

But suppose we lay aside a priori prohibitions against design. In that case, what is wrong with explaining something as designed by an intelligent agent? Certainly there are many everyday occurrences which we explain by appealing to design. Moreover, in our workaday lives it is absolutely crucial to distinguish accident from design. We demand answers to such questions as, Did she fall or was she pushed? Did someone die accidentally or commit suicide? Was this song conceived independently or was it plagiarized? Did someone just get lucky on the stock market or was there insider trading?

Not only do we demand answers to such questions, but entire industries are devoted to drawing the distinction between accident and design. Here we can include forensic science, intellectual property law, insurance claims investigation, cryptography, and random number generation—to name but a few. Science itself needs to draw this distinction to keep itself honest. As a January 1998 issue of *Science* made clear, plagiarism and data falsification are far more common in science than we would like to admit.[11] What keeps these abuses in check is our ability to detect them.

If design is so readily detectable outside science, and if its detectability is one of the key factors keeping scientists honest, why should design be barred from the actual *content* of science? There is a worry here. The worry is that when we leave the constricted domain of human artifacts and enter the unbounded domain of natural objects, the distinction between design and nondesign cannot be reliably drawn. Consider, for instance, the following remark by Darwin in the concluding chapter of his *Origin of Species*:

> Several eminent naturalists have of late published their belief that a multitude of
> reputed species in each genus are not real species; but that other species are real,

that is, have been independently created. . . . Nevertheless they do not pretend that they can define, or even conjecture, which are the created forms of life, and which are those produced by secondary laws. They admit variation as a *vera causa* [i.e., true cause] in one case, they arbitrarily reject it in another, without assigning any distinction in the two cases.[12]

Darwin is here criticizing fellow biologists who claim that some species result from purely natural processes but that other species are specially created. According to Darwin these biologists failed to provide any objective method for distinguishing between those forms of life that were specially created and those that resulted from natural processes (or from what Darwin calls "secondary laws"). Yet without such a method for distinguishing the two, how can we be sure that our ascriptions of design are reliable? It is this worry of falsely ascribing something to design (here construed as creation) only to have it overturned later that has prevented design from entering science proper.

This worry, though perhaps justified in the past, can no longer be sustained. There does in fact exist a rigorous criterion for discriminating intelligently caused from unintelligently caused objects. Many special sciences (e.g., forensic science, artificial intelligence, cryptography, archeology, and the Search for Extraterrestrial Intelligence) already use this criterion, though in a pretheoretic form. I call it the *complexity-specification criterion*. When intelligent agents act, they leave behind a characteristic trademark or signature—what I define as *specified complexity*.[13] The complexity-specification criterion detects design by identifying this trademark of designed objects.

## 1.3   The Complexity-Specification Criterion

A detailed explication and justification of the complexity-specification criterion is technical and was the subject of my book *The Design Inference*. In chapter 2 I will summarize *The Design Inference* as well as introduce some useful simplifications and extensions of my work there. For the remainder of this chapter, however, I want simply to motivate and illustrate the complexity-specification criterion. The basic idea is straightforward. Consider how the radio astronomers in the movie *Contact* detected an extraterrestrial intelligence. This movie, based on a novel by Carl Sagan, was an enjoyable piece of propaganda for the SETI research program—the Search for Extraterrestrial Intelligence. To make the movie interesting, the SETI researchers in *Contact* actually did find an extraterrestrial intelligence (the real-life SETI program has yet to be so lucky).

How, then, did the SETI researchers in *Contact* convince themselves that they had found an extraterrestrial intelligence? To increase their chances of finding an extraterrestrial intelligence, SETI researchers monitor millions of radio signals from outer space. Many natural objects in space produce radio waves (e.g., pulsars). Looking for signs of design among all these naturally produced radio signals is like looking for a needle in a haystack. To sift through the haystack, SETI researchers run the signals they monitor through computers programmed with pattern-matchers. So long as a signal does not match one of the preset patterns, it will pass through the pattern-matching sieve (even if it has an intelligent source). If, on the other hand, it does match one of these patterns, then, depending on the pattern matched, the SETI researchers may have cause for celebration.

The SETI researchers in *Contact* did find a signal worthy of celebration, namely the following:

```
11011101111101111110111111111110111111111111111011111111
11111111101111111111111111111101111111111111111111111101
11111111111111111111111111110111111111111111111111111111
11111011111111111111111111111111111111111111101111111111
11111111111111111111111111110111111111111111111111111111
11111111111111111101111111111111111111111111111111111111
11111111111011111111111111111111111111111111111111111111
11111111111011111111111111111111111111111111111111111111
11111111111111111101111111111111111111111111111111111111
11111111111111111111111111111110111111111111111111111111
11111111111111111111111111111111111111111111110111111
11111111111111111111111111111111111111111111111111111111
11111111111101111111111111111111111111111111111111111111
11111111111111111111111111111111111110111111111111111111
11111111111111111111111111111111111111111111111111111111
11111111111011111111111111111111111111111111111111111111
11111111111111111111111111111111111111111111101111111111
11111111111111111111111111111111111111111111111111111111
11111111111111111111111111111110111111111111111111111111
11111111111111111111111111111111111111111111111111111111
111111111111111111111111111
```

The SETI researchers in *Contact* received this signal as a sequence of 1126 beats and pauses, where 1s correspond to beats and 0s to pauses. This sequence represents the prime numbers from 2 to 101, where a given prime number is represented by the corresponding number of beats (i.e., 1s), and the individual prime numbers are separated by pauses (i.e., 0s). Thus the sequence begins with 2 beats, then a pause, 3 beats, then a pause, 5 beats, then a pause, all the way up to 101 beats. The SETI researchers in *Contact* took this signal as decisive confirmation of an extraterrestrial intelligence.

What about this signal indicates design? Whenever we infer design, we must establish three things: *contingency*, *complexity*, and *specification*. Contingency ensures that the object in question is not the result of an automatic and therefore unintelligent process that had no choice in its production. Complexity ensures that the object is not so simple that it can readily be explained by chance. Finally, specification ensures that the object exhibits the type of pattern characteristic of intelligence. Let us examine these three requirements more closely.

In practice, to establish the contingency of an object, event, or structure, one must establish that it is compatible with the regularities involved in its production, but that these regularities also permit any number of alternatives to it. Typically these regularities are conceived as natural laws or algorithms. By being compatible with but not required by the regularities involved in its production, an object, event, or structure becomes irreducible to any underlying physical necessity. Michael Polanyi and Timothy Lenoir have both described this method of establishing contingency.[14] The method applies quite generally: the position of Scrabble pieces on a Scrabble board is irreducible to the natural laws governing the motion of Scrabble pieces; the configuration of ink on a sheet of paper is irreducible to the physics and chemistry of paper and ink; the sequencing of DNA bases is irreducible to the bonding affinities between the bases; and so on. With the radio signal in *Contact*, the sequence of 0s and 1s forming a sequence of prime numbers is irreducible to the laws of physics that govern the transmission of radio signals. We therefore regard the sequence as contingent.

To see next why complexity is crucial for inferring design, consider the following sequence of bits:

110111011111

These are the first twelve bits in the previous sequence representing the prime numbers 2, 3, and 5 respectively. Now, it is a sure bet that no SETI researcher, if confronted with this twelve-bit sequence, is going to contact the science editor at the *New York Times*, hold a press conference, and announce that an extraterrestrial intelligence has been discovered. No headline is going to read, "Aliens Master First Three Prime Numbers!"

The problem is that this sequence is much too short (and thus too simple) to establish that an extraterrestrial intelligence with knowledge of prime numbers produced it. A randomly beating radio source might by chance just happen to output this sequence. A sequence of 1126 bits representing the prime numbers from 2 to 101, however, is a different story. Here the sequence is sufficiently long (and therefore sufficiently complex) that only an extraterrestrial intelligence could have produced it.

Complexity as I am describing it here is a form of probability. To see the connection between complexity and probability, consider a combination lock. The more possible combinations of the lock, the more complex the mechanism and correspondingly the more improbable that the mechanism can be opened by chance. A combination lock whose dial is numbered from 0 to 39 and which must be turned in three alternating directions will have 64,000 ( = 40 × 40 × 40) possible combinations and thus a 1/64,000 probability of being opened by chance. A more complicated combination lock whose dial is numbered from 0 to 99 and which must be turned in five alternating directions will have 10,000,000,000 ( = 100 × 100 × 100 × 100 × 100) possible combinations and thus a 1/10,000,000,000 probability of being opened by chance. Complexity and probability therefore vary inversely: the greater the complexity, the smaller the probability. Thus to determine whether something is sufficiently complex to underwrite a design inference is to determine whether it has sufficiently small probability.

Even so, complexity (or improbability) is not enough to eliminate chance and establish design. If I flip a coin 1000 times, I will participate in a highly complex (i.e., highly improbable) event. Indeed, the sequence I end up flipping will be one in a trillion trillion trillion . . . , where the ellipsis needs twenty-two more "trillions." This sequence of coin tosses will not, however, trigger a design inference. Though complex, this sequence will not exhibit a suitable pattern. Contrast this with the previous sequence representing the prime numbers from 2 to 101. Not only is this sequence complex, but it also embodies a suitable pattern. The SETI researcher who in the movie *Contact* discovered this sequence put it this way: "This isn't noise, this has structure."

What is a *suitable* pattern for inferring design? Not just any pattern will do. Some patterns can legitimately be employed to infer design whereas others cannot. The intuition underlying the distinction between patterns that alternately succeed or fail to implicate design is, however, easily motivated. Consider the case of an archer. Suppose an archer stands fifty meters from a large wall with bow and arrow in hand. The wall, let us say, is sufficiently large that the archer cannot help but hit it. Now suppose each time the archer shoots an arrow at the wall, the archer paints a target around the arrow so that the arrow sits squarely in the bull's-eye. What can be concluded from this scenario? Absolutely nothing about the archer's ability as an archer. Yes, a pattern is being matched; but it is a pattern fixed only after the arrow has been shot. The pattern is thus purely *ad hoc.*

But suppose instead the archer paints a fixed target on the wall and then shoots at it. Suppose the archer shoots 100 arrows, and each time hits a perfect bull's-eye. What can be concluded from this second scenario? Confronted with this occurrence, we are obligated to infer that here is a world-class archer, one whose shots cannot legitimately be attributed to luck but rather to the archer's skill and mastery. Skill and mastery are, of course, instances of design.

The archer example introduces three elements that are essential for inferring design:

1. A reference class of possible events (here the arrow hitting the wall at some unspecified place);
2. A pattern that restricts the reference class of possible events (here a target on the wall); and
3. The precise event that has occurred (here the arrow hitting the wall at some precise location).

In a design inference, reference class, pattern, and event are linked, with the pattern mediating between event and reference class and helping to decide whether the event is due to chance or design (figure 1.1 illustrates the connections). Note that in determining whether an event is sufficiently improbable or complex to implicate design, the relevant probability is not that of the event itself. In the archery example, that probability corresponds to the size of the arrowhead point in relation to the size of the wall and will be minuscule regardless whether a target is painted on the wall. Rather, the relevant probability is that of hitting the target. In the archery example, that probability corresponds to the size of the target in relation to the size of the wall and can take any value between zero and one. The bigger the target, the easier it is to hit it by chance and thus apart from design. The smaller the target, the harder it is to hit it by chance and thus apart from design. The crucial probability, then, is the probability of the target with respect to the reference class of possible events (see sections 3.4 and 3.5).

The type of pattern where an archer fixes a target first and then shoots at it is common to statistics, where it is known as setting a *rejection region* or *critical region* prior to an experiment.[15] In statistics, if the outcome of an experiment falls within a rejection region, the chance hypothesis supposedly responsible for the outcome is rejected. The reason for setting a rejection region prior to an experiment is to forestall what statisticians call "data

## Ω: **Reference Class of Possibilities**

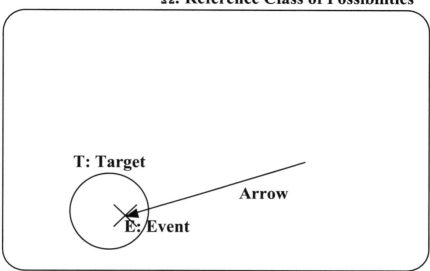

**Figure 1.1.** **The Reference Class–Target–Event Triple.**

snooping" or "cherry picking." Just about any data set will contain strange and improbable patterns if we look hard enough. By forcing experimenters to set their rejection regions prior to an experiment, the statistician protects the experiment from spurious patterns that could just as well result from chance.

Now, a little reflection makes clear that a pattern need not be given prior to an event to eliminate chance and implicate design. Consider the following cipher text:

```
nfuijolt ju jt mjlf b xfbtfm
```

Initially this looks like a random sequence of letters and spaces; initially you lack any pattern for rejecting chance and inferring design.

But suppose next that someone comes along and tells you to treat this sequence as a Caesar cipher,[16] moving each letter one notch down the alphabet. Now the sequence reads,

```
methinks it is like a weasel
```

Even though the pattern (in this case, the cryptographic key) is given after the fact, it is still the right sort of pattern for eliminating chance and inferring design. In contrast to statistics, which always identifies its patterns before an experiment is performed, cryptanalysis must discover its patterns after the fact. In both instances, however, the patterns are suitable for inferring design.

Patterns thus divide into two types, those that in the presence of complexity warrant a design inference and those that despite the presence of complexity do not warrant a design inference. The first type of pattern I call a *specification*, the second a *fabrication*. Specifications are the non-*ad hoc* patterns that can legitimately be used to eliminate chance and warrant a design inference. In contrast, fabrications are the *ad hoc* patterns that cannot legitimately be used to warrant a design inference. As we shall see in chapter 2, this distinction between specifications and fabrications can be made with full statistical rigor.

To sum up, the complexity-specification criterion detects design by establishing three things: contingency, complexity, and specification. When called to explain an event, object, or structure, we have a decision to make—are we going to attribute it to *necessity*, *chance*, or *design*? According to the complexity-specification criterion, to answer this question is to answer three simpler questions: Is it contingent? Is it complex? Is it specified? Consequently, the complexity-specification criterion can be represented as a flowchart with three decision nodes. I call this flowchart the Explanatory Filter (see figure 1.2).

The Explanatory Filter has come under considerable fire both in print and on the Internet, so it is worth it at this early stage in the discussion to flag a few common objections. One concern is that the filter assigns merely improbable events to design. But this is clearly not the case since, in addition to complexity or improbability, the filter needs to assess specification before attributing design. Another concern is that the filter will assign to design regular geometric objects like the star-shaped ice crystals that form on a cold window. This criticism fails because such shapes form as a matter of physical necessity simply in virtue of the properties of water (the filter will therefore assign the crystals to necessity and not to design). Similar considerations apply to self-organizing systems generally (see chapter 3).

According to Gert Korthof, the filter mistakenly attributes design to certain regular arithmetic progressions that arise in the growth and development of biological systems—progressions that instead ought to be attributed to natural necessities.[17] For instance, Fibonacci sequences (for which each number in the sequence is the sum of the two previous numbers) characterize the arrangement of leaves on the stems of certain plants.[18] Accordingly, such Fibonacci sequences, like the sequence of prime numbers considered earlier in this section, would have to be attributed to design. Yet, as Korthof writes,

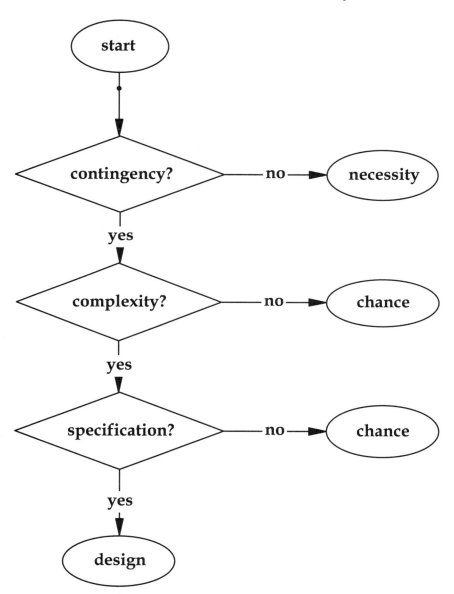

**Figure 1.2.   The Explanatory Filter.**

"the arrangement of leaves on the stem of a plant is a perfectly natural process."[19] But natural in what sense? Korthof fails to appreciate that the design of the biological systems that give rise to Fibonacci sequences is itself in question. Korthof's example is logically equivalent to a computer being programmed to generate Fibonacci sequences. Once programmed, the computer will as a matter of necessity (cf. the necessity node on the filter) output Fibonacci sequences. But whence the computer that runs the program? And whence the program? All the computer hardware and software in our ordinary experience is properly referred not to necessity but to design.

Still another concern is that the filter will miss certain obvious cases of design. Consider an embossed sign that reads "Eat at Frank's" and that falls over in a snow storm, leaving the mirror image of "Eat at Frank's" embedded in the snow.[20] Granted, the sign fell over as a result of undirected natural forces and on that basis the impression the sign left in the snow would not be attributed to design by the filter. Nonetheless, there is a relevant event whose design needs to be assessed, namely, the structuring of the embossed image (whether in the snow or on the sign). This event of structuring the embossed image must be referred back to the activity of the sign's maker and is properly ascribed to design by the Explanatory Filter. Natural forces can serve as conduits of design. As a result, a simple inspection of those natural forces may turn up no evidence of design. Often one must look deeper. To use the Explanatory Filter to identify design requires inputting the right events, objects, and structures into the filter. Just because one input does not turn up design does not mean that another more appropriately chosen input will fail.[21]

John Wilkins and Wesley Elsberry attempt to offer a general argument for why the filter is not a reliable indicator of design. Central to their argument is that if we incorrectly characterize the natural necessities and chance processes that might have been operating to account for a phenomenon, we may omit an undirected natural cause that renders the phenomenon likely and that thereby adequately accounts for it in terms other than design. Granted, this is a danger for the Explanatory Filter. But it is a danger endemic to all of scientific inquiry. Indeed, it is merely a restatement of the problem of induction—to wit, that we may be wrong about the regularities (be they probabilistic or necessitarian) that have operated in the past and are applicable to the present (for more in this vein see section 6.7).[22]

Finally, there is the criticism that in distinguishing chance and necessity, the filter fails to account for the joint action of chance and necessity, especially as they play out in the Darwinian mechanism of natural selection (the necessity component) and random variation (the chance component).[23] In particular, the Darwinian mechanism is said to be capable of delivering all the biological complexity that the filter attributes to design. If correct, this objection would ruin the Explanatory Filter. But it is not correct. I approach chance and necessity as a probabilist for whom necessity is a species of chance in which probabilities collapse to zero and one. Chance as I characterize it thus includes necessity, chance (as it is ordinarily used), and their combination. The robustness of the filter and its applicability for critiquing Darwinism become clear in sections 1.8–1.10 and chapter 4. Suffice it to say, there is no easy refutation of the Explanatory Filter.

## 1.4 Specification

Because specification is so central to identifying intelligence, some further elaboration is appropriate. For a pattern to count as a specification, the important thing is not when it was identified but whether in a certain well-defined sense it is *independent* of the event it describes. Drawing a target around an arrow already embedded in a wall is not independent of the arrow's trajectory. Consequently, such a pattern cannot be used to attribute the arrow's trajectory to design. Patterns that are specifications cannot simply be read off the events whose design is in question—in other words, it is not enough to identify a pattern simply by inspecting an event and noting (i.e., "reading off") its features. Rather, to count as specifications, patterns must be suitably independent of events. I refer to this relation of independence as *detachability* and say that a pattern is *detachable* if and only if it satisfies that relation (see section 2.5).

Detachability can be understood as asking the following question: Given an event whose design is in question and a pattern describing it, would we be able to explicitly identify or exhibit that pattern if we had no knowledge which event occurred? Here is the idea. An event has occurred. A pattern describing the event is given. The event is one from a range of possible events. If all we knew was the range of possible events without any specifics about which event actually occurred (e.g., we know that tomorrow's weather will be rain or shine, but we do not know which), could we still identify the pattern describing the event? If so, the pattern is detachable from the event.

To see what is at stake, consider the following example. It was this example that finally clarified for me what transforms a pattern simpliciter into a pattern qua specification. Consider, therefore, the following event E, an event that to all appearances was obtained by flipping a fair coin 100 times:

THTTTHHTHHTTTTTHTHTTHHHTT
HTHHHTHHHTTTTTTTHTTHTTTHH
THTTTHTHTHHTTHHHHTTTHTTHH
THTHTHHHHTTHHTHHHHTHHHHTT                                    E

Is E the product of chance or not? A standard trick of statistics professors
with an introductory statistics class is to divide the class in two, having
students in one half of the class each flip a coin 100 times, writing down the
sequence of heads and tails on a slip of paper, and having students in the
other half each generate purely with their minds a "random looking" string
of coin tosses that mimics the tossing of a coin 100 times, also writing down
the sequence of heads and tails on a slip of paper. When the students then
hand in their slips of paper, it is the professor's job to sort the papers into
two piles, those generated by flipping a fair coin, and those concocted in the
students' heads. To the amazement of the students, the statistics professor is
typically able to sort the papers with 100 percent accuracy.

There is no mystery here. The statistics professor simply looks for a repetition of six or seven heads or tails in a row to distinguish the truly random
from the pseudorandom sequences. In a hundred coin flips, one is quite likely
to see six or seven such repetitions.[24] On the other hand, people concocting
pseudorandom sequences with their minds tend to alternate between heads
and tails too frequently. Whereas with a truly random sequence of coin tosses
there is a 50 percent chance that one toss will differ from the next, as a
matter of human psychology people expect that one toss will differ from the
next around 70 percent of the time.

How, then, will our statistics professor fare when confronted with E, the
event described above? Will E be attributed to chance or to the musings of
someone trying to mimic chance? According to the professor's crude randomness checker, E would be assigned to the pile of sequences presumed to
be truly random, for E contains a repetition of seven tails in a row. Everything that at first blush would lead us to regard E as truly random checks out.
There are exactly 50 alternations between heads and tails (as opposed to the
70 that would be expected from humans trying to mimic chance). What's
more, the relative frequencies of heads and tails check out: there were 49
heads and 51 tails. Thus it is not as though the coin supposedly responsible
for generating E was heavily biased in favor of one side versus the other.

Suppose, however, that our statistics professor suspects she is not up
against a neophyte statistics student but instead against a fellow statistician
trying to put one over on her. To help organize her problem, study it more
carefully, and enter it into a computer, she will find it convenient to let
strings of 0s and 1s represent the outcomes of coin flips, with 1 corresponding

to heads and 0 to tails. In that case the following pattern D will correspond to the event E:

```
01000110110000010100111001
10111011100000001001000111
01000101011001111000100011
01010111100110111101111100
```
D

Now, the mere fact that the event E conforms to the pattern D is no reason to think that E did not occur by chance. As things stand, the pattern D has simply been read off the event E.

But D need not have been read off of E. Indeed, D could have been constructed without recourse to E. To see this, let us rewrite D as follows:

```
0
1
00
01
10
11
000
001
010
011
100
101
110
111
0000
0001
0010
0011
0100
0101
0110
0111
1000
1001
1010
1011
1100
1101
1110
1111
00
```
D

By viewing D this way, anyone with the least exposure to binary arithmetic immediately recognizes that D was constructed simply by writing binary numbers in ascending order, starting with the one-digit binary numbers (i.e., 0 and 1), proceeding then to the two-digit binary numbers (i.e., 00, 01, 10, and 11), and continuing on until 100 digits were recorded. It is therefore intuitively clear that D does not describe a truly random event (i.e., an event obtained by tossing a fair coin), but rather a pseudorandom event concocted by doing a little binary arithmetic.

Although it is now intuitively clear why chance cannot properly explain E, let us review why this is so. We started with a putative chance event E, supposedly obtained by flipping a fair coin 100 times. Since heads and tails each have probability 1/2 and since this probability gets multiplied for each flip of the coin, it follows that the probability of E is 1 in $2^{100}$, or approximately 1 in $10^{30}$ (i.e., one in a thousand billion billion billion). In addition, we constructed a pattern D to which E conforms. Initially D proved insufficient to eliminate chance as the explanation of E, since in its construction D was simply read off of E. Thus, to eliminate chance we had also to recognize that D exhibited a pattern independent of E. And for this it was enough to see that D could have been readily identified by performing some simple arithmetic operations with binary numbers. Thus, to eliminate chance we needed to consult our knowledge of binary arithmetic. This item of background knowledge detached the pattern D from the event E and thereby rendered D a specification.

For detachability to hold, an item of background knowledge must enable us to identify the pattern to which an event conforms, yet without recourse to the actual event. This is the crucial insight. Because the item in our background knowledge is conditionally and therefore epistemically independent of the event, any pattern identified from it is obtained without recourse to the event. In this way any pattern identified from this item of background knowledge avoids the charge of being *ad hoc*. Such patterns, then, are the detachable patterns and therefore the specifications. We will return to detachability in chapter 2.

## 1.5 Probabilistic Resources

To round out this preliminary discussion of the complexity-specification criterion, we need to address one more question, namely, What degree of complexity is needed for the complexity-specification criterion to implicate design reliably? Since complexity and probability are correlative notions (i.e., higher complexity corresponds to smaller probability), this question can be reformulated probabilistically: How small does a probability have to be so that in the presence of a specification it reliably implicates design?

To answer this question we will need to introduce the concept of a *probabilistic resource*. A probability is never small in isolation but only in relation to a set of probabilistic resources that describe the number of relevant ways an event might occur. There are two types of probabilistic resources, *replicational* and *specificational*. To understand replicational resources, imagine that a massive revision of the criminal justice system has just taken place. Henceforth a convicted criminal is sentenced to serve time in prison until he flips k heads in a row. The number k is thus selected according to the severity of the offense (the worse the offense, the bigger k). We assume that all coin flips are fair and duly recorded—no cheating is possible.

Thus for a 10-year prison sentence, if we assume the prisoner can flip a coin once every five seconds (this seems reasonable), the prisoner will perform 12 tosses per minute or 720 per hour or 5,760 in an eight-hour work day or 34,560 in a six-day work week or 1,797,120 in a year or 17,971,200 in ten years. Of these, on average half will be heads. Of these on average half will be followed by heads. Of these in turn on average half will be followed by heads. Continuing in this vein we find that a sequence of 17,971,200 coin tosses yields 23 heads in a row roughly half the time and strictly less than 23 heads in a row the other half (note that $2^{23}$ is about 9,000,000, which is about half of the total number of coin tosses). Thus if we required a prisoner to flip 23 heads in a row before being released, we would on average expect to see him out within approximately 10 years. Of course specific instances will vary—some prisoners being released after only a short stay, others never recording the elusive 23 heads!

Flipping 23 heads in a row is unpleasant but doable. On the other hand, flipping 100 heads in a row is effectively impossible. The probability of getting 100 heads in a row on a given trial is so small that the prisoner has no practical hope of getting out of prison, even if his life expectancy and coin-tossing ability were dramatically increased. If he could, for instance, make 10 billion attempts each year to obtain 100 heads in a row (this is coin-flipping at a rate of over 500 coin flips per second every hour of every day for a full year), then he stands only an even chance of getting out of prison in $10^{20}$ years (i.e., a hundred billion billion years). His probabilistic resources are so inadequate for obtaining the desired 100 heads that it is pointless for him to entertain hopes of freedom.

Let us now turn to the other type of probabilistic resource. In the preceding example probabilistic resources consisted of the number of opportunities for a certain event to occur. We called this type of probabilistic resource a *replicational resource* (i.e.,the number of trials or replications for a given event to occur). Replicational resources are not the only type of probabilistic resource, however. Probabilistic resources can also assume another form in which the key question is not how many opportunities there are for a given event to *occur*, but rather how many opportunities there are to *specify* an as-yet undetermined event. We will call this other type of probabilistic resource a *specificational resource*. Because lotteries provide the perfect illustration of specificational resources, we consider next a lottery example.

To eliminate the national debt, suppose the federal government decides to hold a national lottery in which the grand prize is the nation's gold reserves at Fort Knox. The federal government has constructed the lottery so that the probability of any one ticket winning is 1 in $2^{100}$, or approximately 1 in $10^{30}$. Specifically, the lottery has been constructed so that to buy a ticket, the lottery player pays a fixed price, in this case ten dollars, and then records a bit string of length 100—whichever string she chooses so long as it does not match a string that has already been chosen. She is permitted to purchase as many tickets as she wishes, subject only to her financial resources and the time it takes to record bit strings of length 100. The lottery is to be drawn at a special meeting of the United States Senate: in alphabetical order each senator is to flip a single coin once and record the resulting coin toss (heads corresponding to 1 and tails to 0).

Suppose now that the fateful day has arrived. A trillion tickets have been sold at ten dollars apiece. To prevent cheating Congress has enlisted the National Academy of Sciences. In accord with the NAS's recommendation, each ticket holder's name is duly entered onto a secure database, together with the tickets purchased and the ticket numbers (i.e., the bit strings relevant to deciding the winner). All this information is now in place. After much fanfare the senators start flipping their coins. As soon as the last senator's toss is announced, the database is consulted to determine whether the lottery has a winner. Lo and behold, the lottery does indeed have a winner, whose name is _____.

From a probabilist's perspective there is one overriding implausibility to this example. The implausibility consists not in the federal government sponsoring a lottery to eliminate the national debt, or in the choice of prize, or in the way the lottery is decided at a special meeting of the Senate, or even in the fantastically poor odds of anyone winning the lottery. The implausibility rests with the lottery having a winner. By all means the federal government should institute such a lottery if it seemed likely to redress the national debt since it is obvious that if the lottery is run fairly, there will be no winner—the odds are simply too much against it. Suppose, for instance, that a trillion tickets are sold at 10 dollars apiece. An elementary calculation shows that the probability that even one of these tickets (qua specifications) will match the winning string of 0s and 1s drawn by the Senate cannot exceed 1 in $10^{18}$ (i.e., one in a billion billion). Even if we increase the number of lottery tickets sold by several orders of magnitude, there still will not be sufficiently many tickets sold for the lottery to stand a reasonable chance of having a winner.

Sometimes it is necessary to consider both types of probabilistic resources together, those depending on the number of opportunities for an event to occur (i.e., replicational) as well as those depending on the number of opportunities to specify a given event (i.e., specificational). Suppose, for instance, that in the preceding lottery the senate will hold up to a thousand drawings to determine a winner. Thus instead of having senators flip their pennies in succession just once, we have them repeat this process up to a thousand times, stopping short of the thousand repetitions in case there happens to be a winner. If we now assume as before that a trillion tickets have been sold, then for this probabilistic setup the probabilistic resources include both a trillion specifications and a thousand possible replications. An elementary calculation now shows that the probability of this modified lottery having a winner is no greater than 1 in $10^{15}$ (i.e., one in a thousand trillion). It therefore remains highly unlikely that this modified lottery, despite the increase in probabilistic resources, will have a winner.

Probabilistic resources comprise the relevant ways an event can occur (replicational resources) and be specified (specificational resources). The important question therefore is not What is the probability of the event in question? but rather What does its probability become after all the relevant probabilistic resources have been factored in? Probabilities can never be considered in isolation, but must always be referred to a relevant reference class of possible replications and specifications. A seemingly improbable event can become quite probable when placed within the appropriate reference class of probabilistic resources. On the other hand, it may remain improbable even after all the relevant probabilistic resources have been factored in. If it remains improbable (and therefore complex) and if the event is also specified, then the complexity-specification criterion is satisfied.

One final point about probabilistic resources is important here to note. In the observable universe, probabilistic resources come in very limited supplies. Within the known physical universe there are estimated around $10^{80}$ elementary particles. Moreover, the properties of matter are such that transitions from one physical state to another cannot occur at a rate faster than $10^{45}$ times per second. This frequency corresponds to the Planck time, which constitutes the smallest physically meaningful unit of time.[25] Finally, the universe itself is about a billion times younger than $10^{25}$ seconds (assuming the universe is between ten and twenty billion years old). If we now assume that any specification of an event within the known physical universe requires at least one elementary particle to specify it and cannot be generated any faster than the Planck time, then these cosmological constraints imply that the total number of specified events throughout cosmic history cannot exceed

$$10^{80} \times 10^{45} \times 10^{25} = 10^{150}.$$

It follows that any specified event of probability less than 1 in $10^{150}$ will remain improbable even after all conceivable probabilistic resources from the observable universe have been factored in. A probability of 1 in $10^{150}$ is therefore a *universal probability bound*.[26] A universal probability bound is impervious to all available probabilistic resources that may be brought against it. Indeed, all the probabilistic resources in the known physical world cannot conspire to render remotely probable an event whose probability is less than this universal probability bound. The universal probability bound of 1 in $10^{150}$ is the most conservative in the literature. The French mathematician Emile Borel proposed 1 in $10^{50}$ as a universal probability bound below which chance could definitively be precluded (i.e., any specified event as improbable as this could never be attributed to chance).[27] Cryptographers assess the security of cryptosystems in terms of a brute force attack that employs as many probabilistic resources as are available in the universe to break a cryptosystem by chance. In its report on the role of cryptography in securing the information society, the National Research Council set 1 in $10^{94}$ as its universal probability bound to ensure the security of cryptosystems against chance-based attacks.[28] As we shall see in chapter 5, such levels of improbability are easily attained by real physical systems. It follows that if such systems are also specified, then they are designed.

## 1.6   False Negatives and False Positives

As with any criterion, we need to make sure that the judgments of the complexity-specification criterion agree with reality. Consider medical tests. Any medical test is a criterion. A perfectly reliable medical test would detect the presence of a disease whenever it is indeed present, and fail to detect the disease whenever it is absent. Unfortunately, no medical test is perfectly reliable, and so the best we can do is keep the proportion of false positives and false negatives as low as possible.[29]

All criteria, and not just medical tests, face the problem of false positives and false negatives. A criterion attempts to classify individuals with respect to a target group (in the case of medical tests, those who have a certain disease). When the criterion places in the target group an individual who should not be there, it commits a false positive. Alternatively, when the criterion fails to place in the target group an individual who should be there, it commits a false negative.

Let us now apply these observations to the complexity-specification criterion. This criterion purports to detect design. Is it a reliable criterion? The target group for this criterion comprises all things intelligently caused. How accurate is this criterion at correctly assigning things to this target group and correctly omitting things from it? The things we are trying to explain have causal stories. In some of those causal stories intelligent causation is indispensable whereas in others it is dispensable. An inkblot can be explained without appealing to intelligent causation; ink arranged to form meaningful text cannot. When the complexity-specification criterion assigns something to the target group, can we be confident that it actually is intelligently caused? If not, we have a problem with false positives. On the other hand, when this criterion fails to assign something to the target group, can we be confident that no intelligent cause underlies it? If not, we have a problem with false negatives.

Consider first the problem of false negatives. When the complexity-specification criterion fails to detect design in a thing, can we be sure that no intelligent cause underlies it? No, we cannot. To determine that something is not designed, this criterion is not reliable. False negatives are a problem for it. This problem of false negatives, however, is endemic to detecting intelligent causes. One difficulty is that intelligent causes can mimic necessity and chance, thereby rendering their actions indistinguishable from such unintelligent causes. A bottle of ink may fall off a cupboard and spill onto a sheet of paper. Alternatively, a human agent may deliberately take a bottle of ink and pour it over a sheet of paper. The resulting inkblot may look identical in both instances, but in the one case results by chance, in the other by design. Another difficulty is that detecting intelligent causes requires background knowledge on our part. It takes an intelligent cause to recognize an intelligent cause. But if we do not know enough, we will miss it. Consider a spy listening in on a communication channel whose messages are encrypted. Unless the spy knows how to break the cryptosystem used by the parties on whom she is eavesdropping (i.e., knows the cryptographic key), any messages passing the communication channel will be unintelligible and might in fact be meaningless.

The problem of false negatives therefore arises either when an intelligent agent has acted (whether consciously or unconsciously) to conceal one's actions, or when an intelligent agent in trying to detect design has insufficient background knowledge to determine whether design actually is present. Detectives face this problem all the time. A detective confronted with a murder needs first to determine whether a murder has indeed been committed. If the murderer was clever and made it appear that the victim died by accident, then the detective will mistake the murder for an accident. So too, if the detective is stupid and misses certain obvious clues, the detective will mistake the murder for an accident. In doing so, the detective commits a false negative. Contrast this, however, with a detective facing a murderer intent on revenge and who wants to leave no doubt that the victim was intended to die. In that case the problem of false negatives is unlikely to arise.

Intelligent causes can do things that unintelligent causes cannot and can make their actions evident. When for whatever reason an intelligent cause fails to make its actions evident, we may miss it. But when an intelligent cause succeeds in making its actions evident, we take notice. This is why false negatives do not invalidate the complexity-specification criterion. This criterion is fully capable of detecting intelligent causes intent on making their presence evident. Masters of stealth intent on concealing their actions may successfully evade the criterion. But masters of self-promotion bank on the complexity-specification criterion to make sure their intellectual property gets properly attributed. Indeed, intellectual property law would be impossible without this criterion.

And this brings us to the problem of false positives. Even though specified complexity is not a reliable criterion for *eliminating* design, it is, I shall argue, a reliable criterion for *detecting* design. The complexity-specification criterion is a net. Things that are designed will occasionally slip past the net. We would prefer that the net catch more than it does, omitting nothing due to design. But given the ability of design to mimic unintelligent causes and the possibility of ignorance causing us to pass over things that are designed, this problem cannot be remedied. Nevertheless, we want to be very sure that whatever the net does catch includes only what we intend it to catch—to wit, things that are designed. Only things that are designed had better end up in the net. If this is the case, we can have confidence that whatever the complexity-specification criterion attributes to design is indeed designed. On the other hand, if things end up in the net that are not designed, the criterion is vitiated.

I want, then, to argue that specified complexity is a reliable criterion for detecting design. Alternatively, I want to argue that the complexity-specification criterion successfully avoids false positives—in other words, whenever it attributes design, it does so correctly. Let us now see why this is the case. I offer two arguments. The first is a straightforward inductive argument: In every instance where the complexity-specification criterion attributes design and where the underlying causal story is known (i.e., where we are not just dealing with circumstantial evidence, but where, as it were, the video camera is running and any putative designer would be caught red-handed), it turns out design actually is present; therefore, design actually is present whenever the complexity-specification criterion attributes design. The conclusion of this argument is a straightforward inductive generalization. It has the same logical force as concluding that all ravens are black given that all ravens observed to date have been found to be black.

How well does this inductive justification of the complexity-specification criterion hold up? I am arguing inductively that this criterion reliably detects design. The conclusion of this argument is that whenever the criterion attributes design, design actually is present. The premise of this argument is that whenever the criterion attributes design and the underlying causal story can be verified, design actually is present. Now, even though the conclusion follows as an inductive generalization from the premise, the premise itself seems false. There are a lot of coincidences out there that seem best explained without invoking design. Consider, for instance, the Shoemaker-Levy comet. The Shoemaker-Levy comet crashed into Jupiter exactly twenty-five years to the day after the Apollo 11 moon landing. What are we to make of this coincidence? Do we really want to explain it in terms of design? What if we submitted this coincidence to the complexity-specification criterion and out popped design? Our intuitions strongly suggest that the comet's trajectory and NASA's space program were operating independently, and that at best this coincidence should be referred to chance and not design.

This objection is readily met. The fact is that the complexity-specification criterion does not yield design all that easily, especially if the complexities are kept high (or correspondingly, the probabilities are kept small). It is simply not the case that unusual and striking coincidences automatically yield design. Martin Gardner is correct when he notes, "The number of events in which you participate for a month, or even a week, is so huge that the probability of noticing a startling correlation is quite high, especially if you keep a sharp outlook."[30] The implication he means to draw, however, is incorrect, namely, that therefore startling correlations or coincidences may be uniformly relegated to chance. Yes, the fact that the Shoemaker-Levy comet crashed into Jupiter exactly twenty-five years to the day after the Apollo 11 moon landing is a coincidence best referred to chance. But the fact that Mary Baker Eddy's writings on Christian Science bear a remarkable resemblance to Phineas Parkhurst Quimby's writings on mental healing is a coincidence that cannot be explained by chance and is properly explained by positing Quimby as a source for Eddy.[31]

The complexity-specification criterion is robust and easily resists counterexamples of the Shoemaker-Levy variety. Assuming, for instance, that the Apollo 11 moon landing serves as a specification for the crash of Shoemaker-Levy into Jupiter (a generous concession at that), that the comet could have crashed at any time within a period of a year, and that the comet crashed to the very second precisely twenty-five years after the moon landing, then a straightforward probability calculation indicates that the probability of this coincidence is no smaller than 1 in $10^8$. This simply is not all that small a probability (i.e., high complexity), especially when considered in relation to all the events astronomers are observing in the solar system (i.e., in relation to the probabilistic resources relevant to astronomy—see section 1.5). Certainly this probability is nowhere near the universal probability bound of 1 in $10^{150}$ proposed in the last section.

Is there a convincing application of the complexity-specification criterion that employs this universal probability bound and in which a coincidence better explained by undirected natural causes gets attributed to design? Perhaps the best candidate is the natural nuclear reactors at the Oklo uranium mine in Gabon, West Africa.[32] Producing a nuclear chain reaction these days is not simple. It requires getting the right proportion of uranium 235 and uranium 238 and placing sufficient quantities in a controlled environment. Indeed, all the nuclear chain reactions on earth these days result from human design. Nonetheless, two billion years ago nature produced its own nuclear chain reaction at the Oklo mine. What's more, it was a controlled reaction, with water facilitating the reaction and continually cooling the uranium to prevent an explosion—just as in today's power plants.

Many highly specific and seemingly unlikely conditions had to be satisfied for the Oklo reactors to form. For instance, because the half-life of uranium 235 is about six times as short as that of uranium 238, the proportion of uranium 235 to uranium 238 on earth is nowadays much less than required for the enriched uranium that powers artificial nuclear reactors. Nevertheless, two billion years ago, the proportions would have been just right, and thus a necessary condition for nature to produce nuclear chain reactions would have been satisfied back then. Likewise other highly specific conditions had to be satisfied for Oklo to produce natural nuclear chain reactions. The Oklo reactors are remarkable. To date no other natural nuclear reactors have been identified.[33]

What are we to make of the Oklo reactors? My own view is that although the conditions these natural nuclear reactors had to satisfy were highly specific, they were not that improbable. For instance, with respect to the proportions of uranium 235 and 238, given their differing decay rates, there was bound to come a time when the proportions would be ideal for a nuclear chain reaction. That the time should be two billion years ago, coinciding with the Oklo reactors, is therefore not the sort of coincidence that triggers a design inference. Rather, it lands on the necessity node of the Explanatory Filter. Other conditions that needed to be satisfied, even when considered jointly, do not appear to trigger a design inference either. Although these conditions are specified, they do not appear that improbable. The precise probabilities for such conditions have yet to be ascertained. Consequently, it is not possible at this time to decide whether the Oklo reactors satisfy the complexity-specification criterion. My own very strong suspicion, however, is that should such probabilities be ascertained, the complexity-specification criterion would not be satisfied.

But suppose the Oklo reactors ended up satisfying this criterion after all. Would this vitiate the complexity-specification criterion? Not at all. At worst it would indicate that certain naturally occurring events or objects that we initially expected to involve no design actually do involve design. This would no doubt seem counterintuitive in light of the naturalism that currently dominates science, but it would not provide a counterexample to the complexity-specification criterion. Rather, it would constitute a borderline case, one that without the complexity-specification criterion would be consigned to the category of unexplained coincidences, but with it would be attributed to design. If design is as pervasive in nature as design theorists claim, then we certainly need to find it in such traditional repositories of design like biology, but we should also not be surprised if we find it in unexpected places.

The complexity-specification criterion is inductively sound and is not vitiated by finding specified complexity in unexpected places. Indeed, the only way to refute this criterion is on a case-by-case basis by showing that it fails to yield design for some object in question either because the relevant probability cannot be calculated, or because the relevant pattern cannot be shown to be suitably independent and therefore a specification, or because the relevant probability, though calculable, was in fact miscalculated. In the last case, what typically undermines the complexity-specification criterion is that the relevant probability was underestimated. Dawkins's book *Climbing Mount Improbable* is exclusively devoted to the theme that biological systems are not nearly as improbable as they seem.[34] For Dawkins, highly improbable (i.e., complex) specified events require explanation. A proper explanation of such events for Dawkins requires showing that the improbabilities are not as bad as initially suspected. Thus in biology the Darwinian mechanism of natural selection and random variation is supposed to wash away all vast improbabilities associated with biological complexity by breaking these improbabilities into a series of more manageable probabilities. The usual course in mitigating the force of the complexity-specification criterion for natural objects is to argue on a case-by-case basis that the complexities (i.e., improbabilities) that seemingly implicate design are in fact not complex enough to satisfy the complexity-specification criterion and thereby warrant a design inference. As we shall see in chapter 5, there is no reason to think that all putative instances of specified complexity in nature submit to this divide-and-conquer approach.

## 1.7   Why the Criterion Works

My second argument for showing that specified complexity reliably detects design considers the nature of intelligent agency and, specifically, what it is about intelligent agents that makes them detectable. Even though induction confirms that specified complexity is a reliable criterion for detecting design, induction does not explain why this criterion works. To see why the complexity-specification criterion is exactly the right instrument for detecting design, we need to understand what it is about intelligent agents that makes them detectable in the first place. The principal characteristic of intelligent agency is *choice*. Even the etymology of the word "intelligent" makes this clear. "Intelligent" derives from two Latin words, the preposition *inter*, meaning between, and the verb *lego*, meaning to choose or select. Thus, according to its etymology, intelligence consists in *choosing between*. For an intelligent agent to act is therefore to choose from a range of competing possibilities.

This is true not just of humans but of animals as well as of extraterrestrial intelligences. A rat navigating a maze must choose whether to go right or left at various points in the maze. When SETI researchers attempt to discover intelligence in the extraterrestrial radio transmissions they are monitoring, they assume an extraterrestrial intelligence could have chosen any number of possible radio transmissions and then attempt to match the transmissions they observe with certain patterns as opposed to others. Whenever a human being utters meaningful speech, a choice is made from a range of possible sound-combinations that might have been uttered. Intelligent agency always entails discrimination, choosing certain things, ruling out others.

Given this characterization of intelligent agency, the crucial question is how to recognize it. Intelligent agents act by making a choice. How, then, do we recognize that an intelligent agent has made a choice? A bottle of ink spills accidentally onto a sheet of paper; someone takes a fountain pen and writes a message on a sheet of paper. In both instances ink is applied to paper. In both instances one among an almost infinite set of possibilities is realized. In both instances a contingency is actualized and others are ruled out. Yet in one instance we ascribe agency, in the other chance.

What is the relevant difference? Not only do we need to observe that a contingency was actualized, but we ourselves need also to be able to specify that contingency. The contingency must conform to an independently given pattern, and we must be able independently to construct that pattern. A random ink blot is unspecified; a message written with ink on paper is specified. To be sure, the exact message recorded may not be specified. But orthographic, syntactic, and semantic constraints will nonetheless specify it.

Actualizing one among several competing possibilities, ruling out the rest, and specifying the one that was actualized encapsulates how we recognize intelligent agency or, equivalently, how we detect design. Experimental psychologists who study animal learning and behavior have known this all along. To learn a task an animal must acquire the ability to actualize behaviors suitable for the task as well as the ability to rule out behaviors unsuitable for the task. Moreover, for a psychologist to recognize that an animal has learned a task, it is necessary not only to observe the animal making the appropriate discrimination but also to specify the discrimination.

Thus, to recognize whether a rat has successfully learned how to traverse a maze, a psychologist must first specify which sequence of right and left turns conducts the rat out of the maze. No doubt, a rat randomly wandering a maze also discriminates a sequence of right and left turns. But by randomly wandering the maze, the rat gives no indication that it can discriminate the appropriate sequence of right and left turns for exiting the maze. Consequently, the psychologist studying the rat will have no reason to think the rat has learned how to traverse the maze.

Only if the rat executes the sequence of right and left turns specified by the psychologist will the psychologist recognize that the rat has learned how to traverse the maze. Now it is precisely the learned behaviors we regard as intelligent in animals. Hence it is no surprise that the same scheme for recognizing animal learning recurs for recognizing intelligent agency generally—to wit, actualizing one among several competing possibilities, ruling out the others, and specifying the one actualized.

Note that complexity is implicit here as well. To see this, consider again a rat traversing a maze, but now take a very simple maze in which two right turns conduct the rat out of the maze. How will a psychologist studying the rat determine whether it has learned to exit the maze? Just putting the rat in the maze will not be enough. Because the maze is so simple, the rat could by chance just happen to take two right turns, and thereby exit the maze. The psychologist will therefore be uncertain whether the rat actually learned to exit this maze or whether the rat just got lucky.

But contrast this with a complicated maze in which a rat must take just the right sequence of left and right turns to exit the maze. Suppose the rat must take 100 appropriate right and left turns and that any mistake will prevent the rat from exiting the maze. A psychologist who sees the rat take no erroneous turns and in short order exit the maze will be convinced that the rat has indeed learned how to exit the maze and that this was not dumb luck.

This general scheme for recognizing intelligent agency is but a thinly disguised form of the complexity-specification criterion. In general, to recognize intelligent agency we must observe an actualization of one among several competing possibilities, note which possibilities were ruled out, and then be able to specify the possibility that was actualized. What's more, the competing possibilities that were ruled out must be live possibilities and sufficiently numerous so that specifying the possibility that was actualized cannot be attributed to chance. In terms of complexity, this is just another way of saying that the range of possibilities is complex. In terms of probability, this is just another way of saying that the possibility that was actualized has small probability.

All the elements in this general scheme for recognizing intelligent agency (i.e., actualizing, ruling out, and specifying) find their counterpart in the complexity-specification criterion. Consequently, this criterion makes precise what we have been doing right along when we recognize intelligent agency. The complexity-specification criterion pinpoints how we detect design.

## 1.8 The Darwinian Challenge to Design

Darwin would not have agreed. Indeed, he would have seen his mechanism of random variation and natural selection as purchasing the very specified complexity that I am attributing exclusively to intelligence. Darwin's foil was ever design. In formulating his theory, Darwin was responding to the tradition of British natural theology embodied in William Paley's *Natural Theology*, subtitled *Evidences of the Existence and Attributes of the Deity, Collected from the Appearances of Nature*.[35] The subtitle is revealing. Paley's project was to examine features of the natural world ("appearances of nature") and therewith draw conclusions about a designing intelligence responsible for those features (whom Paley identified with the God of Christianity). Paley is best remembered for his famous watchmaker analogy. According to Paley, if we find a watch in a field, the watch's adaptation of parts to telling time ensures that it is the product of an intelligence. So too, according to Paley, the marvelous adaptations of means to ends in organisms ensure that organisms are the product of an intelligence.

Paley's *Natural Theology* is one of the great unread books. If it is read at all today, it is read in philosophy of religion courses, where it is presented as a foil not to Charles Darwin but to David Hume. In his *Dialogues Concerning Natural Religion* Hume had criticized the design argument as a feeble argument from analogy with no probative force.[36] It is this criticism that for many philosophers of religion remains decisive against design. Schematically, an argument from analogy takes the following form: we are given two objects, U and V, which share certain properties, call them A, B, C, and D. U and V are therefore similar with respect to A, B, C, and D. Now, suppose we know that U has some property Q, and suppose further that we want to determine whether V also has property Q. An argument from analogy then warrants that V has property Q because U and V share properties A, B, C, and D, and U has property Q. In terms of premises and conclusion, the argument from analogy therefore looks as follows:

> U has property Q.
> U and V share properties A, B, C, and D.
> Therefore V also has property Q.

In the case of Paley's watchmaker argument, U is a watch, V is an organism, and the property Q is that something is intelligently designed. For the watch there is no question that it actually is intelligently designed. For the organism, on the other hand, this is not so immediately clear. Yet because the watch and the organism share several features in common, call them A, B, C, and D (like functional interdependence of parts, self-propulsion, etc.), we are, according to the argument from analogy, warranted in concluding that organisms are also intelligently designed. In terms of premises and conclusion, the argument looks as follows:

> Watches are intelligently designed.
> Watches and organisms are similar.
> Therefore organisms are also intelligently designed.

Although arguments from analogy can be intuitively appealing, they are not valid deductive arguments for which the truth of the premises guarantees the truth of the conclusion. Sometimes an argument from analogy leads us to the right conclusion:

> In human beings, the blood circulates.
> Human beings and dogs are similar.
>
> Therefore in dogs, the blood circulates.

But at other times arguing by analogy leads us astray:

> In human beings, the blood circulates.
> Human beings and plants are similar.
>
> Therefore in plants, the blood circulates.

The chief difficulty with arguments from analogy is that they are always also arguments from disanalogy. If U and V were identical, there would be no question about V having property Q if U has that property. The reason there is a question about V having property Q is because U and V are not identical. What this means schematically is that there are properties I, J, K, and L that U possesses but which V does not possess. U has properties A, B, C, and D, which V shares, but also properties I, J, K, and L, which V does not share. Moreover, U has property Q. The big question, therefore, is whether Q is a property like A, B, C, and D, which V shares with U, or whether Q is a property like I, J, K, and L, which V does not share with U. Without additional information, the argument from analogy has no way of deciding this question.

By itself, therefore, the argument from analogy provides no compelling support for its conclusion. The property that stands in question, here Q, might just as well be part of the disanalogy as part of the analogy. At best, therefore, the argument from analogy gives us reason to suspect that two objects that share similarities might share still an additional similarity. Analogies may thus point us to further analogies. Yet without additional information we can draw no definite conclusion.

There is, however, a way to strengthen the argument from analogy, and that is to make its conclusion follow as an inductive generalization: When objects U and V both possess properties A, B, C, and D, and when U also possesses property Q, the conclusion that V possesses Q follows inductively if in every instance where an object possesses A, B, C, and D, and where it can be determined whether the object also possesses Q, the object actually does possess Q. In other words, in this strengthened form of the argument from analogy, the appearance of A, B, C, and D has yet to be divorced from the appearance of Q. This strengthened form of the argument therefore has an additional premise and can be formulated as follows:

U has property Q.
U and V share properties A, B, C, and D.
There is no known instance where A, B, C, and D occur without Q.
Therefore, V has property Q.

Granted, this is still not a deductive argument. But for the conclusion to fail, V would have be the first known instance of an object that possesses A, B, C, and D without possessing Q. It is possible to formulate Paley's watchmaker argument as such a strengthened argument from analogy. So formulated, it constitutes a valid inductive argument (though whether the argument is sound—i.e., whether all the premises are also true—is another matter). Thus U would correspond to the watch; V to an organism; A, B, C, and D to such properties as functional integration of parts, storage of information, processing of energy, and self-propulsion; and Q to the property of being designed by an intelligence. Such a revamped argument from analogy, to my mind, would go a long way toward addressing Hume's objections to design.

Nevertheless, my point here is not to rebut Hume, but to show how Darwin unseated even this strengthened form of Paley's design argument. Paley had claimed that certain structural features of objects reliably correlate with a designing intelligence (the principal features being functional integration of parts and adaptation of means to ends). Instead of raising philosophical doubts about this claim, as David Hume had, Darwin took Paley on his own terms and proposed a natural mechanism to account for precisely those features that Paley thought were inexplicable apart from design. Darwin's refutation of Paley was thus far bolder than any refutation inherent in Hume's *Dialogues*. Hume was a skeptic and caviller. Darwin was a destroyer and rebuilder. Hume found logical flaws in the design argument. Darwin showed not that design was philosophically substandard but that it was scientifically redundant—that a purely natural mechanism could produce the appearance of design in nature. Darwin claimed to discover this mechanism. It is this discovery that Daniel Dennett regards as the pinnacle of human intellectual achievement:

> If I were to give an award for the single best idea anyone has ever had, I'd give it to Darwin, ahead of Newton and Einstein and everyone else. In a single stroke, the idea of evolution by natural selection unifies the realm of life, meaning, and

purpose with the realm of space and time, cause and effect, mechanism and physical law.[37]

## 1.9 The Constraining of Contingency

Darwin fundamentally transformed the debate about design. Before Darwin our explanatory options were those described in the earlier sections of this chapter, namely, necessity, chance, and design. Thus, explanations were divided neatly into necessity and contingency. Moreover, because contingency could be undirected or directed, contingency was in turn neatly divided into chance and design (see section 1.1). Accordingly, the anatomy of explanation prior to Darwin looked as in figure 1.3.

Darwin challenged this account of explanation. For him, an undirected contingency, though not guided by intelligence, could nonetheless be *constrained* by natural laws and thereby mimic the effects of intelligence. The problem with the pre-Darwinian anatomy of explanation, therefore, was that it failed adequately to distinguish undirected contingencies that were wholly unconstrained (i.e., pure chance) from undirected contingencies that were

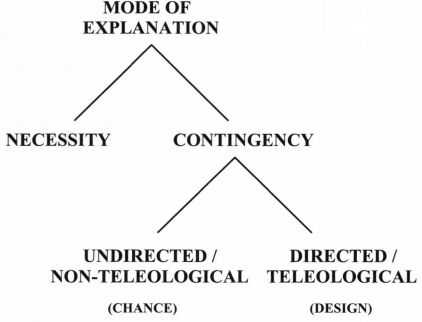

**MODE OF EXPLANATION**

**NECESSITY**        **CONTINGENCY**

**UNDIRECTED /
NON-TELEOLOGICAL**        **DIRECTED /
TELEOLOGICAL**

**(CHANCE)**        **(DESIGN)**

**Figure 1.3.  Anatomy of Explanation (Pre-Darwinian).**

constrained by natural laws (i.e., chance modified by law). Within the category of constrained but undirected contingency Darwin claimed to find all the resources to explain what for Paley had required directed contingency (i.e., design). Accordingly, the anatomy of explanation after Darwin looked as in figure 1.4.

In place of the three traditional modes of explanation, Darwin introduced a further subdivision and gave us four. Necessity and design remained unchanged, but chance was subdivided into pure chance and chance modified by law. For Darwin this last mode of explanation offered to supplant design in nature. How could it do that? To be sure, neither Darwin nor anyone else

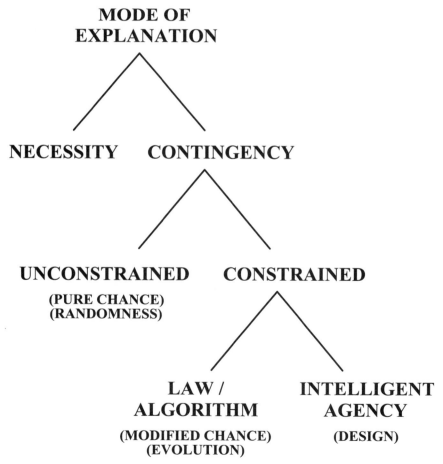

**Figure 1.4.  Anatomy of Explanation (Post-Darwinian).**

for that matter ever laid out the precise causal pathway by which chance modified by law could produce, say, a mammalian eye. But shorter causal pathways leading to less marvelous but still ordered structures seemed well within the reach of chance modified by law. And since chance modified by law could operate over vast time scales, this was sufficient justification for Darwin to extrapolate from processes we can observe creating ordered structures to those we cannot.

Examples of chance modified by law creating novel ordered structures abound. Consider an agitator that sifts rocks according to size. An agitator is a container into which one places rocks and then shakes them up. This shaking (or "agitating") constitutes a chance process that, apart from any further constraints, would yield a totally disordered configuration of rocks. Here on earth, however, there is a further constraint, namely, gravity. Agitating the container in the presence of a downward-directed gravitational force ensures that the biggest rocks will rise to the top and that the smallest will proceed to the bottom. Of course, precisely where at a given level a rock of the same size as others will land will still be random. Thus randomness or chance still operates horizontally. But vertically there is a clear order imposed on rocks agitated in the presence of gravity.

Darwin's great insight was to see that the same sort of constraining of contingency (or alternatively modification of chance by law) occurs in biology. He opens his *Origin of Species* by describing animal breeding experiments. The offspring of animals, though similar to their parents, nonetheless differ from them. These differences for all we know are controlled by chance. (Darwin referred to these differences as variations; we now think of them as arising from mutations in DNA, which are much more clearly chance driven. Darwin did not speculate about how variations arise.)

Now, if animal breeders flipped a coin whenever they decided which of their animals could reproduce, then any chance difference between parents and offspring would remain unconstrained. But if, on the other hand, animal breeders monitored their animals and permitted only those to reproduce which, albeit by chance, exhibited certain desirable characteristics, then over time those desirable characteristics would become intensified and entrenched in the population. Thus chance, in the form of variation between parent and offspring, would be shaped or constrained by the animal breeder to yield those characteristics that the animal breeder prefers. But note that the animal breeder is limited to those characteristics that chance produces. A breeder of furry animals, for instance, will not be able to obtain smooth-skinned animals if all the animals bred stay consistently furry.

Human animal breeders are of course absent from most of the history of life. What, then, constrains the reproduction of living things in nature? Clearly, nature herself. Nature, as it were, selects those organisms that will reproduce and eliminates those that will not. Since without constraints the reproduction of organisms would proceed exponentially, and since nature clearly does not have the resources to accommodate such exponential growth, only a small proportion of organisms will in any generation have the opportunity to reproduce. Those whose characteristics best facilitate reproduction will therefore be selected to leave offspring whereas the rest will die without leaving them. Thus, according to Darwin, nature herself constitutes the supreme animal breeder that constrains the history of life.

## 1.10   The Darwinian Extrapolation

But is nature, acting through the Darwinian mechanism of chance variations and natural selection, capable of accounting for the full diversity of life? Experimental and observational evidence for the Darwinian mechanism is necessarily limited to the brief time spans allotted to human investigators and thus inevitably reveals only limited variation within seemingly fixed boundaries (e.g., insects may develop resistance to insecticides, but they remain insects). Fossil evidence, on the other hand, spans vast time scales but is typically incomplete so that one cannot construct detailed evolutionary pathways connecting highly disparate organisms. According to the Darwinian theory, organisms possess unlimited plasticity to diversify across all boundaries; moreover, natural selection is said to have the capability of exploiting that plasticity and thereby delivering the spectacular diversity of living forms that we see.

Such a theory, however, necessarily commits an extrapolation. And as with all extrapolations, there is always the worry that what we are confident about in a limited domain may not hold more generally outside that domain. This is always a danger in science—to think that one's theory encompasses a far bigger domain than it actually does. In the early heady days of Newtonian mechanics, physicists thought Newton's laws gave a total account of the constitution and dynamics of the universe. Maxwell, Einstein, and Heisenberg each showed that the proper domain of Newtonian mechanics was far more constricted. It is therefore fair to ask whether the Darwinian mechanism may not face similar limitations.

In chapters 4 and 5 I shall examine whether the Darwinian extrapolation is indeed justified. For the moment, however, I want provisionally to accept the Darwinian extrapolation and show why it constitutes such a bold stroke in the history of ideas. Examining this extrapolation makes clear why Darwinists like Daniel Dennett feel justified calling evolution by natural selection "the single best idea anyone has ever had." Clearly, if the Darwinian mechanism produced no more than the changes we see organisms undergo in the lab or field, there would be far less interest in Darwin's theory. Which is not to say the theory would not be important—for instance, the development of antibiotic resistance by pathogens via the Darwinian mechanism is experimentally verified and rightly of great concern to the medical field. But the almost alchemical transformation of organisms having vastly different morphologies is a direct consequence of Darwin's theory, and it is this genealogical melding of what prima facie appear utterly distinct organisms that makes the Darwinian extrapolation so remarkable.

Most extrapolations are, after all, unremarkable. Given a curve that fits a data set, we can ask what shape the curve would take for data points outside the data set. With a well-defined curve this is perfectly straightforward— simply a matter of inputting hitherto untried data points and outputting their values. Here the relation between input and output is fully explicit. Moreover, if the inputs and outputs become experimentally accessible, then it becomes possible to confirm or disconfirm the extrapolation, testing whether it holds for data points not in the original data set.

But that is not how the Darwinian extrapolation works. Not with all the experimental evidence in the world for what the Darwinian mechanism can accomplish within investigator-controlled settings will it be possible to determine how the Darwinian mechanism actually transformed, say, a reptile into a mammal over the course of natural history. There are simply too many historical contingencies and too many missing data to form an accurate picture of precisely what happened. Fossil evidence may confirm that the transformation was indeed effected over the course of natural history, and the Darwinian mechanism may forever constitute the best scientific explanation of how that transformation was effected. But in extrapolating the Darwinian theory from experimentally controlled settings to nature at large, we are ceding to nature all the hard work of extrapolation.

What do I mean by "the hard work of extrapolation"? Consider, by contrast, the statistician who fits a curve to a data set. The statistician specifies the curve and knows how to determine the shape of the curve for data points outside the original data set. Here it is the statistician who does all the hard work of extrapolation: The statistician specifies a curve, identifies data points of interest, and then evaluates the curve at those data points. But in the Darwinian theory the biologist, unlike the statistician, cannot specify what structures might evolve outside the biologist's domain of experimental control short of looking to nature and seeing what structures actually did evolve.

Independent of what actually happens in nature, the statistician is able to specify what nature would do given that a statistical extrapolation is correct. Not so the biologist. The biologist must consult nature to determine what nature has actually done before describing the course of Darwinian evolution.

This is why Darwinian explanation so often takes the form of a just-so story. The term *just-so story* comes from Rudyard Kipling, who told fantastic tales about how animals obtained their various features (How did the giraffe get its neck? How did the elephant get its trunk? etc.). If the Darwinian mechanism is indeed how animals obtained their various features, then suitably refined just-so stories are about the best one can do to account for those features. For instance, how did the mammalian eye evolve? Richard Dawkins offers the following just-so story:

> Not only is it clear that part of an eye is better than no eye at all. We also can find a plausible series of intermediates among modern animals. This doesn't mean, of course, that these modern intermediates really represent ancestral types [N.B.]. But it does show that intermediate designs are capable of working.
>
> Some single-celled animals have a light-sensitive spot with a little pigment screen behind it. The screen shields it from light coming from one direction, which gives it some "idea" of where the light is coming from. Among many-celled animals, various types of worm and some shellfish have a similar arrangement, but the pigment-backed light-sensitive cells are set in a little cup. This gives slightly better direction-finding capability, since each cell is selectively shielded from light rays coming into the cup from its own side. In a continuous series from flat sheet of light-sensitive cells, through shallow cup to deep cup, each step in the series, however small (or large) the step, would be an optical improvement. Now, if you make a cup very deep and turn the sides over, you eventually make a lensless pinhole camera. . . .
>
> When you have a cup for an eye, almost any vaguely convex, vaguely transparent or even translucent material over its opening will constitute an improvement, because of its slight lens-like properties. It collects light over its area and concentrates it on a smaller area of retina. Once such a crude proto-lens is there, there is a continuously graded series of improvements, thickening it and making it more transparent and less distorting, the trend culminating in what we would recognize as a true lens.[38]

Thus the Darwinian mechanism delivers up a mammalian eye. This account of the mammalian eye is both revealing and frustrating. It is frustrating because all the historical and biological details in the eye's construction are lost. How exactly did a lens form within a pinhole camera? How did a spot become innervated and thereby light-sensitive? With respect to embryology, what developmental changes are required to go from a light-sensitive sheet to a light-sensitive cup? None of these questions receives an answer in purely Darwinian terms. Instead—and this is what is revealing about Dawkins's account of the eye—the Darwinian mechanism provides a plausible description of various adaptive changes that might have led from one structure to another. This perforce entails a certain indeterminacy, but it is the best that the Darwinian mechanism can do in the absence of precise historical and biological detail, and such detail tends unfortunately to be overwhelmingly absent for the history of life.

The Darwinian mechanism therefore constitutes at once a lazy and a profound solution to the immensely difficult problem of biological complexity and diversity. The laziness here is ours and the profundity is nature's. Accordingly, nature is a creative force that we with our recently evolved intellects cannot hope to rival. Like the ancients in the presence of the numinous, all we can do is stand in awe before the power of selection to produce the marvels of nature. To be sure, we can offer as myths just-so stories that for the moment seem plausible enough. What's more, now and again we may even catch a glimpse of how the Darwinian mechanism actually did effect some transformation (e.g., a minor adaptive change like finch-beak variation). But in general we must content ourselves with recording what nature has wrought without being able to trace nature's hand or reproduce her steps.

This disparity between the nature's power in implementing the Darwinian mechanism and our own inability to track nature's implementation of the Darwinian mechanism is illumined with a concept from the philosophy of mathematics—*surveyability*. According to Ludwig Wittgenstein, a mathematical proof or calculation is surveyable if one can command a clear view of it.[39] For instance, one can command a clear view of simple mathematical proofs, like the proof from Peano's axioms showing that $2 + 2 = 4$. On the other hand, one cannot command a clear view of the classification theorem for finite simple groups. Though the statement of the theorem requires but half a page, its proof required 10,000 pages and employed the joint effort of hundreds of mathematicians spanning several decades.[40] The mathematical community considers the theorem proven, but no one mathematician is able to survey the entire proof.

Even more acute is the problem of surveyability in the proof of the four-color problem (the problem of showing that no more than four colors are needed to color a map so that no two adjacent countries have identical colors). Not only can no single human being survey its proof, but even the task of constructing its proof—unlike the classification theorem for finite simple groups—did not fall entirely to humans. Instead, the proof required the assistance of a computer.[41] The computer performed a set of calculations so arduous that no team of humans, given only pencil and paper, could perform them, much less hope to perform them accurately. The four-color problem, though now considered decisively settled, is therefore not surveyable.

Wittgenstein's concept of surveyability applies no less to biology than to mathematics. The biological community does not command a clear view of how the Darwinian mechanism shaped the history of life. The history of life is fraught with countless contingencies that natural selection is said to have exploited but that shall never be repeated. These contingencies are behind the veil of history and cannot be reconstructed except in broad strokes. Yes, there may be good evidence that a meteor slamming into the earth was responsible for the extinction of the dinosaurs.[42] But the millions of contingencies that via selection are said to have turned a reptile into a mammal are forever lost.

The Darwinian mechanism tells us the logic by which the history of life was shaped, but is silent about the details. This is the genius of Darwinism but also the source of continued skepticism about the theory. If the Darwinian mechanism is indeed the engine that drives evolution, then Daniel Dennett is not far from the truth when he attributes to Darwin the greatest idea ever thought. But the logically prior question of whether the Darwinian mechanism can in principle adequately account for the complexity and diversity of living things needs first to be answered. Closer inspection of the actual mechanism reveals a significant limitation, namely, it is incapable of generating specified complexity. To see this conclusively requires fleshing out the theoretical apparatus for detecting design sketched in this introductory chapter. The next three chapters are devoted to this task.

## Notes

1. Moses Maimonides, *The Guide for the Perplexed*, trans. M. Friedländer (New York: Dover, 1956), 188.

2. Ibid.

3. Isaac Newton, *Mathematical Principles of Natural Philosophy*, trans. A. Motte, ed. F. Cajori (Berkeley, Calif.: University of California Press, 1978), 543–544.

4. Pierre Simon de Laplace, *Celestial Mechanics*, 4 vols., trans. N. Bowditch (New York: Chelsea, 1966).

5. See the introduction to Pierre Simon de Laplace, *A Philosophical Essay on Probabilities*, trans. F. W. Truscott and F. L. Emory (New York: Dover, 1996).

6. See John S. Bell, *Speakable and Unspeakable in Quantum Mechanics* (Cambridge: Cambridge University Press, 1987).

7. See Aristotle, *Metaphysics*, Book 5, ch. 2, in *The Basic Works of Aristotle*, ed. R. McKeon (New York: Random House, 1941), 752.

8. Francis Bacon, *The Advancement of Learning*, in *Great Books of the Western World*, vol. 30, ed. R. M. Hutchins (Chicago: Encyclopedia Britannica, 1952).

9. Jacques Monod, *Chance and Necessity* (New York: Vintage, 1972), 21.

10. Stanley Jaki, *Chesterton, A Seer of Science* (Urbana, Ill.: University of Illinois Press, 1986), 139–140, n. 2.

11. Eliot Marshall, "Medline Searches Turn Up Cases of Suspected Plagiarism," *Science* 279 (1998): 473–474.

12. Charles Darwin, *On the Origin of Species*, facsimile 1st ed. (1859; reprinted Cambridge, Mass.: Harvard University Press, 1964), 482.

13. The term "specified complexity" goes back at least to 1973, when Leslie Orgel used it in connection with origins-of-life research: "Living organisms are distinguished by their specified complexity. Crystals such as granite fail to qualify as living because they lack complexity; mixtures of random polymers fail to qualify because they lack specificity." See Orgel, *The Origins of Life* (New York: Wiley, 1973), 189. The challenge of specified complexity to nonteleological accounts of life's origin continues to loom large. Thus according to Paul Davies, "Living organisms are mysterious not for their complexity *per se*, but for their tightly specified complexity." See Paul Davies, *The Fifth Miracle* (New York: Simon & Schuster, 1999), 112.

14. Michael Polanyi, "Life Transcending Physics and Chemistry," *Chemical and Engineering News* (21 August 1967): 54–66; Michael Polanyi, "Life's Irreducible Structure," *Science* 113 (1968): 1308–1312; Timothy Lenoir, *The Strategy of Life: Teleology and Mechanics in Nineteenth Century German Biology* (Dordrecht: Reidel, 1982), 7–8. See also Hubert Yockey, *Information Theory and Molecular Biology* (Cambridge: Cambridge University Press, 1992), 335.

15. See Ian Hacking, *Logic of Statistical Inference* (Cambridge: Cambridge University Press, 1965), 82.

16. Simon Singh, *The Code Book: The Evolution of Secrecy from Mary Queen of Scots to Quantum Cryptography* (New York: Doubleday, 1999), 10–13.

17. See http://home.wxs.nl/~gkorthof/kortho44.htm (last accessed 7 June 2001).

18. Ian Stewart, *Life's Other Secret: The New Mathematics of the Living World* (New York: John Wiley, 1998), ch. 6.

19. http://home.wxs.nl/~gkorthof/kortho44.htm.

20. I am indebted to John Mark Reynolds for this example.

21. Del Ratzsch seems to have missed this point in arguing against the Explanatory Filter—see his book, *Nature, Design and Science: The Status of Design in Natural Science* (Albany, N.Y.: SUNY Press, 2001), 164.

22. For the critique by John Wilkins and Wesley Elsberry see their article "The Advantages of Theft over Toil: The Design Inference and Arguing from Ignorance," *Biology and Philosophy*, forthcoming.

23. See Taner Edis, "Darwin in Mind: 'Intelligent Design' Meets Artificial Intelligence," *Skeptical Inquirer* 25(2) (March/April 2001): 35–39.

24. The proof is quite simple: In 100 coin tosses, on average half will repeat the previous toss, implying about 50 two-repetitions. Of these 50 two-repetitions, on average half will repeat the previous toss, implying about 25 three-repetitions. Continuing in this vein, we find on average 12 four-repetitions, 6 five-repetitions, 3 six-repetitions, and 1 seven-repetition. See Ivars Peterson, *The Jungles of Randomness: A Mathematical Safari* (New York: Wiley, 1998), 5.

25. See David Halliday and Robert Resnick, *Fundamentals of Physics*, 3rd ed. extended (New York: Wiley, 1988), 544. Note that universal time bounds for electronic computers have clock speeds between ten and twenty magnitudes slower than the Planck time—see Ingo Wegener, *The Complexity of Boolean Functions* (Stuttgart: Wiley-Teubner, 1987), 2.

26. For the details justifying this universal probability bound, see William A. Dembski, *Design Inference: Eliminating Chance through Small Probabilities* (Cambridge: Cambridge University Press, 1998), sec. 6.5.

27. Emile Borel, *Probabilities and Life*, trans. M. Baudin (New York: Dover, 1962), 28. See also Eberhard Knobloch, "Emile Borel as a Probabilist," 215–233 in *The Probabilistic Revolution*, vol. 1, eds. L. Krüger, L. J. Daston, and M. Heidelberger (Cambridge, Mass.: MIT Press, 1987), 228.

28. Kenneth W. Dam and Herbert S. Lin, eds., *Cryptography's Role in Securing the Information Society* (Washington, D.C.: National Academy Press, 1996), 380, n. 17. See also Singh, *The Code Book*, which is filled with arguments that tacitly appeal to universal probability bounds. For instance, Singh quotes William Crowell, deputy director of the National Security Agency: "If all the personal computers in the world—approximately 260 million computers—were to be put to work on a single PGP encrypted message, it would take on average an estimated 12 million times the age of the universe to break a single message" (317).

29. For the statistics behind medical tests, see Charles H. Hennekens and Julie E. Buring, *Epidemiology in Medicine* (Boston: Little, Brown and Company), ch. 13.

30. Martin Gardner, "Arthur Koestler: Neoplatonism Rides Again," *World* (1 August 1972): 87–89.

31. Walter Martin, *The Kingdom of the Cults*, rev. ed. (Minneapolis: Bethany House, 1985), 127–130.

32. I am indebted to Del Ratzsch for this example. See his book *Nature, Design and Science: The Status of Design in Natural Science* (Albany, N.Y.: SUNY Press, 2001), 12–13, 66–69.

33. For an introduction to the Oklo reactors see Paul Kuroda, *The Origin of Chemical Elements and the Oklo Phenomenon* (New York: Springer-Verlag, 1982). Information on the Oklo reactors is readily available on the web. See for instance http://www.curtin.edu.au/curtin/centre/waisrc/OKLO/index.shtml (last accessed 7 June 2001).

34. Richard Dawkins, *Climbing Mount Improbable* (New York: Norton, 1996).

35. William Paley, *Natural Theology: Or Evidences of the Existence and Attributes of the Deity Collected from the Appearances of Nature* (1802; reprinted Boston: Gould and Lincoln, 1852).

36. David Hume, *Dialogues Concerning Natural Religion* (1779; reprinted Buffalo, N.Y.: Prometheus Books, 1989).

37. Daniel Dennett, *Darwin's Dangerous Idea* (New York: Simon & Schuster, 1995), 21.

38. Richard Dawkins, *The Blind Watchmaker* (New York: Norton, 1987), 85–86.

39. Ludwig Wittgenstein, *Remarks on the Foundations of Mathematics*, eds. G. H. von Wright, R. Rhees, and G. E. M. Anscombe, trans. G. E. M. Anscombe, rev. ed. (Cambridge, Mass.: MIT Press, 1983), 145.

40. See Daniel Gorenstein, *The Classification of Finite Simple Groups*, vol. 1 (New York: Plenum, 1983), as well as Daniel Gorenstein, "Classifying the Finite Simple Groups," *Bulletin of the American Mathematical Society* 14 (1986): 1–98.

41. K. Appel, W. Haken, and J. Koch, "Every Planar Map Is Four Colorable," *Illinois Journal of Mathematics* 21 (1977): 429–567.

42. L. Alvarez, W. Alvarez, F. Asaro, and H. Michel, "Extraterrestrial Cause for the Cretaceous-Tertiary Extinction," *Science* 208 (1980): 1095–1108.

# CHAPTER TWO

~

# Another Way to Detect Design?

## 2.1  Fisher's Approach to Eliminating Chance

In Ronald Fisher's approach to hypothesis testing that he developed in the 1920s, one is justified rejecting or eliminating a chance hypothesis if a sample falls within a prespecified *rejection region* (also known as a *critical region*).[1] For example, suppose one's chance hypothesis is that a coin is fair. To test whether the coin is biased in favor of heads, and thus not fair, one can set a rejection region of ten heads in a row and then flip the coin ten times. In Fisher's approach, if the coin lands ten heads in a row, then one is justified rejecting the chance hypothesis. As a criterion for detecting design, specified complexity extends Fisher's approach to hypothesis testing in two ways: First, it generalizes the types of rejection regions by which chance is eliminated (the generalized rejection regions being what I call *specifications*). Second, it eliminates all relevant chance hypotheses that could characterize an event rather than just a single chance hypothesis.

Fisher's approach to hypothesis testing is the one most widely used in the applied statistics literature and certainly the first one taught in introductory statistics courses. Nevertheless, in its original formulation Fisher's approach is problematic. The problem is this. For a rejection region to warrant rejecting a chance hypothesis, the rejection region must have sufficiently small probability. But how small is small enough? Given a chance hypothesis and a rejection region, how small does the probability of the rejection region have to be so that if a sample falls within it, then the chance hypothesis can legitimately be rejected? The problem here is to justify what is called a *significance level* such that whenever the sample falls within the rejection region and the probability of the rejection region given the chance hypothesis is less than the significance level, then the chance hypothesis can be legitimately rejected.

More formally, the problem is to justify a significance level $\alpha$ (always a positive real number less than one) such that whenever the sample (an event we will call E) falls within the rejection region (call it R) and the probability of the rejection region given the chance hypothesis (call it **H**) is less than $\alpha$ (i.e., $P(R|H) < \alpha$), then the chance hypothesis **H** can be legitimately rejected as the explanation of the sample. In the applied statistics literature it is common to see significance levels of .05 and .01. The problem to date has been that any such proposed significance levels have seemed arbitrary, lacking what Howson and Urbach call "a rational foundation."[2]

To see what is at stake in setting a rationally compelling significance level, let us contrast eliminating a chance hypothesis via a significance test with eliminating a hypothesis (chance or otherwise) via ordinary logic. Ordinary logic, which is nonstatistical, has a straightforward method for eliminating a hypothesis, namely this: Assume the hypothesis and see if it leads to a logical contradiction. If so, reject the hypothesis. This method for eliminating hypotheses is known as *reductio ad absurdum* and is a staple of logicians and mathematicians.[3] The success of *reductio ad absurdum* depends on being able to demonstrate contradictions, which are defined as statements of the form E & ~E (i.e., E conjoined with its negation). Now, from the vantage of probabilities, contradictions are impossible events of probability zero. Indeed, the axioms of probability theory guarantee that any statement of the form E & ~E has probability zero.[4]

Consequently, a probabilistic rendering of *reductio ad absurdum* looks as follows: If a rejection region R has probability zero with respect to a hypothesis **H**, and if an event E occurs and lands in R, then **H** is to be rejected. Note that this probabilistic rendering of *reductio ad absurdum* is not logically equivalent to ordinary *reductio ad absurdum*. Ordinary *reductio ad absurdum* trades in contradictions; probabilistic *reductio ad absurdum* trades in events or statements of probability zero. To be sure, contradictions have probability zero. Events of probability zero, however, need not be contradictions. If, for instance, one randomly samples a real number from the unit interval, then any number picked will have probability zero. The very act of randomly sampling the unit interval ensures that some number or other will be picked, and the picking of such a number need involve no contradiction. Even so, if we specify a rejection region of probability zero in this interval, sample randomly from it, and then find that the number picked falls in the rejection region, we will reject the hypothesis that the sample was chosen randomly (i.e., that it was chosen with respect to a uniform probability on the unit interval).

This probabilistic rendering of *reductio ad absurdum* is uncontested throughout the statistical literature. Indeed, hypotheses that confer zero probability are invariably rejected or passed over in favor of other hypotheses that confer nonzero probability. This is true not only of Fisher's approach to hypothesis testing but also of the others (e.g., Neyman-Pearson, likelihood, and Bayesian approaches). Hypotheses are never confirmed by conferring zero probability. Indeed, hypotheses can only be confirmed by conferring nonzero probability. But that raises the question whether hypotheses can also be disconfirmed by conferring nonzero probability. Fisher says yes, allowing a chance hypothesis to be disconfirmed simply in virtue of the small nonzero probability it confers on a given event. The other approaches also say yes but qualify their yes. The other approaches are essentially comparative: they require more than one chance hypothesis, pit competing chance hypotheses against each other, and then see which confers the greater probability on some region of probability space. Those hypotheses conferring greater probability are confirmed; those conferring less are disconfirmed.

I will discuss comparative approaches to hypothesis testing later in this chapter. For now, however, I want to summarize my work in *The Design Inference* and show how it places Fisher's approach to hypothesis testing on a rational foundation. In *The Design Inference* I argue that significance levels cannot be set in isolation but must always be set in relation to the probabilistic resources relevant to an event's occurrence.[5] Critics of Fisher's approach to hypothesis testing like Howson and Urbach are therefore correct in claiming that significance levels of .05, .01, and the like that regularly appear in the applied statistics literature are arbitrary. A significance level is supposed to provide an upper bound on the probability of a rejection region and thereby ensure that the rejection region is sufficiently small to justify the elimination of a chance hypothesis (essentially, the idea is to make a target so small that an archer is highly unlikely to hit it by chance). Rejection regions eliminate chance hypotheses when events putatively characterized by those hypotheses fall within the rejection regions. The problem with significance levels like .05 and .01 is that typically they are instituted without any reference to the probabilistic resources relevant to controlling for false positives. (A false positive here is the error of eliminating a chance hypothesis as the explanation of an event when the chance hypothesis actually is operative. Statisticians refer to such false positives as "type I errors.")

Suppose, for instance, an academic journal in the social sciences institutes a significance level α of .01 to control for false positives. In that case, articles that record what would be an interesting theoretical result so long as the result is not due to chance are accepted for publication only if the result falls inside a rejection region whose probability is .01 or less. Any number of journals in experimental psychology, for instance, require an α-level of .01 for submission of manuscripts. But what does such an α-level accomplish? In general, for every hundred experiments conducted by researchers in the journal's field of specialization and in which the chance hypothesis actually was operating, on average one experiment will satisfy an α-level of .01. Thus for a journal requiring an α-level of .01, on average one in a hundred of such experiments will slip through the cracks and form the basis of an article acceptable to that journal.

Does an α-level of .01 therefore provide stringent enough controls on false positives for such a journal? The answer depends on the number of experiments conducted by researchers in the field who submit their findings to the journal. As this number increases, the number of experiments in which chance actually was operating, but for which chance will be (falsely) eliminated, will increase, thus increasing the number of false positives that could conceivably slip into the journal. A journal requiring an α-level of .01 will on average allow one in a hundred of such experiments into its pages. More generally, a journal requiring a significance level α will on average allow the proportion α of such experiments into its pages. The more experimental research there is in the journal's field of specialization, the more likely the journal is to include false positives. The relevant probabilistic resource here is therefore the replicational resource consisting of the number N of separate experiments performed by researchers in the journal's field of specialization (see section 1.5).[6]

Although the amount of research activity in the journal's field of specialization determines the number of false positives that could conceivably slip into the pages of the journal, in practice the precise α-level a journal sets to control for this error will depend on the needs and interests of the editors of the journal. If the field to which the journal is directed is good about correcting past mistakes, there may be no problem setting the α-level high (e.g., α = .05). On the other hand, if the field to which the journal is directed is bad about correcting past mistakes, it may be a good idea to set the α-level low (e.g., α = .0001) so that what errors do make it into the journal will be few. Nevertheless, the choice of any such α-level need not be arbitrary but obtains definite meaning by reference to the probabilistic resources relevant to precluding false positives (i.e., relevant to preventing chance from being mistakenly eliminated).[7] Suffice it to say, there exists a non-question-begging account of statistical significance testing.

## 2.2  Generalizing Fisher's Approach

I want next to show how specified complexity, as a criterion for detecting design, generalizes Fisher's approach to hypothesis testing. Fisher's approach has the following logical structure: We are given a reference class of possibilities $\Omega$ and a chance hypothesis $H$ that induces a probability measure $P$ defined on $\Omega$ (i.e., $P(\cdot|H)$ is defined for subsets of $\Omega$). What's more, we are given a significance level $\alpha$ (always a positive real number less than one) and a rejection region R whose probability is less than $\alpha$ (i.e., $P(R|H) < \alpha$). Having specified R in advance of an experiment, we now perform the experiment and observe an event E, which we call the sample. If the sample E falls within the rejection region R (logically this means that E entails R), and provided that the significance level $\alpha$ factors in the probabilistic resources that in the context of inquiry are relevant to precluding false positives, then the chance hypothesis $H$ can legitimately be rejected as inadequate to explain the sample E.

In generalizing Fisher's approach to hypothesis testing, I now need to do two things: (1) generalize the types of rejection regions capable of eliminating chance and (2) show what it means to eliminate all relevant chance hypotheses that could characterize an event. Let us begin with generalizing the rejection regions capable of eliminating chance. As I have laid out the logic of Fisher's approach, for the chance occurrence of a sample E to be legitimately rejected because E falls in a rejection region R, R had to be identified prior to the occurrence of E. This is to avoid the familiar problem known among statisticians as "data snooping" or "cherry picking," in which a pattern (in this case R) is imposed on an event (in this case the sample E) after the fact. Requiring the rejection region to be set prior to the occurrence of E safeguards against attributing patterns to E that are factitious and that do not properly preclude the occurrence of E by chance. This safeguard, however, is unduly restrictive. Indeed, even within Fisher's approach to eliminating chance hypotheses this safeguard is routinely relaxed.

To see this, consider that the reference class of possibilities $\Omega$, against which patterns and events are defined, usually comes with some additional geometric structure together with a privileged (probability) measure that preserves that geometric structure. In case $\Omega$ is finite or countably infinite, this measure is typically just the counting measure (i.e., it counts the number of elements in a given subset of $\Omega$; note that normalizing this measure on finite sets yields a uniform probability). In case $\Omega$ is uncountable but suitably bounded (or, as topologists would say, compact), this privileged measure is a uniform probability. In case $\Omega$ is uncountable and unbounded, this privileged measure becomes a uniform probability when restricted to and normalized with respect to the suitably bounded subsets of $\Omega$. Let us refer to this privileged measure as $U$.[8]

Now the interesting thing about $U$ in reference to Fisher's approach to eliminating chance hypotheses is that it allows probabilities of the form $P(\cdot|H)$ (i.e., those defined explicitly with respect to a chance hypothesis $H$) to be represented as probabilities of the form f·dU (i.e., the product of a nonnegative function f, known as a *probability density function*, and the measure $U$). The "d" in front of $U$ here signifies that to evaluate this probability requires integrating f with respect to $U$.[9] The technical details here are not important. What is important is that within Fisher's approach to hypothesis testing the probability density function f is used to identify rejection regions that in turn are used to eliminate chance. These rejection regions take the form of *extremal sets*, which we can represent as follows ($\gamma$ and $\delta$ are real numbers):

$$T^\gamma = \{\omega \in \Omega \mid f(\omega) \geq \gamma\},$$
$$T_\delta = \{\omega \in \Omega \mid f(\omega) \leq \delta\}.$$

$T^\gamma$ consists of all possibilities in the reference class $\Omega$ for which the density function f is at least $\gamma$. Likewise $T_\delta$ consists of all possibilities in the reference class $\Omega$ for which the density function f is no more than $\delta$. Although this way of characterizing rejection regions may seem terribly abstract, what is actually going on here is quite simple. Any probability density function f is a nonnegative real-valued function defined on $\Omega$. As such, f induces what might be called a *probability landscape* (think of $\Omega$ as a giant plane and f as describing the elevation of a landscape over $\Omega$). Where f is high corresponds to where the probability measure f·dU concentrates a lot of probability. Where f is low corresponds to where the probability measure f·dU concentrates little probability. Since f cannot fall below zero, we can think of the landscape as never dipping below sea-level. $T_\delta$ then corresponds to those places in $\Omega$ where the probability landscape induced by f is no higher than $\delta$ whereas $T^\gamma$ corresponds to those places in $\Omega$ where probability landscape is at least as high as $\gamma$. For a bell-shaped curve, $T^\gamma$ corresponds to the region under the curve where it approaches a maximum, and $T_\delta$ corresponds to the tails of the distribution.

Since this way of characterizing rejection regions may still seem overly abstract, let us apply it to some concrete examples. Suppose, for instance, that $\Omega$ is the real line and that the hypothesis **H** describes a normal distribution with, let us say, mean zero and variance one. In that case, for the probability measure f·d**U** that corresponds to **P**(·|**H**), the density function f has the form

$$f(x) \; = \; \frac{1}{\sqrt{2\pi}} \, e^{-\frac{x^2}{2}}.$$

This function is everywhere positive and attains its maximum at the mean (i.e., at x = 0, where f takes the value $1/\sqrt{2\pi}$, or approximately .399). At two standard deviations from the mean (i.e., for the absolute value of x at least 2), this function attains but does not exceed .054. Thus for $\delta$ = .054, $T_\delta$ corresponds to two tails of the normal probability distribution (see figure 2.1). Moreover, since those tails are at least two standard deviations from the mean, it follows that **P**($T_\delta$|**H**) < .05. Thus for a significance level $\alpha$ = .05 and a sample E that falls within $T_\delta$, Fisher's approach rejects attributing E to the chance hypothesis **H**.

Now the important thing to note in this example is not the precise level of significance that was set (here $\alpha$ = .05) or the precise form of the probability distribution under consideration (here a normal distribution with mean zero and variance one). These were simply given for concreteness. Rather, the important thing here is that the temporal ordering of rejection region and sample, where a rejection region (here $T_\delta$) is first specified and a

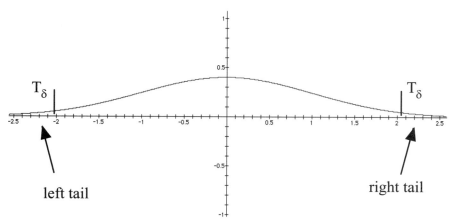

**Figure 2.1.   The Tails of a Bell Curve.**

sample (here E) is then taken, simply was not an issue. For a statistician to eliminate the chance hypothesis **H** as an explanation of E, it is not necessary for the statistician first to identify $T_8$ explicitly, then perform an experiment to elicit E, and finally determine whether E falls within $T_8$. **H**, by inducing a probability density function f, automatically also induces the extremal set $T_8$, which, given the significance level $\alpha = .05$, is adequate for eliminating **H** once a sample E falls within $T_8$. If you will, the rejection region $T_8$ comes automatically with the chance hypothesis **H**, making it unnecessary to identify it explicitly prior to the sample E. Eliminating chance when samples fall in such extremal sets is standard statistical practice. Thus, in the case at hand, if a sample falls sufficiently far out in the tails of a normal distribution, the distribution is rejected as inadequate to account for the sample.

In this last example we considered extremal sets of the form $T_8$ at which the probability density function concentrates minimal probability. Nonetheless, extremal sets of the form $T^\gamma$ at which the probability density function concentrates maximal probability can also serve as rejection regions within Fisher's approach, and this holds even if such regions are not explicitly identified prior to an experiment. Consider, for instance, the following example. Imagine that a die is to be thrown 6,000,000 times. Within Fisher's approach the "null" hypothesis (i.e., the chance hypothesis most naturally associated with this probabilistic set-up) is a chance hypothesis **H** according to which the die is fair (i.e., each face has probability 1/6) and the die rolls are stochastically independent (i.e., one roll does not affect the others). Consider now the reference class of possibilities $\Omega$ consisting of all 6-tuples of nonnegative integers that sum to 6,000,000. The 6-tuple (6,000,000, 0, 0, 0, 0, 0) would, for instance, correspond to tossing the die 6,000,000 times with each time the die landing "1." The 6-tuple (1,000,000, 1,000,000, 1,000,000, 1,000,000, 1,000,000, 1,000,000), on the other hand, would correspond to tossing the die 6,000,000 times with each face landing exactly as often as the others. Both these 6-tuples represent possible outcomes for tossing the die 6,000,000 times, and both belong to $\Omega$.

Suppose now that the die is thrown 6,000,000 times and that each face appears *exactly* 1,000,000 times (as in the previous 6-tuple). Even if the die is fair, there is something anomalous about getting *exactly* one million appearances of each face of the die. What's more, Fisher's approach is able to make sense of this anomaly. The probability distribution that **H** induces on $\Omega$ is known as the multinomial distribution,[10] and for each 6-tuple $(x_1, x_2, x_3, x_4, x_5, x_6)$ in $\Omega$, it assigns a probability of

$$f(x_1, x_2, x_3, x_4, x_5, x_6) = \frac{6{,}000{,}000!}{x_1! \cdot x_2! \cdot x_3! \cdot x_4! \cdot x_5! \cdot x_6!}(1/6)^{6{,}000{,}000}.$$

Although the combinatorics involved with the multinomial distribution are rather complicated (hence the common practice of approximating it with continuous probability distributions like the chi-square distribution), the reference class of possibilities $\Omega$, though large, is finite, and the cardinality of $\Omega$ (i.e., the number of elements in $\Omega$), denoted by $|\Omega|$, is well-defined (its order of magnitude is around $10^{33}$).

The function f defines not only a probability for individual elements $(x_1,x_2,x_3,x_4,x_5,x_6)$ of $\Omega$ but also a probability density with respect to the counting measure $\mathbf{U}$ (i.e., $\mathbf{P}(\cdot|\mathbf{H}) = \mathrm{f}\cdot\mathrm{d}\mathbf{U}$), and attains its maximum at the 6-tuple for which each face of the die appears exactly 1,000,000 times. Let us now take $\gamma$ to be the value of $\mathrm{f}(x_1,x_2,x_3,x_4,x_5,x_6)$ at $x_1 = x_2 = x_3 = x_4 = x_5 = x_6 = 1,000,000$ ($\gamma$ is approximately $2.475 \times 10^{-17}$). Then $\mathrm{T}^\gamma$ defines a rejection region corresponding to those elements of $\Omega$ where f attains at least the value $\gamma$. Since $\gamma$ is the maximum value that f attains and since there is only one 6-tuple where this value is attained (i.e., at $x_1 = x_2 = x_3 = x_4 = x_5 = x_6 = 1,000,000$), it follows that $\mathrm{T}^\gamma$ contains only one member of $\Omega$ and that the probability of $\mathrm{T}^\gamma$ is $\mathbf{P}(\mathrm{T}^\gamma|\mathbf{H}) = \mathrm{f}(1,000,000, 1,000,000, 1,000,000, 1,000,000, 1,000,000, 1,000,000)$, which is approximately $2.475 \times 10^{-17}$. This rejection region is therefore highly improbable and will in most practical applications be far more extreme than any significance level we happen to set. Thus, if we observed exactly 1,000,000 appearances of each face of the die, we would according to Fisher's approach conclude that the die was not thrown by chance. Unlike the previous example, in which by falling in the tails of the normal distribution a sample deviated from expectation by too much, the problem here is that the sample matches expectation too closely. In general, samples that fall in rejection regions of the form $\mathrm{T}_\delta$ deviate from expectation too much whereas samples that fall in rejection regions of the form $\mathrm{T}^\gamma$ match expectation too closely.

The problem of matching expectation too closely is as easily handled within Fisher's approach as the problem of deviating from expectation too much. Fisher himself recognized as much and saw rejection regions of the form $\mathrm{T}^\gamma$ as the key to uncovering data falsification. For instance, Fisher uncovered a classic case of data falsification in analyzing Gregor Mendel's data on peas. Fisher inferred that "Mendel's data were massaged," as one statistics text puts it, because Mendel's data matched his theory too closely.[11] The match that elicited this charge of data falsification was a specified event whose probability was no more extreme than one in a hundred thousand (a probability that is huge compared to the 1 in $10^{17}$ probability computed in the preceding example). Fisher concluded by charging Mendel's gardening assistant with deception.

Given the chance hypothesis **H**, the extremal sets $T^\gamma$ and $T_\delta$ were defined with respect to the probability density function f where $P(\cdot|H) = f \cdot dU$. Moreover, given a significance level $\alpha$ and a sample E that falls within either of these extremal sets, Fisher's approach eliminates **H** provided that the extremal sets have probability less than $\alpha$. An obvious question now arises: In using either of the extremal sets $T^\gamma$ or $T_\delta$ to eliminate the chance hypothesis **H**, what is so special about basing these extremal sets on the probability density function f associated with the probability measure $P(\cdot|H)$ ( $= f \cdot dU$)? Why not let f be an arbitrary real-valued function?

Clearly, some restrictions must apply to f. If f can be arbitrary, then for any subset A of the reference class $\Omega$, we can define f to be the indicator function $1_A$ (by definition this function equals 1 whenever an element of $\Omega$ is in A and 0 otherwise).[12] For such an f, if $\gamma = 1$, then $T^\gamma = \{\omega \in \Omega \mid f(\omega) \geq \gamma\} = A$, and thus any subset of $\Omega$ becomes an extremal set for some function f. But this would allow Fisher's approach to eliminate any chance hypothesis whatsoever: For any sample E (of sufficiently small probability), find a subset A of $\Omega$ that includes E and such that $P(A|H)$ is less than whatever significance level $\alpha$ was decided. Then for $f = 1_A$ and $\gamma = 1$, $T^\gamma = A$. Consequently, if Fisher's approach extends to arbitrary f, then **H** would have to be rejected. But by thus eliminating all chance hypotheses, Fisher's approach becomes useless.

What restrictions on f will therefore safeguard it from frivolously disqualifying chance hypotheses? Recall that initially it was enough for rejection regions to be given in advance of an experiment. Then we noted that the rejection regions corresponding to extremal sets associated with the probability density function also worked (regardless of when they were identified—in particular, they did not have to be identified in advance of an experiment). Finally, we saw that placing no restrictions on the function f eliminated chance willy-nilly and thus could not legitimately be used to eliminate chance. But why was the complete lack of restrictions inadequate for eliminating chance? The problem—and this is why statisticians are so obsessive about making sure hypotheses are formulated in advance of experiments—is that with no restriction on the function f, f can be *tailored* to the sample E and thus perforce eliminate **H** even if **H** obtains. What needs to be precluded, then, is the tailoring of f to E. Alternatively, f needs to be independent (in some appropriate sense) of the sample E.

Let us now return to the question we left hanging: In eliminating **H**, what is so special about basing the extremal sets $T^\gamma$ and $T_\delta$ on the probability density function f associated with the chance hypothesis **H** (i.e., **H** induces the probability measure $P(\cdot|H)$ which can be represented as $f \cdot dU$)? Answer:

There is nothing special about f being the probability density function associated with **H**; instead, what is important is that f be capable of being defined independently of E, the event or sample that is observed. And indeed, Fisher's approach to eliminating chance hypotheses has already been extended in this way, though the extension thus far has mainly been tacit rather than explicit.

## 2.3  Case Study: Nicholas Caputo

To see that Fisher's approach to chance elimination does not depend solely on the extremal sets associated with probability density functions, I want next to consider two examples: first, the case of Nicholas Caputo; second, the compressibility of bit strings. Let us then begin with Caputo. In the summer of 1985 the *New York Times* reported a trial that has since made it into the statistics textbooks:[13]

> TRENTON, July 22—The New Jersey Supreme Court today caught up with the "man with the golden arm," Nicholas Caputo, the Essex County Clerk and a Democrat who has conducted drawings for decades that have given Democrats the top ballot line in the county 40 out of 41 times. . . . The court noted that the chances of picking the same name 40 out of 41 times were less than 1 in 50 billion. It said that "confronted with these odds, few persons of reason will accept the explanation of blind chance." [The court] believed that election officials have a duty to strengthen public confidence in the election process after such a string of "coincidences," [and thus suggested] changes in the way Mr. Caputo conducts the drawings to stem "further loss of public confidence in the integrity of the electoral process."

Nicholas Caputo was brought before the New Jersey Supreme Court because the Republican party had filed suit against him, claiming Caputo had consistently rigged the ballot line in the New Jersey county where he was county clerk. It is a known fact that first position on a ballot increases one's chances of winning an election. Since in every instance but one Caputo positioned the Democrats first on the ballot line, the Republicans argued that in selecting the order of ballots Caputo had favored his own Democratic party. In short, the Republicans claimed that Caputo had cheated.

The question therefore before the New Jersey Supreme Court was, Did Caputo actually rig the order, or was it without malice and forethought on his part that the Democrats happened 40 out of 41 times to appear first on the ballot? Since Caputo denied any wrongdoing, and since he conducted the drawing of ballots so that witnesses were unable to observe how he actually did draw the ballots, determining whether Caputo did in fact rig the order of ballots becomes a matter of evaluating the circumstantial evidence connected with this case. How, then, is this evidence to be evaluated?

In determining how to explain the remarkable coincidence of Nicholas Caputo selecting the Democrats 40 out of 41 times to head the ballot line, the court quickly dispensed with the possibility that Caputo chose poorly his procedure for selecting ballot lines so that instead of genuinely randomizing the ballot order, it just kept putting the Democrats on top. Caputo claimed to have placed capsules designating the various political parties running in New Jersey into a container and then to have swished them around. Caputo was therefore implicitly claiming to use an urn model. Since urn models are among the most reliable randomization techniques available, there was no reason for the court to think that Caputo's randomization procedure was at fault.

The key question therefore before the court was whether Caputo actually put this procedure into practice when he made the ballot selections, or whether he purposely circumvented this procedure in order that the Democrats would consistently come out on top. And since Caputo's actual drawing of the capsules was obscured to witnesses, it was this question that the court had to answer. From a statistical vantage, the court's task was therefore to determine whether chance could reasonably be invoked to explain Caputo's ballot line selections. The court reasoned as follows: Having noted that the chances of picking the same political party at least 40 out of 41 times were less than 1 in 50 billion, the court concluded that "confronted with these odds, few persons of reason will accept the explanation of blind chance."

This certainly seems right. Nevertheless, a bit more needs to be said about the court's reasoning. After all, highly improbable events happen by chance all the time. What is it about giving the Democrats the top ballot line 40 out of 41 times that makes chance an unreasonable explanation? To answer this question, I want next to offer a rational reconstruction of the court's decision that makes explicit its use of Fisher's approach to chance elimination.

Let $\Omega$ be all sequences of Ds and Rs that are 41 characters in length ("D" for Democrat first on the ballot line and "R" for Republican first). A generic element of $\Omega$ therefore looks as follows:

DRRDRDRRDDRDRDDRDRRDRRDRRRDRRRDRDDDRDRDD.

Thus for this possible sequence of ballot line selections, Caputo would have given the top ballot line to the Democrats the first time he assigned ballot lines, then to the Republicans the next two times, then to the Democrats, and so on.

The cardinality of $\Omega$ is $2^{41}$, or approximately $2 \times 10^{12}$ (i.e., $\Omega$ has approximately 2 trillion such sequences). The chance hypothesis **H** before the court was that Democrats and Republicans should both have had probability 1/2 of obtaining the top ballot line on any given assignment of ballot lines and that these assignments were probabilistically independent (i.e., one assignment should not affect the others). Consequently, the probability of any sequence in $\Omega$ should have probability 1 in $2^{41}$.

For concreteness, let us now suppose Caputo's actual selection of ballot lines was the following event E:

E           DDDDDDDDDDDDDDDDDDDDDDRDDDDDDDDDDDDDDDDDDD .

Thus we suppose that for the initial 22 times, Caputo chose the Democrats to head the ballot line; then at the 23rd time, he chose the Republicans; after which, for the remaining times, he chose the Democrats. (The *New York Times* article cited above did not explicitly mention the one place where Caputo gave the top ballot line to the Republicans. For the sake of this discussion I will therefore assume E is what actually happened.)

The probability of E given **H** (i.e., $P(E|H)$) is $2^{-41}$ or approximately 1 in 2 trillion. Indeed, any sequence in $\Omega$ whatsoever has this same small probability, and the probability density function corresponding to the probability measure $P(\cdot|H)$ assumes this same value for every sequence in $\Omega$ (i.e., the probability density function is constant and takes only the value $2^{-41}$).[14] Consequently, there is no way to get any interesting extremal sets out of the probability density function (the extremal sets of a constant function comprise the universe of possibilities $\Omega$ and the empty set—nothing else).

How then did the New Jersey Supreme Court arrive at a rejection region for E? The answer occurs in the following statement from the *New York Times* article: "The court noted that the chances of picking the same name 40 out of 41 times were less than 1 in 50 billion." Leaving aside for the moment how the "1 in 50 billion" probability was calculated, it is clear that the court was counting the number of times the Democrats appeared first on the ballot line. In other words, they were considering a real-valued function f that for every sequence in $\Omega$ calculates the number of Ds that occur. For the sequence E (which we are assuming is Caputo's actual sequence of ballot line selections), f(E) = 40. The function f is integer-valued and takes values ranging from 0 to 41.

Now it is clear that f can be identified independently of E. Indeed, there is nothing to connect f to E. Given a sequence of diverse elements, humans have for millennia been categorizing such elements and then counting the number of elements in each category. That is all f is doing. One did not need to witness E to identify f. The function f is not tailored to the event E. It is this function f, then, that the court tacitly employed to formulate its rejection region and eliminate the chance hypothesis **H**.

A rational reconstruction of the court's reasoning therefore looks as follows. Given $\Omega$, the chance hypothesis **H**, and the event E (which we are assuming corresponds to Caputo's actual ballot line selections), the court defined a function f that could be identified independently of E. This function counts the number of Ds in a generic element of $\Omega$. Given this function, the court next set a lower bound of $\gamma = 40$ and defined the extremal set $T^\gamma$ based on the function f:

$$T^\gamma = \{\omega \in \Omega \mid f(\omega) \geqslant \gamma\}.$$

$T^\gamma$ is the rejection region implicitly used by the New Jersey Supreme Court to defeat chance as the explanation of Caputo's ballot line selections. $T^\gamma$ corresponds to an event consisting of 42 possible outcomes: the Democrats attaining priority all 41 times (1 possible outcome) and the Democrats attaining priority exactly 40 times (41 possible outcomes). It follows that $P(T^\gamma|H) = 42 \times 2^{-41}$, which is the probability of "1 in 50 billion" computed by the court. Included among the possible outcomes in $T^\gamma$ is of course the event E, which for the sake of this discussion we are assuming is the event that actually occurred. Provided the significance level decided on by the court was no less than 1 in 50 billion, E's occurrence and inclusion within $T^\gamma$ is, on Fisher's approach, enough to warrant eliminating the chance hypothesis **H**.

## 2.4  Case Study: The Compressibility of Bit Strings

We turn next to the work of Gregory Chaitin, Andrei Kolmogorov, and Ray Solomonoff for another example of extremal sets that, though not induced by probability density functions, nonetheless successfully eliminate chance within Fisher's approach to hypothesis testing. In the 1960s, Chaitin, Kolmogorov, and Solomonoff investigated what makes a sequence of coin flips random.[15] Also known as algorithmic information theory (see section 3.2), the Chaitin-Kolmogorov-Solomonoff theory began by noting that conventional probability theory is incapable of distinguishing bit strings of identical length (if we think of bit strings as sequences of coin tosses, then any sequence of n flips has probability 1 in $2^n$).

Consider a concrete case. If we flip a fair coin and note the occurrences of heads and tails in order, denoting heads by 1 and tails by 0, then a sequence of 100 coin flips looks as follows:

(R)  1100001101011000110111111101000110001101100110111  
0001100100001011110111011001111101001010010101011110.

This is in fact a sequence I obtained by flipping a coin 100 times. Now the problem algorithmic information theory seeks to resolve is this: Given probability theory and its usual way of calculating probabilities for coin tosses, how is it possible to distinguish these sequences in terms of their degree of randomness? Probability theory alone is not enough. For instance, instead of flipping (R) I might just as well have flipped the following sequence:

(N)  11111111111111111111111111111111111111111111111111  
11111111111111111111111111111111111111111111111111.

Sequences (R) and (N) have been labeled suggestively, R for "random," N for "nonrandom." Chaitin, Kolmogorov, and Solomonoff wanted to say that (R) was "more random" than (N). But given the usual way of computing probabilities, all one could say was that each of these sequences had the same small probability of occurring, namely 1 in $2^{100}$, or approximately 1 in $10^{30}$. Indeed, every sequence of 100 coin tosses has exactly this same small probability of occurring.

To get around this difficulty Chaitin, Kolmogorov, and Solomonoff supplemented conventional probability theory with some ideas from recursion theory, a subfield of mathematical logic that provides the theoretical underpinnings for computation and generally is considered quite far removed from probability theory.[16] What they said was that a string of 0s and 1s becomes increasingly random as the shortest computer program that generates the string increases in length. For the moment, we can think of a computer program as a shorthand description of a sequence of coin tosses. Thus the sequence (N) is not very random because it has a very short description, namely,

repeat '1' a hundred times.

Note that we are interested in the shortest descriptions since any sequence can always be described in terms of itself. Thus (N) has the longer description

copy  '11111111111111111111111111111111111111111111111111  
11111111111111111111111111111111111111111111111111'.

But this description holds no interest since there is one so much shorter.
 The sequence

(H) 11111111111111111111111111111111111111111111111111
00000000000000000000000000000000000000000000000000

is slightly more random than (N) since it requires a longer description, for example,

repeat '1' fifty times, then repeat '0' fifty times.

So too the sequence

(A) 10101010101010101010101010101010101010101010101010
10101010101010101010101010101010101010101010101010

has a short description,

repeat '10' fifty times.

The sequence (R), on the other hand, has no short and neat description (at least none that has yet been discovered). For this reason, algorithmic information theory assigns it a higher degree of randomness than the sequences (N), (H), and (A).

Since one can always describe a sequence in terms of itself, (R) has the description

copy '11000011010110001101111111010001100011011001110111
00011001000010111101110110011111010010100101011110'.

Because (R) was constructed by flipping a coin, it is very likely that this is the shortest description of (R). It is a combinatorial fact that the vast majority of sequences of 0s and 1s have as their shortest description just the sequence itself. In other words, most sequences are random in the sense of being algorithmically incompressible. It follows that the collection of nonrandom sequences has small probability among the totality of sequences so that observing a nonrandom sequence is reason to look for explanations other than chance.[17]

 Let us now reconceptualize this algorithmic approach to randomness within Fisher's approach to chance elimination. For definiteness, let us assume we have a computer with separate input and output memory-registers each consisting of N bits of information (N being large, let us say at least a billion bits). Each bit in the output memory is initially set to zero. Each initial sequence of bits in the input memory is broken into bytes, interpreted as ASCII characters, and treated as a Fortran program that records its results in the output memory. Next we define the length of a bit sequence u in input memory as N minus the number of uninterrupted 0s at the end of u. Thus, if u consists entirely of 0s, it has length 0. On the other hand, if u has a 1 in its very last memory location, u has length N.

Given this computational set-up, there exists a function $\varphi$ that for each input sequence u treats it as a Fortran program, executes the program, and then, if the program halts (i.e., does not loop endlessly or freeze), delivers an output sequence v in the output memory-register. $\varphi$ is therefore a partial function (i.e., it is not defined for all sequences of N bits but only for those that can be interpreted as well-defined Fortran programs and that halt when executed). Given $\varphi$, we now define the following function f on the output memory-register: f(v) for v an output sequence is defined as the length of the shortest program u (i.e., input sequence) such that $\varphi(u) = v$ (recall that the length of u is N minus the number of uninterrupted 0s at the end of u); if no such u exists (i.e., there is no u that $\varphi$ maps onto v), then we define f(v) = N. The function f is integer-valued and ranges between 0 and N. What's more, given a real number $\delta$, it induces the following extremal set (the reference class $\Omega$ here comprises all possible sequences of N bits in the output memory-register):

$$T_\delta = \{v \in \Omega \mid f(v) \leq \delta\}.$$

As a matter of simple combinatorics it now follows that for $\delta$ an integer between 0 and N the cardinality of $T_\delta$ (i.e., the number of elements in $T_\delta$) is no greater than $2^{\delta+1}$. If we denote the cardinality of $T_\delta$ by $|T_\delta|$, this means that $|T_\delta| \leq 2^{\delta+1}$. The argument demonstrating this claim is straightforward. The function $\varphi$ that maps input sequences to output sequences associates at most one output sequence to any given input sequence (which is not to say that the same output sequence may not be mapped onto by many input sequences). Since for any integer $\delta$ between 0 and N, there are at most 1 input sequence of length 0, 2 input sequences of length 1, 4 input sequences of length 2, . . . and $2^\delta$ input sequences of length $\delta$, it follows that $|T_\delta| \leq 1 + 2 + 4 + \ldots + 2^\delta = 2^{\delta+1} - 1 < 2^{\delta+1}$.

Suppose now that **H** is a chance hypothesis characterizing the tossing of a fair coin. Any output sequence v in the reference class $\Omega$ will therefore have probability $2^{-N}$. Moreover, since the extremal set $T_\delta$ contains at most $2^{\delta+1}$ elements of $\Omega$, it follows that the probability of the extremal set $T_\delta$ conditional on **H** will be bounded as follows:

$$P(T_\delta|\mathbf{H}) \leq 2^{\delta+1}/2^N = 2^{\delta+1-N}.$$

For N large and $\delta$ small, this probability will be minuscule, and certainly smaller than any significance level we might happen to set. Consequently, for the sequences with short programs (i.e., those whose programs have length no greater than δ), Fisher's approach will warrant eliminating the chance hypothesis **H**. And indeed, this is exactly the conclusion reached by Chaitin, Kolmogorov, and Solomonoff. Kolmogorov even used the language of statistical mechanics to describe this result, calling the random sequences high entropy sequences and the nonrandom sequence low entropy sequences.[18] To sum up, the collection of algorithmically compressible (and therefore nonrandom) sequences has small probability among the totality of sequences, so that observing such a sequence is reason to look for explanations other than chance.

## 2.5 Detachability

Given these two case studies, let us now pick up our story about how to generalize Fisher's approach to hypothesis testing. If we allow as rejection regions the extremal sets associated with an independently given function f, then Fisher's approach to hypothesis testing assumes the following logical form: Suppose we are given a reference class of possibilities $\Omega$ and a chance hypothesis **H** that induces a probability measure **P** defined on $\Omega$ (i.e., **P**($\cdot$|**H**) is defined for subsets of $\Omega$). Suppose further we are given a significance level $\alpha$ (always a positive real number less than one). We now perform an experiment and observe an event E, which we call the sample. What's more, independently of E we are able to identify a real-valued function f on $\Omega$. Ordinarily, probabilists refer to f simply as a *random variable*.[19] When f is used to induce rejection regions via its extremal sets, statisticians also refer to f as a *test statistic*.[20] Yet to make explicit f's role in inducing rejection regions, I will refer to f as a *rejection function*. Thus with the rejection function f in hand, we next determine a rejection region R such that: (1) R is an extremal set of the form $T^{\gamma} = \{\omega \in \Omega \mid f(\omega) \geq \gamma\}$ or $T_{\delta} = \{\omega \in \Omega \mid f(\omega) \leq \delta\}$ ($\gamma$ and $\delta$ are real numbers); (2) **P**(R|**H**) < $\alpha$; and (3) E falls within R. Finally, provided that the significance level $\alpha$ factors in the probabilistic resources that in the context of inquiry are relevant to precluding false positives, we are then justified eliminating the chance hypothesis **H** as inadequate to explain the sample E.

The only point at issue in this generalization of Fisher's approach to hypothesis testing is to make precise what it means for a rejection function f to be given independently of the sample E. The following definition provides the answer: Given a reference class of possibilities $\Omega$, a chance hypothesis **H**, a probability measure induced by **H** and defined on $\Omega$ (i.e., **P**($\cdot$|**H**)), and an event/sample E from $\Omega$; a rejection function f is *detachable* from E if and only if a subject possesses background knowledge $K$ that is conditionally independent of E (i.e., $P(E|H\&K) = P(E|H)$) and such that $K$ explicitly and univocally identifies the function f. Any rejection region R of the form $T^\gamma = \{\omega \in \Omega \mid f(\omega) \geq \gamma\}$ or $T_\delta = \{\omega \in \Omega \mid f(\omega) \leq \delta\}$ is then said to be *detachable* from E as well. Furthermore, R is then called a *specification* of E, and E is said to be *specified*.

Several remarks about this definition are in order.

**Remark 2.5.1**. Detachability is always relativized to a subject or subjects possessing certain background knowledge. For instance, the SETI researchers in section 1.3 who inferred that an extraterrestrial intelligence was communicating with them needed to know about prime numbers before they could ascertain that an incoming sequence of radio signals representing an ascending sequence of prime numbers was in fact just that and therefore due to an intelligence. A design inference is one intelligence inferring the activity of another intelligence by the judicious exploitation of background knowledge. The background knowledge acts as a sieve that collects artifacts due to design. The more complete the background knowledge and the better it is utilized, the more refined the sieve.

**Remark 2.5.2**. The requirement that background knowledge $K$ be conditionally independent of the sample E employs a well-defined notion from probability theory. Conditional independence means that the probability of E does not change once the background knowledge is taken into account. Conditional independence is the standard probabilistic way of unpacking epistemic independence. Two things are epistemically independent if knowledge about one thing (in this case the background knowledge $K$) does not affect our knowledge about the other (in this case the probability of the occurrence of E).

**Remark 2.5.3**. In case $P(E|H)$ equals zero, satisfying the condition $P(E|H\&K) = P(E|H)$ ($= 0$) tends to be uninformative. This is because $K$ may tell us much about E, yet still not confer on E a nonzero probability. For instance, if E is a randomly sampled real number from the unit interval with $H$ specifying the uniform probability on that interval and if $K$, based on direct observation of E, asserts that the real number just sampled from the unit interval is less than 1/2, then $P(E|H\&K) = P(E|H) = 0$. Nonetheless, the probability distribution involved in E's selection has now radically changed given $K$. In case $P(E|H)$ equals zero, we therefore need to require a stronger version of conditional independence, namely, $P(\cdot|H) = P(\cdot|H\&K)$. In other words, in case $P(E|H)$ equals zero, the probability distribution induced by $H$ must remain unchanged by factoring in our knowledge $K$. All the same, for discrete distributions where every elementary event or outcome

E has positive probability, $P(E|H\&K) = P(E|H)$ ($\neq 0$) works just fine as the defining condition for detachability.

**Remark 2.5.4.** To say that the background knowledge **K** explicitly and univocally identifies the rejection function f is simply to say that **K** uniquely describes f. Indeed, in the conditional independence condition we might just as well have written $P(E|H\&f) = P(E|H)$. The only reason we did not is because it would involve an abuse of notation: chance hypotheses and background knowledge are properly represented as statements whereas functions are mathematical objects and therefore distinct from statements (though describable by statements). Consequently, we should think of **K** as a description of f that leaves no doubt about f's identity. Even so, **K** need not exhaustively describe f in the sense of enabling us to evaluate f at every element of $\Omega$. For instance, it is common for differential equations to admit an existence and uniqueness proof so that we know that there is one and only one function (known as the solution) satisfying a given differential equation with certain initial conditions. Given such a differential equation, the unique function that solves it is fully specified. Nevertheless, we may be unable to evaluate that function precisely but must instead content ourselves with merely being able to approximate it (cf. the three-body problem in classical mechanics).[21]

**Remark 2.5.5.** Often the rejection function f has been identified explicitly before the sample E is taken. In that case the conditional independence between the background knowledge **K** describing f and the sample E is immediate. Recall the sequence of 100 bits in section 1.4 beginning `0100011011000001` . . . and which we referred to as D. By dividing this sequence as `0|1|00|01|10|11|000|001|` . . . it became evident that this sequence was constructed simply by writing binary numbers in ascending lexicographic order, starting with the one-digit binary numbers (i.e., 0 and 1), proceeding to the two-digit binary numbers (i.e., 00, 01, 10, and 11), and continuing until 100 digits were recorded. Letting **K** correspond to our knowledge of binary arithmetic and lexicographic orderings induces $f = 1_D$, the indicator function of D. One of f's extremal sets is D and **K** is conditionally independent of E. Even so, D (and hence $1_D$) was explicitly identified well before I ever attempted this sort of detachability analysis. The sequence beginning `0100011011000001` . . . is known as the Champernowne sequence and has the property that any N-digit combination of bits appears in this sequence with limiting frequency $2^{-N}$. D. G. Champernowne identified this sequence back in 1933.[22]

**Remark 2.5.6.** The rejection functions in sections 2.3 and 2.4 are detachable in the sense defined here. This is immediately evident in the Caputo case, where the rejection function f merely counts the number of Ds in a sequence of 41 Rs and Ds. Here **K** is our knowledge about counting elements of a given type from a finite collection of elements distinguishable by type. This knowledge has no way of altering the probabilities associated with random ballot-line selections. As for the compressibility of bit strings, the rejection function that evaluates the shortest computer program that generates a given bit string is defined solely in recursion-theoretic terms. This recursion-theoretic knowledge has no way of altering the probabilities associated with randomly generating a bit string by tossing a fair coin (heads = 1, tails = 0).

**Remark 2.5.7.** As a mathematical function that assigns real numbers to elements from a reference class of possibilities $\Omega$, the rejection function f is an abstract mathematical object. It follows that f is describable purely in terms of a priori mathematical knowledge. But if so, how can the knowledge **K** that identifies f fail to be probabilistically independent of E? Indeed, how could purely a priori knowledge of an abstract mathematical object like f possibility affect or be affected by the probability of an empirical observation? It seems, therefore, that I have failed to provide a principled way to distinguish specifications from nonspecifications (or fabrications as I called them in section 1.3). This objection fails to grasp that the knowledge that renders a rejection function detachable and induces a specification derives not from what is knowable in general (be it purely on mathematical grounds or also with reference to empirical factors) but from what subjects like us who are drawing design inferences actually do know. There is no reason to think that our knowledge of rejection functions is sufficient to exhaust all possible rejection regions and thus able to collapse the distinction between specifications and nonspecifications. Our ability to identify rejection functions solely on the basis of our knowledge of mathematics is quite limited. Nor does our knowledge of empirical factors that enter rejection functions change that. (As an example of a rejection function incorporating empirical factors, consider a normal density with a given mean and variance derived from empirical observation.)

**Remark 2.5.8.** In section 1.4 I characterized detachability and specification as follows: Given an event whose design is in question and a pattern describing it, would we be able to explicitly identify or exhibit that pattern if we had no knowledge which event occurred? If so, the pattern is detachable from the event and constitutes a specification. Inherent in this pretheoretic definition of detachability and specification is the ability to identify a pattern of the right sort—such patterns now being defined as rejection regions given by detachable rejection functions. Specifically, a rejection region R induced by a rejection function f is a specification provided that a subject

S possesses knowledge K and with that knowledge is able explicitly and univocally to identify f. We might therefore want to say that the rejection region R is specifiable or that the sample E falling within R is specifiable. Nonetheless, referring to something as specifiable can be misleading. Anything might be specifiable. But whether it is depends on whether some subject possesses the background knowledge to specify it. Nothing is ever specifiable as such but only in relation to a subject that does the specifying. And for that subject to specify something requires that the subject make explicit the knowledge that specifies it.[23]

**Remark 2.5.9.** Specification depends on the knowledge of subjects. Is specification therefore subjective? Yes, but not in a way that limits specification's usefulness for science. It is important here to grasp John Searle's distinction between ontological subjectivity and epistemic objectivity.[24] Money, for instance, is ontologically subjective in that it depends on the social conventions of human subjects. Nonetheless, money is epistemically objective—any dispute about the paper in my wallet being money can be objectively settled and is not like the dispute whether Mozart was a better composer than Beethoven (matters of taste being epistemically subjective). Specifications are ontologically subjective but epistemically objective.

**Remark 2.5.10.** Readers of *The Design Inference* will note that the characterization of detachability and specification given here differs from the one given there.[25] Specifically, I have retained the conditional independence condition but removed the tractability condition. What happened to the tractability condition? The tractability condition was always concerned with a subject's epistemic resources for generating specifications. The greater those epistemic resources, the more specifications a subject can generate and the more difficult for that subject to eliminate chance and infer design. The tractability condition was therefore not so much an essential part of the definition of detachability and specification as a way of disciplining our use of detachability and specification in drawing design inferences. The tractability condition has therefore been moved to the Generic Chance Elimination Argument (see section 2.7 and specifically step #5). The Generic Chance Elimination Argument describes the logic of design inferences. Within the context of that logic, the tractability condition's role becomes clear in a way that it was not when it was part of the definition of detachability. Consequently, once embedded in the logic of the Generic Chance Elimination Argument, the tractability condition no longer requires an explicit statement (though its intent is still realized). For the sake of continuity with *The Design Inference* I will flag the tractability condition's role in the Generic

Chance Elimination Argument when I discuss this argument schema in section 2.7.

## 2.6   Sweeping the Field of Chance Hypotheses

Earlier I indicated that as a criterion for detecting design, specified complexity generalizes Fisher's approach to hypothesis testing in two ways: (1) it generalizes the types of rejection regions capable of eliminating chance and (2) it eliminates all relevant chance hypotheses that could characterize an event (as opposed to just a single chance hypothesis as in Fisher's approach). The first of these points has now been adequately addressed: the generalized rejection regions capable of eliminating chance are the specifications, and these are induced by detachable rejection functions. Let us therefore turn to the second of these points, namely, the elimination of all relevant chance hypotheses that could characterize an event. I refer to this second point as "sweeping the field of chance hypotheses."

In Fisher's approach to hypothesis testing, eliminating one chance hypothesis typically opens the door to others. By contrast, specified complexity precludes chance decisively. Consequently, for an event E to exhibit specified complexity, it is not enough to know that E is specified by some rejection region R and that R has small probability (i.e., has probability less than some significance level $\alpha$) with respect to some probability distribution or other. Rather, we must know that whatever probability distribution may have been responsible for E, R is a detachable rejection region for all those probability distributions and that additionally the probability of R with respect to all those probability distributions is less than $\alpha$.

Thus, unlike the statistician, who typically operates from a position of ignorance in trying to determine what probability distributions might be responsible for the samples he or she observes, before we even begin to assess whether E exhibits specified complexity, we need to be reasonably confident about what probability distribution(s), if any, were operating to produce the event. There is thus a crucial difference between the way statistics eliminates chance and the way the design inference eliminates chance. When statistics eliminates chance, it is always a particular probability distribution (or sometimes set of probability distributions) that gets eliminated, with the question remaining what alternative probability distributions might be operating in its place. On the other hand, the design inference eliminates chance by decisively precluding alternative probability distributions.

Design inferences therefore eliminate chance in the global sense of closing the door to all relevant chance explanations. To be sure, this cannot be done with absolute finality since there is always the possibility that some crucial probability distribution was missed. Nonetheless, it is not enough for the design skeptic merely to note that adding a new chance explanation to the mix can upset a design inference. Instead, the design skeptic needs to explicitly propose a new chance explanation and argue for its relevance to the case at hand. Design inferences are falsifiable in the sense that if we are wrong in our assessment of the probability distributions that might characterize an event, then we can be wrong in eliminating chance and inferring design for that event. But the mere possibility of falsifiability is never enough to falsify a claim. Nor does feeding false claims into an argument invalidate that argument. An argument's validity depends on its logical structure and not on the truth or falsehood of its premises (for the validity of the design inference see section 2.7).

To clarify the distinction between statistical inferences and design inferences, consider an example of each. When a statistician wants to determine whether a certain fertilizer will help improve the yield of a certain crop, she assumes initially that the fertilizer will have no effect. Her research presumably is going to be used by farmers to decide whether to use this fertilizer. Because fertilizer costs money, she does not want to recommend the fertilizer if in fact it has no positive effect on the crop. Initially she will therefore assume that the probability distribution governing crop yield with the fertilizer is the same as without (this is the null hypothesis).

Having made this assumption, she will now compare how the crop fares with both fertilized and unfertilized soil. She will therefore conduct an experiment, raising, let us say, one acre of the crop in fertilized soil and another in unfertilized soil. In raising the crop, she will try to keep all sources of variation other than the fertilizer constant (e.g., water and sunshine). Our statistician does not know in advance how the crop will fare in the fertilized soil. Nevertheless, it is virtually certain she will not observe exactly the same yields from both acres.

The statistician's task, therefore, is to determine whether any difference in yield is due to the intrinsic effectiveness of the fertilizer or to chance fluctuations. Let us say that the fertilized crop yielded more than the unfertilized crop. Her task then is to determine whether this difference is significant enough to overturn her initial assumption that the fertilizer has no effect on the crop. If there is only a 1 percent increase in the fertilized crop over the unfertilized crop, this may not be enough to overthrow her initial assumption. On the other hand, a 100 percent increase in the fertilized crop might well overthrow the statistician's initial assumption, providing evidence that the fertilizer is highly effective.

The point then is this. Our statistician started by assuming the probability distribution characterizing crop yield for fertilized soil is the same as the probability distribution characterizing crop yield for unfertilized soil. She made this assumption because without prior knowledge about the effectiveness of the fertilizer, it was the safest assumption to make. Indeed, she does not want to recommend that farmers buy this fertilizer unless it significantly improves their crop yield. At least initially she will therefore assume the fertilizer has no effect on crops and that differences in crop yield are due to random fluctuations. The truth of this initial assumption, however, remains uncertain until she performs her experiment. If the experiment indicates little or no difference in crop yield, she will stick with her initial assumption. But if there is a big difference, she will discard her initial assumption and revise her probability distribution. Note, however, that in discarding her initial assumption and revising her probability distribution, our statistician has not eliminated chance but merely exchanged one chance hypothesis for another. Probabilities continue to characterize crop yield for the fertilized soil.

Let us now contrast the probabilistic reasoning of this statistician with the probabilistic reasoning of a design theorist who, let us say, must explain why a certain bank's safe that was closed earlier now happens to be open. Let us suppose the safe has a combination lock that is marked with a hundred numbers ranging from 00 to 99 and for which five turns in alternating directions are required to open the lock. We assume that precisely one sequence of alternating turns is capable of opening the lock (e.g., 43-89-52-90-17). There are thus 10 billion possible combinations, of which precisely one opens the lock. Random twirling of the combination lock's dial is therefore exceedingly unlikely to open it. The probability is 1 in 10 billion, and we may assume this probability is more extreme than the significance level $\alpha$ that we happened to set. What's more, opening the lock is specified. Indeed, the very construction of the lock's tumblers specifies which one of the 10 billion combinations opens the lock (this is easily formalizable in terms of the apparatus of section 2.5).

Now the crucial thing to observe here is that the probability distribution(s) associated with opening the lock is/are not open to question in the way the probability distribution(s) associated with the effectiveness of fertilizers was/were open to question in the last example. The design theorist's initial assessment of probability for the combination lock is stable; the statistician's initial assessment about the effectiveness of a given fertilizer is not. We have a great deal of prior knowledge about locks in general, and combination locks in particular, before the specific combination lock we are considering crosses our path. On the other hand, we know virtually nothing about a new fertilizer and its effect on crop yield until after we perform an experiment with it. Whereas the statistician's initial assessment of probability is likely to change, the design theorist's initial assessment of probability tends to be stable. The statistician wants to exclude one chance explanation only to replace it with another. The design theorist, on the other hand, wants to exclude the only available chance explanation(s), replacing it/them with a completely different mode of explanation, namely, a design explanation.

The combination lock example presents the simplest case of a design inference. Here there is only one chance hypothesis to consider, and when it is eliminated, any appeal to chance is effectively dead. But design inferences in which multiple chance hypotheses have to be considered and then eliminated can arise as well. We might, for instance, imagine explaining the occurrence of the sequence of prime numbers in the movie *Contact* (i.e., 110111011111011111110 . . .—see section 1.3) by repeatedly flipping a coin (1 for heads and 0 for tails) where the coin is either fair or weighted in favor of heads with probability .75. To eliminate chance and infer design we would now have to eliminate two chance hypotheses, one where the probability of heads is .5 (i.e., the coin is fair) and the other where the probability of heads is .75. To do this we would have to make sure that for both probability distributions this sequence of prime numbers is highly improbable (i.e., has probability less than some significance level $\alpha$), and then show that this sequence is also specified (neither of which is a problem). In case still more chance hypotheses might be operating, a design inference obtains only if each of these additional chance hypotheses get eliminated as well, which means that the sequence has to be statistically significant with respect to all the relevant chance hypotheses and in each case be specified as well.

But what is this set consisting of "all the relevant chance hypotheses"? Clearly this set cannot contain all possible chance hypotheses that might account for an event. Rather this set consists of all chance hypotheses that *in light of our context of inquiry* (which includes our background knowledge, epistemic values, and local circumstances) might plausibly account for the event in question. For any event whatsoever, there exists a probability distribution that concentrates all probability on that event and thus assigns it a probability of one. It therefore makes no sense to criticize my generalization of Fisher's approach to hypothesis testing for failing to consider all possible chance hypotheses. Statistical hypothesis testing must necessarily limit the set of hypotheses to be tested. This is not a fault of statistical theory. To be sure, it means that sweeping the field of chance hypotheses is falsifiable in the sense that we might have omitted a crucial chance hypothesis with respect to which an event, though previously exhibiting specified complexity, no longer does so. But the mere possibility of falsifiability is no reason to dismiss the design inference. Archeologists infer that certain chunks of rocks are arrowheads. Detectives infer that certain deaths were deliberate. Cryptographers infer that certain random looking symbol strings are actually encrypted messages. In every case they might be wrong, and further knowledge might reveal a plausible chance process behind what originally appeared to be designed. But such sheer possibilities by themselves do nothing to overturn our confidence in design inferences.

In concluding this section, I want to connect sweeping the field of chance hypotheses with the Explanatory Filter of section 1.3. Throughout this section I have focused exclusively on chance hypotheses and their elimination in establishing design. Nonetheless, in section 1.3 not only chance but also necessity had to be eliminated before design could be inferred. As it turns out, necessity can be viewed as a special case of chance in which the probability distribution governing necessity collapses all probabilities either to zero or one. For instance, for me to release a solid metal sphere near the surface of the earth will, in the absence of forces other than the earth's gravity, cause the sphere to fall with probability one. Nevertheless, even though necessity can be assimilated to chance in this way, in practice we often establish contingency (the opposite of necessity) directly by recognizing that the laws involved in an event's production do not uniquely constrain the event but also permit any number of alternatives to it (this was discussed in greater detail in section 1.3). The combination lock example considered earlier in this section illustrates this method of eliminating necessity. The laws of physics prescribe two possible motions of the combination lock, namely, clockwise and counterclockwise turns. Dialing any particular combination is compatible with these possible motions of the lock, but in no way dictated by these motions. To open the lock by hitting the right combination is therefore irreducible to these possible motions of the lock. In sum, sweeping the field of chance hypotheses corresponds to eliminating both the chance and the necessity nodes on the Explanatory Filter.

## 2.7   Justifying the Generalization

The preceding generalization of Fisher's approach to chance-elimination arguments now admits the following codification. I refer to this codification as the Generic Chance Elimination Argument. The Generic Chance Elimination Argument is an argument schema that encapsulates the common pattern of reasoning that underlies chance-elimination arguments generally. Here is the argument schema:

#1    A subject S learns that an event E has occurred and notes that E belongs to a reference class of possible events $\Omega$.

#2    By examining the circumstances under which E occurred (and thus relative to S's context of inquiry), S determines that only those chance processes characterized by chance hypotheses in the set $\{H_i\}_{i \in I}$ could have been operating to produce E (I here is an index set). This is a relevance condition: the set $\{H_i\}_{i \in I}$ consists of all chance hypotheses relevant to the production of E, not all chance hypotheses *überhaupt*.

#3    S identifies a rejection function f and therewith a rejection region R that includes E and that is an extremal set of f. R is therefore of the form $T^\gamma = \{\omega \in \Omega \mid f(\omega) \geq \gamma\}$ or $T_\delta = \{\omega \in \Omega \mid f(\omega) \leq \delta\}$ where $\gamma$ and $\delta$ are real numbers. Typically $\gamma$ is chosen as large as possible and $\delta$ as small as possible so that $T^\gamma$ or $T_\delta$ still includes E. The idea is to make the target around E as tight as possible. Thus, the rejection region R ($= T^\gamma$ or $T_\delta$) includes E but also has the smallest probability of the rejection regions derived from f that include E (see step #7).

#4    S identifies background knowledge **K** that explicitly and univocally identifies the rejection function f. Moreover, S confirms that **K** satisfies the conditional independence condition for each chance hypothesis in $\{H_i\}_{i \in I}$, i.e., $P(E|H_i \& K) = P(E|H_i)$ for all i in the index set I (in other words, the rejection function f is detachable for each of the chance hypotheses in $\{H_i\}_{i \in I}$). Note that if any $P(E|H_i)$ equals zero, S needs instead to confirm the stronger version of conditional independence defined in section 2.5, namely, $P(\cdot|H_i) = P(\cdot|H_i \& K)$ (see remark 2.5.3).

#5    Depending on how important it is for S to avoid a false positive (i.e., to avoid attributing E to something other than the chance hypotheses in $\{H_i\}_{i \in I}$ provided one of these hypotheses actually was responsible for E), S identifies a relevant set of probabilistic resources, which we will denote **ProbRes**, characterizing the different opportunities for E to occur and be specified relative to S's context of inquiry.

#6    On the basis of this set of probabilistic resources, S fixes a significance level $\alpha$ having the following property: An event of probability less than $\alpha$ remains improbable with respect to each of the chance hypotheses in $\{H_i\}_{i\in I}$ even after all the probabilistic resources in **ProbRes** are factored in.

#7    S calculates the probability of the rejection region R conditional on each of the chance hypotheses in $\{H_i\}_{i\in I}$ and determines that $P(R|H_i) < \alpha$ for all i in the index set I.

---

#8    S is warranted in inferring that E did not occur according to any of the chance hypotheses in $\{H_i\}_{i\in I}$ and therefore that E exhibits specified complexity.

I want to devote the remainder of this section to justifying the Generic Chance Elimination Argument. In doing so, I will not address whether the Generic Chance Elimination Argument reliably detects design. Specified complexity, the conclusion of the Generic Chance Elimination Argument, is a statistical notion. Design is a causal notion that refers to the activity of an intelligent agent. The connection between the two was addressed in sections 1.6 and 1.7 and will be taken up in section 2.10 as well as in chapter 5. Section 2.10 is of especial interest because it addresses the legitimacy of inferring design via an eliminative argument (the concern being that design inferences are arguments from ignorance—as we shall see, they are not). Nonetheless, for now I want simply to focus on whether the Generic Chance Elimination Argument is a valid argument schema for eliminating a set of chance hypotheses $\{H_i\}_{i\in I}$ (setting aside whether some crucial chance hypothesis that overturns design was omitted from this set as well as whether specified complexity is a reliable indicator of design).

Let us start with a simplification: Since the Generic Chance Elimination Argument eliminates a set of chance hypotheses $\{H_i\}_{i\in I}$ by eliminating each hypothesis in this set individually, to justify this argument schema it is enough to justify it for a single chance hypothesis **H**. Further, I am going to assume that this argument schema makes good sense for the special case that Fisher considered in which the rejection function f is just the probability density function associated with **H** (i.e., f satisfies $P(\cdot|H) = f\cdot dU$ where **U** is a uniform probability or a variant of it—see section 2.2). Readers unconvinced by Fisher's approach to hypothesis testing (despite my refurbishing of statistical significance levels using probabilistic resources) will need to read sections 2.9 and 2.10 to appreciate that alternatives to Fisher's approach unwittingly presuppose it despite ostensibly denying it.

Justifying the Generic Chance Elimination Argument now centers on one main issue. In Fisher's approach to hypothesis testing, the rejection function f is the probability density uniquely associated with the chance hypothesis

H. Consequently, when an event E falls within a rejection region R induced by such a uniquely given rejection function, R can rightly be regarded as independent of and therefore not tailored to E. But in the more general case, since background knowledge **K** that induces a rejection function f is merely constrained to be conditionally independent of E given **H**, there is no longer a unique rejection function associated with **H** whose rejection regions are then used to eliminate **H**. Rather, depending on the extent of our background knowledge, we may be able to identify many different items of background knowledge, each of which induces a detachable rejection function, and each of which in turn induces multiple rejection regions.

Now it turns out that practically speaking a given rejection function induces only two rejection regions. This is because for a rejection function f, its extremal sets have the following property: For $\gamma_1 < \gamma_2$, $T^{\gamma_1} \supset T^{\gamma_2}$ (i.e., $\{\omega \in \Omega \mid f(\omega) \geq \gamma_1\}$ includes $\{\omega \in \Omega \mid f(\omega) \geq \gamma_2\}$) and for $\delta_1 < \delta_2$, $T_{\delta_1} \subset T_{\delta_2}$ (i.e., $\{\omega \in \Omega \mid f(\omega) \leq \delta_1\}$ is included in $\{\omega \in \Omega \mid f(\omega) \leq \delta_2\}$). Thus, in choosing a rejection region R associated with the rejection function f, it is enough to consider extremal sets $T^\gamma$ and $T_\delta$ for which $P(T^\gamma|H)$ and $P(T_\delta|H)$ are close to but just less than the significance level $\alpha$. It follows that a given rejection function does not induce many different rejection regions but essentially just two.

That still leaves the problem, however, of a subject identifying numerous distinct items of background knowledge each of which is conditionally independent of E given **H**. Suppose each such item of background knowledge induces a rejection function that in turn induces a pair of rejection regions. Then each such rejection region is potentially capable of eliminating **H**. Hence by identifying sufficiently many items of such background knowledge, a subject can in principle run through one rejection region after another until hitting one that eliminates the chance hypothesis **H**. The worry, then, is that the Generic Chance Elimination Argument might eliminate any chance hypothesis whatsoever.

Think of it this way: A subject S identifies items of background knowledge in a set $\{K_j\}_{j \in J}$ (J is an indexing set) each of which is conditionally independent of E given **H**. Next, S uses each $K_j$ to identify a detachable rejection function $f_j$ and then successively runs through rejection regions $R_{j+}$ and $R_{j-}$ where $R_{j+}$ is of the form $T^\gamma = \{\omega \in \Omega \mid f_j(\omega) \geq \gamma\}$ and $R_{j-}$ is of the form $T_\delta = \{\omega \in \Omega \mid f_j(\omega) \leq \delta\}$ such that $P(R_{j+}|H)$ and $P(R_{j-}|H)$ are close to but just less than the significance level $\alpha$ (note that $\gamma$ and $\delta$ will vary with the index j). Given sufficiently many such items of background knowledge and thus given sufficiently many such rejection regions, one can exhaust the reference class of possible events $\Omega$ and subsume any event E whatsoever within such a rejection region. In effect, the Generic Chance Elimination Argument seems to allow for just the sort of tailoring of rejection regions to events that it was designed to forestall.

The problem with this criticism is that if the significance level α is small enough, there will not be enough items in our background knowledge that are conditionally independent of E and that enable us—in case E is due to chance—to recover E within a detachable rejection region R of probability less than α. Essentially the reason the Generic Chance Elimination Argument works is because we do not know enough.[26] Indeed, we neither know enough nor can know enough to defeat this argument schema when the significance level α is sufficiently small and when the event E is in fact due to the chance hypothesis **H**. Thus, if we identify multiple detachable rejection regions each of probability less than α and if after successive tries one of these regions happens to include the event E, we are justified in eliminating **H** provided α is small enough. The presumption here is that if a subject S can figure out an independently given pattern to which E conforms (i.e., a detachable rejection region of probability less than α that includes E), then so could someone else. Indeed, the presumption is that someone else used that very same item of background knowledge—the one used by S to eliminate **H**—to bring about E in the first place.

To illustrate that this is exactly right, consider the case of cryptography. Suppose that Alice and Bob are trading signals across a communication channel and that Eve is listening in. Let us assume the signals are all bit strings. The signals look random to Eve, that is, they look as though they could have been generated by flipping a fair coin. Indeed, when Eve runs the signals through her various tests for randomness, she finds that the signals pass all the tests. Are the signals in fact random and therefore the result of chance? Eve suspects that Alice and Bob are communicating meaningful messages that only appear random because they have been encrypted. Eve therefore runs through various decryption schemes that could have been derived independently of the bit strings she has collected eavesdropping on Alice and Bob. It is important that these decryption schemes be independent (i.e., detachable) since otherwise Eve could make the bit strings she has collected say anything she wants them to say. Once Eve finds such a decryption scheme and it transforms the bit strings she has collected into meaningful messages, there is no longer any question but that these bit strings are nonrandom.

More is true—Eve is convinced she has broken the very cryptosystem that Alice and Bob are using. It is conceivable that Alice and Bob are using one decryption scheme and Eve another, so that the messages Eve attributes to

Alice and Bob are different from the ones they are actually exchanging. In other words, it is conceivable that there is a problem of *underdetermination* here, with Eve making sense of the bit strings she has collected in one way and Alice and Bob making sense of them in another. Though in principle underdetermination might be a problem for cryptography, in practice it never is. If a proposed decryption scheme is successful at coherently interpreting numerous ciphertext transmissions, the cryptosystem is considered broken. Breaking a cryptosystem is like finding a key that turns a lock. Once the lock turns, we are confident the locking mechanism is disabled. True, there is always the possibility that the lock will turn without the locking mechanism being disabled. But as a proposed decryption scheme assigns a coherent sense not only to prior transmissions but also to incoming transmissions of ciphertext, any doubts about the correctness of the decryption scheme disappear.[27]

Given an event E whose chance occurrence the Generic Chance Elimination Argument is trying to decide, a subject S needs to factor in the opportunities for E to occur (i.e., the replicational resources, which we will denote by **ReplRes**) and the opportunities for E to be specified (i.e., the specificational resources, which we will denote by **SpecRes**). Together **ReplRes** and **SpecRes** constitute the probabilistic resources relevant for assessing whether the probability of a rejection region (i.e., specification) is sufficiently small to warrant the elimination of chance. We denote the combination of **ReplRes** and **SpecRes** by **ProbRes**. By factoring in all the relevant probabilistic resources, specification effectively precludes unbridled tailoring of events to rejection regions—each attempt to specify an event incurs a probabilistic cost that has to be counterbalanced by a probabilistic resource that takes that attempt into account.

Whereas enumerating replicational resources is usually straightforward, enumerating specificational resources tends to require some care. In *The Design Inference* I formalized how to enumerate specificational resources in what I called "the tractability condition" (a condition that in *The Design Inference* I assimilated to the definition of detachability but that properly belongs to the logic of inferring design).[28] The tractability condition employs a complexity measure $\varphi$ that characterizes the complexity of patterns relative to S's background knowledge and abilities as a cognizer to perceive and generate patterns. Such a measure is objectively given (relative to S) and determined up to monotonic transformations.[29] For a given specification R, S can therefore evaluate $\varphi(R)$. Moreover, S can use $\varphi$ to determine the number of specifications that have complexity no more than $\varphi(R)$ and whose probability is no more than $P(R|H)$. These, roughly, constitute the relevant specificational resources.

Actually, this is one place where we have to be careful about including the entire set of relevant chance hypotheses $\{H_i\}_{i \in I}$. Thus, formally, we define **SpecRes** as follows: **SpecRes** is the set of all specifications T such that (1) T is a subset of $\Omega$, (2) T is detachable from E for S with respect to all the chance hypotheses $\{H_i\}_{i \in I}$ (i.e., T is a specification qua rejection region for each of the chance hypotheses in $\{H_i\}_{i \in I}$), (3) $\max_{i \in I} P(T|H_i) \le \max_{i \in I} P(R|H_i)$, (4) $\varphi(T) \le \varphi(R)$, and (5) T is set-theoretically maximal with respect to conditions (1) to (4) (i.e., T is not a proper subset of any set that satisfies conditions (1) to (4)). I will refer to condition (5) as the *maximality condition*. Note that in all practical applications R is a member of **SpecRes**; moreover, when it is not, the maximal completion of R is. Any specification satisfying these five conditions is by definition said to satisfy the *tractability condition*. Specifications satisfying the tractability condition constitute S's relevant set of specificational resources for determining whether E happened by chance. Thus, by definition, **SpecRes** is the set of all specifications satisfying the tractability condition.

A word about the motivation underlying this definition of **SpecRes** is in order. Specificational resources are like lottery tickets (indeed, lottery tickets at a properly conducted lottery are specifications). The more lottery tickets that are sold, the more likely that someone is to win the lottery by chance. Unfortunately, in most instances the specifications relevant to deciding whether there are enough of them to render chance plausible are not nearly as straightforwardly identified as in a lottery, where one merely counts the number of tickets sold. Yet clearly, any specification has to identify some portion of the reference class of possibilities. Hence (1). Moreover, like lottery tickets, they actually have to be specifications and not fabrications. Hence (2).

Next, if chance is to be precluded in explaining E, then the specifications relevant to deciding E's chance occurrence must all have small probability. In the lottery analogy, we cannot have lottery tickets with extremely small probability of winning and then an exception with, say, probability 1/2 of winning. Lottery tickets must each have identical small probability of winning. Condition (3) attempts as far as possible to model this feature of lotteries. The condition $\max_{i \in I} P(T|H_i) \le \max_{i \in I} P(R|H_i)$ guarantees that all the specifications that factor into S's specificational resources have probability no greater than R with respect to all the relevant chance hypotheses. In effect, the specifications considered in deciding E's chance occurrence cannot nibble away more probability from the reference class $\Omega$ than the rejection region of interest R. Hence (3).

With enough background knowledge and cognitive/computational resources, one can identify as a specification any subset of $\Omega$ whatsoever. Specifications, however, are not all created equal, even those of small probability as constrained by condition (3). Some specifications are more complex than others. In deciding what specificational resources are relevant for deciding whether E occurred by chance, S therefore limits oneself to those specifications that are no more complex than the specification of prime interest, namely R. Hence (4).

Finally, in placing targets on a wall, it makes no sense to place targets within targets if hitting any target yields the same prize. If one target is included within another target and if all that is at issue is whether some target was hit, then all that matters is the biggest targets. The smaller ones that sit inside bigger ones are, in that case, merely along for the ride. The point of the maximality condition is to preclude nested targets from figuring into the enumeration of specificational resources. The specifications within a set of specificational resources are used to eliminate chance when an event falls within any one of those specifications. It is therefore enough only to consider maximal specifications. Permitting nonmaximal specifications inflates specificational resources and makes it harder than necessary to eliminate chance. Hence (5).

To see what is at stake with the complexity measures used in defining specificational resources, consider two cryptosystems: first, an extremely simple one, like the Caesar cipher, in which each letter of the alphabet is moved a fixed number of notches up or down the alphabet; second, an extremely complicated one, like the one-time pad, in which each letter of the alphabet is moved a random number of notches up or down the alphabet and where the alteration of one letter at one location is probabilistically independent of other alterations at other locations. With the Caesar cipher the complexity of the cryptosystem is very low and leads to low specificational resources. On the other hand, with the one-time pad the complexity of the cryptosystem is as high as it can be and leads to specificational resources as numerous as the possible encrypted alphabetic strings. The low complexity of the Caesar cipher enables it to be easily broken. The high complexity of the one-time pad secures it against cryptographic attacks. The low complexity of the Caesar cipher removes any doubts about underdetermination in the breaking of the cryptosystem and thus whether a decrypted string contains a meaningful message and is therefore designed. The high complexity of the one-time pad guarantees that underdetermination cannot be avoided and that we can never be sure that an alphabetic string is other than random. The huge specificational resources associated with the one-time pad mean we can never draw a design inference for its encrypted messages.

In assessing whether a significance level $\alpha$ is small enough to warrant the elimination of chance, it is necessary to factor in both replicational and specificational resources. Imagine a large wall with N identically-sized non-overlapping targets painted on it and M arrows in your quiver. Let us say that your probability of hitting any one of these targets, taken individually, with a single arrow by chance is p. Then the probability of hitting any one of these N targets, taken collectively, with a single arrow by chance is bounded by Np, and the probability of hitting any of these N targets with at least one of your M arrows by chance is bounded by MNp. We therefore take **ProbRes** (i.e., the probabilistic resources) as the Cartesian product of **ReplRes** (i.e, the replicational resources) and **SpecRes** (i.e., the specificational resources). Thus, if the number of replicational resources in **ReplRes** is M and the number of specificational resources in **SpecRes** is N, the number of probabilistic resources in **ProbRes** is MN. It follows that a detachable rejection region R has probability small enough to warrant the elimination of chance relative to the probabilistic resources **ProbRes** provided that the probability of R is less than $\alpha$ for $\alpha$ satisfying $MN\alpha < 1/2$.

Whence the number 1/2? A significance level $\alpha$ needs to be small enough so that a rejection region of probability less than $\alpha$ remains improbable even after all the probabilistic resources in **ProbRes** have been taken into account. In general, if a set of probabilistic resources contains N resources (replicational, specificational, or their combination as a Cartesian product), then an event of probability less than $1/(2N)$ has probability less than 1/2 once all N probabilistic resources have been taken into account. A significance level $\alpha$ is therefore said to characterize a small probability event provided that the probability $N\alpha$ that is revised in light of the relevant probabilistic resources is less than 1/2. The rationale here is that since factoring in all relevant probabilistic resources leaves us with an event of probability less than 1/2, that event is less probable than not, and consequently we should favor the opposite event, which is more probable than not and precludes it. Full details for converting probabilistic resources into statistical significance levels are given in chapter 6 of *The Design Inference*.

To bring this discussion down to earth, let us consider a concrete example, namely, the Caputo case of section 2.3. I am going to run this example through the Generic Chance Elimination Argument and thereby do a rational reconstruction of the Caputo case. Thus we are to imagine that the

New Jersey Supreme Court was following the steps of the Generic Chance Elimination Argument during Caputo's trial.

In step #1, a subject S, here the New Jersey Supreme Court, learns that the following event E has occurred:

E    DDDDDDDDDDDDDDDDDDDDDDRDDDDDDDDDDDDDDDDDD.

This event, we assume, constitutes Nicholas Caputo's actual ballot-line selections in 41 elections as Essex County Clerk. Thus in the initial 22 elections Caputo chose the Democrats to head the ballot line, then in the 23rd election he chose the Republicans, and then in the remaining elections he chose the Democrats again. The reference class of possible events $\Omega$ here consists of all sequences of Ds and Rs of length 41.

In step #2, the New Jersey Supreme Court, by examining the circumstances under which E occurred, determines that the only chance process that could have been operating to produce E was one in which Ds and Rs are equiprobable and probabilistically independent of one another. The court comes to this determination because Caputo himself was responsible for the ballot selections and claims to have used this chance process. The collection of chance hypotheses potentially capable of characterizing the chance process responsible for E thus consists of only one chance hypothesis, which we will denote by **H**. **H** is essentially a coin-tossing hypothesis that assigns to each sequence of Ds and Rs in $\Omega$ the probability $2^{-41}$ or approximately 1 in 2 trillion.

In step #3, the New Jersey Supreme Court identifies a rejection function f that counts the number of Democrats given the top ballot line in any sequence in $\Omega$. This function induces a rejection region R that is an extremal set of f of the form $T^\gamma = \{\omega \in \Omega \mid f(\omega) \geq \gamma\}$ where $\gamma = 40$. In other words $R = T^{40} = \{\omega \in \Omega \mid f(\omega) \geq 40\}$. The rejection region R includes E. (Note that R, the rejection region, needs to be distinguished from the Rs that appear in the sequences of $\Omega$.)

In step #4, the New Jersey Supreme Court determines that the rejection function f that counts the number of Democrats heading the ballot line for sequences in $\Omega$ is detachable. As indicated in section 2.3, given a diverse collection of elements, humans have for millennia been categorizing such elements and then counting the number in each category. The New Jersey Supreme Court's background knowledge **K** about counting and sorting is therefore conditionally independent of the event E given **H**. That knowledge directly induces f, which in turn induces the rejection region R.

In step #5, the New Jersey Supreme Court identifies a set of probabilistic resources **ProbRes** that gives Nicholas Caputo every opportunity to account for E in terms of **H**. The court is generous in handing out probabilistic resources. The court imagines that each state in the United States has c = 500 counties (an exaggeration), that each county has e = 5 elections per year (another exaggeration), that there were s = 100 states (we imagine rampant American imperialism doubling the number of states in the union), and that the present form of government endures y = 500 years (over double the current total). The product of c times e times s times y equals 125 million and signifies an upper bound on the total number of elections that might reasonably be expected to occur throughout U.S. history. These constitute the relevant replicational resources, which we denote by **ReplRes**, to account for Caputo giving Democrats the top ballot line 40 out of 41 times.

The court also identifies the specificational resources relevant to the Caputo case. Although the court has been apprised of E and determined R with reference to E (though without artificially tailoring R to E since R is detachable), we imagine the court stepping back and enumerating the specifications in $\Omega$ that it might reasonably use to preclude the chance occurrence of E. The court therefore defines **SpecRes** as all specifications belonging to $\Omega$ that are detachable from E, whose probability is no more than $P(R|H)$, whose complexity is no more than the complexity of R, and that are set-theoretically maximal. How complex is R? From the court's vantage, not very complex at all. Here, beginning with R, are some specifications of equal or smaller complexity whose probability is no greater than the probability of R:

1. The sequence of Ds and Rs of length 41 with one or no Republicans.
2. The sequence of Ds and Rs of length 41 consisting of no Republicans.
3. All sequences of Ds and Rs of length 41 with exactly one Republican.
4. All sequences of Ds and Rs of length 41 with one or no Democrats.
5. The sequence of Ds and Rs of length 41 consisting of no Democrats.
6. All sequences of Ds and Rs of length 41 with exactly one Democrat.
7. All sequences of Ds and Rs of length 41 consisting of alternating Democrats and Republicans.

Although there are seven specifications here, in fact only three of them can figure into **SpecRes**. On account of the maximality condition, specifications 2 and 3 need to be assimilated into specification 1 and specifications 5 and 6 need to be assimilated into specification 4. R is among the simplest specifications available to the court, and it is doubtful that there are more than a handful of maximal specifications of complexity no more than R. But let us be generous. Let us assume that the New Jersey Supreme Court decided to play it safe and stipulate that there were 100 specificational resources in **SpecRes** (despite the best indications being that the actual number is in the single digits). It then follows that **ProbRes** is the Cartesian product of **ReplRes** and **SpecRes** and therefore contains 125 million times 100 probabilistic resources. In other words, the number of probabilistic resources in **ProbRes** is 12.5 billion, or $1.25 \times 10^{10}$.

The remaining steps are now straightforward. In step #6, the New Jersey Supreme Court needs to calculate a significance level $\alpha$ such that an event of probability less than $\alpha$ remains improbable once all the probabilistic resources in **ProbRes** have been taken into account. In the Caputo case N equals 12.5 billion. The significance level $\alpha$ can now be read off of N: $\alpha$ needs only to satisfy the inequality $N\alpha < 1/2$. The court can therefore take $\alpha$ to be just less than 1 in 25 billion. For practical purposes, however, the court can take $\alpha$ exactly equal to $1/(2N)$, especially since the court has been generous in allowing probabilistic resources. Thus the court takes $\alpha$ equal to 1 in 25 billion. In step #7, the court computes $P(R|H)$, which not just in this rational reconstruction but also in real life the court computed to be less than 1 in 50 billion. This is strictly less than the $\alpha$-level computed in step #6. R is therefore a specification of probability less than $\alpha$. Step #8 now follows immediately: The New Jersey Supreme Court is warranted in inferring that E did not happen according to the chance hypothesis **H**.

This rational reconstruction of the Caputo case is in my view not only faithful to the reasoning employed by the New Jersey Supreme Court in its deliberations but also normative for how we should conduct chance elimination arguments generally. Philosopher Robin Collins disagrees. He argues that my analysis of the Caputo case is fundamentally flawed. Central to his argument is calling in computers to explicitly record rejection regions. Once computers are called in, they can explicitly list every possible sequence of ballot lines that Caputo might have chosen. According to Collins, computers vastly inflate specificational resources and thus make it impossible in most practical circumstances for the Generic Chance Elimination Argument to eliminate chance even when chance should be eliminated.[30]

But suppose the New Jersey Supreme Court justices had access to a computer with a petabyte hard drive that recorded all possible sequences of Ds and Rs of length 41. To be sure, a computer program generating all these sequences is easy to write and can be written on the basis of background knowledge **K** that is conditionally independent of E given **H**. But **K** does not enable us to identify the rejection region R, much less any other rejection region included in R that in turn includes E. **K** does not enable us to identify a detachable rejection region that includes E. What **K** does is identify an ensemble of possible rejection regions, one of which includes E. Given this ensemble, it still needs to be sorted through with respect to the complexity of the rejection regions (and specifically by employing the tractability condition). For this reason, having computers that willy-nilly generate rejection regions is of no use in eliminating chance. If background knowledge **K** is going to be effective in eliminating chance, it must explicitly and univocally identify a rejection function that in turn explicitly and univocally identifies a rejection region that includes E. An ensemble of rejection regions, one of which includes E, will not do it; and that is all computers have to offer. They can exhaust possibilities. They cannot identify conditionally independent background knowledge that locates a possibility of interest.

In concluding this section, I want to address the possible concern that a subject's context of inquiry determines which probabilistic resources get employed in the Generic Chance Elimination Argument. The concern here is over subjectivism. Subjective factors often do influence the setting of probabilistic resources, and the Generic Chance Elimination Argument faithfully reflects this fact (further confirmation that this argument schema provides a sound rational reconstruction of how we eliminate chance). Nonetheless, as I indicated in section 1.5 and argue at length in chapter 6 of *The Design Inference*, it is also possible to set probabilistic resources objectively and calculate a *universal probability bound*. Such a universal probability bound takes into account all the specificational resources that might ever be encountered in the known physical universe (interestingly, by exhausting the specificational resources of the universe, we also exhaust all the replicational resources that might ever arise[31]). Universal probability bounds come up regularly in cryptographic research where they are used to assess the security of cryptosystems against brute force attacks in which the entire universe is enlisted as a (nonquantum) computer.[32] In *The Design Inference* I compute a universal probability bound of 1 in $10^{150}$ and argue that it is beyond the capacity of the observable universe to generate sufficiently many specifications so that a specified event of that improbability could reasonably be attributed to chance. As a consequence, we never need to consider a statistical significance level less than $10^{-150}$. This is the smallest significance level we ever need.

## 2.8　The Inflation of Probabilistic Resources

Implicit in a universal probability bound such as $10^{-150}$ is that the universe is too small a place to generate specified complexity by sheer exhaustion of possibilities. Stuart Kauffman develops this theme at length in his book

*Investigations.*[33] In one of his examples (and there are many like it throughout the book), he considers the number of possible proteins of length 200 (i.e., $20^{200}$ or approximately $10^{260}$) and the maximum number of pairwise collisions of particles throughout the history of the universe (he estimates $10^{193}$ total collisions supposing the reaction rate for collisions can be measured in femtoseconds). Kauffman concludes: "The known universe has not had time since the big bang to create all possible proteins of length 200 [even] once."[34] To emphasize this point, he notes: "It would take at least 10 to the 67th times the current lifetime of the universe for the universe to manage to make all possible proteins of length 200 at least once."[35]

Kauffman even has a name for numbers that are so big that they are beyond the reach of operations performable by and within the universe—he refers to them as *transfinite*. For instance, in discussing a small discrete dynamical system whose dynamics are nonetheless so complicated that they cannot be computed, he writes: "There is a sense in which the computations are transfinite—not infinite, but so vastly large that they cannot be carried out by any computational system in the universe."[36] Kauffman justifies such proscriptive claims in exactly the same terms that I justified the universal probability bound in section 1.5. Thus as justification he looks to the Planck time, the Planck length, the radius of the universe, the number of particles in the universe, and the rate at which particles can change states.[37] Although Kauffman's idea of transfinite numbers is insightful, the actual term is infelicitous since it already has currency within mathematics, where transfinite numbers are by definition infinite (in fact, the transfinite numbers of transfinite arithmetic can assume any infinite cardinality whatsoever).[38] I therefore propose to call such numbers *hyperfinite numbers.*[39]

Kauffman often writes about the universe being unable to exhaust some set of possibilities. Yet at other times he puts an adjective in front of the word universe, claiming it is the *known* universe that is unable to exhaust some set of possibilities.[40] Is there a difference between the universe (no adjective in front) and the *known* or *observable* universe (adjective in front)? To be sure, there is no empirical difference. Our best scientific observations tell us that the world surrounding us appears quite limited. Indeed, the size, duration, and composition of the known universe are such that $10^{150}$ is a hyperfinite number. For instance, if the universe were a giant computer, it could perform no more than this number of operations (quantum computation, by exploiting superposition of quantum states, enriches the operations performable by an ordinary computer but cannot change their number); if the universe were devoted entirely to generating specifications, this number would set an upper bound; if cryptographers confine themselves to brute-force methods on ordinary computers to test cryptographic keys, the number of keys they can test will always be less than this number.

But what if the universe is in fact much bigger than the known universe? What if the known universe is but an infinitesimal speck within the actual universe? Alternatively, what if the known universe is but one of many possible universes, each of which is as real as the known universe but causally inaccessible to it? If so, are not the probabilistic resources needed to eliminate chance vastly increased and is not the validity of $10^{-150}$ as a universal probability bound thrown into question? This line of reasoning has gained widespread currency among scientists and philosophers in recent years. In this section I want to argue that this line of reasoning is fatally flawed.

Indeed, I shall argue that it is illegitimate to rescue chance by invoking probabilistic resources from outside the known universe. To do so artificially inflates one's probabilistic resources. Only probabilistic resources from the known universe may legitimately be employed in testing chance hypotheses. In particular, probabilistic resources imported from outside the known universe are incapable of overturning the universal probability bound of $10^{-150}$. My basic argument to support this claim is quite simple, though I need to tailor it to some of the specific proposals now current for inflating probabilistic resources. The basic argument is this: It is never enough to postulate probabilistic resources merely to prop an otherwise failing chance hypothesis. Rather, one needs independent evidence whether there really are enough probabilistic resources to render chance plausible.

Consider, for instance, two state lotteries, both of which have printed a million lottery tickets. Let us assume that each ticket has a one in a million probability of winning and that whether one ticket wins is probabilistically independent of whether another wins (multiple winners are therefore a possibility). Suppose now that one of these state lotteries sells the full one million tickets but that the other sells only two tickets. Ostensibly both lotteries have the same number of probabilistic resources—the same number of tickets were printed for each. Nevertheless, the probabilistic resources relevant for deciding whether the first lottery produced a winner by chance greatly exceed those of the second. Probabilistic resources are *opportunities* for an event to happen or be specified. To be relevant to an event, those opportunities need to be actual and not merely possible. Lottery tickets sitting on a shelf collecting dust might just as well never have been printed.

This much is uncontroversial. But let us now turn the situation around. Suppose we know nothing about the number of lottery tickets sold, and are informed simply that the lottery had a winner. Suppose further that the probability of any lottery ticket producing a winner is extremely low. Now what can we conclude? Does it follow that many lottery tickets were sold? Hardly. We are entitled to this conclusion only if we have independent evidence that many lottery tickets were sold. Apart from such evidence we have no way of assessing how many tickets were sold, much less whether the lottery was conducted fairly and whether its outcome was due to chance. It is illegitimate to take an event, decide for whatever reason that it must be due to chance, and then propose numerous probabilistic resources because otherwise chance would be implausible. I call this the *inflationary fallacy*.[41]

Stated thus, the inflationary fallacy is readily rejected as a bogus form of argument. Nevertheless, it can be nuanced so that the problem inherent in it is mitigated (though by no means eliminated). The problem inherent in the inflationary fallacy is always that it multiplies probabilistic resources in the absence of independent evidence that such resources exist. Typically, however, when probabilistic resources get inflated, the rationale for inflating them is not simply to render chance plausible when otherwise it would be implausible. Hardly anyone is so crass as to admit, "I didn't like the alternatives to chance so I simply decided to invent some probabilistic resources." The rationale for inflating probabilistic resources is always more subtle, seeking confirmation in general coherence or consilience considerations even though independent evidence is lacking.

The inflationary fallacy therefore has a crass and a nuanced form. The crass form looks as follows:

| | |
|---|---|
| Premise 1: | Alternatives to chance are for whatever reason unacceptable for explaining some event—call that event X. |
| Premise 2: | With the probabilistic resources available in the known universe, chance is not a reasonable explanation of X. |
| Premise 3: | If probabilistic resources could be expanded, then chance would be a reasonable explanation of X. |
| Premise 4: | <u>Let there be more probabilistic resources.</u> |
| Conclusion: | Chance is now a reasonable explanation of X. |

The problem with this argument is Premise 4 (the "fiat" premise), which creates probabilistic resources ex nihilo simply to ensure that chance becomes a reasonable explanation.

The more nuanced form of the inflationary fallacy is on the surface less objectionable. It looks as follows:

| | |
|---|---|
| Premise 1: | There is an important problem, call it Y, that admits a solution as soon as one is willing to posit some entity, |

process, or stuff outside the known universe. Call whatever this is that resides outside the known universe Z.

Premise 2:   Though not confirmed by any independent evidence, Z is also not inconsistent with any empirical data.

Premise 3:   With the probabilistic resources available in the known universe, chance is not a reasonable explanation of some event—call the event X.

Premise 4:   But when Z is added to the known universe, probabilistic resources are vastly increased and now suffice to account for X by chance.

Conclusion:   Chance is now a reasonable explanation of X.

This nuanced form of the inflationary fallacy appears in various guises and has gained widespread currency. It purports to solve some problem of general interest and importance by introducing what we will call a Z-factor, to wit, some entity, process, or stuff outside the known universe. In addition to solving some problem, this Z-factor has associated with it numerous probabilistic resources that come along for the ride as a by-product. These resources in turn help to shore up chance when otherwise chance would seem unreasonable in explaining some event.

I want therefore next to consider four proposals for Z-factors that purport to resolve important problems and that have gained wide currency. The Z-factors I will consider are these: the bubble universes of Alan Guth's inflationary cosmology, the many worlds of Hugh Everett's interpretation of quantum mechanics, the self-reproducing black holes of Lee Smolin's cosmological natural selection, and the possible worlds of David Lewis's extreme modal realist metaphysics.[42] My choice of proposals, though selective, is representative of the forms that the inflationary fallacy takes. While I readily admit that these Z-factors propose solutions to important problems, I will argue that the costs of these solutions outweigh their benefits. In general, Z-factors that inflate probabilistic resources so that what was unattributable to chance within the known universe now becomes attributable to chance after all are highly problematic and create more difficulties than they solve.

Let us start with Alan Guth's inflationary cosmology. Inflationary cosmology posits a very brief period of hyper-rapid expansion of space just after the Big Bang. Though consistent with general relativity, such expansion is not required. What's more, the expansion has now stopped (at least as far as we can tell within the known universe). Guth introduced inflation to solve such problems in cosmology as the flatness, horizon, and magnetic monopole problems. In standard Big Bang cosmology the first two of these problems seem to require considerable fine-tuning of the initial conditions of the universe whereas the third seems unresolvable if standard Big Bang cosmology is combined with grand unified theories. Inflationary cosmology offers to resolve these problems in one fell swoop. In so doing, however, the known universe becomes a bubble universe within a vast sea of other bubble universes, and the actual universe now constitutes the sea that contains these bubble universes.

Next let us consider Hugh Everett's interpretation of quantum mechanics. Everett's many worlds interpretation of quantum mechanics proposes a radical solution to what in quantum mechanics is known as the measurement problem. The state function of a quantum mechanical system corresponds to a probability distribution that upon measurement assumes a definite value. The problem is that any physical system whatsoever can be conceived as a quantum mechanical system described by a state function. Now what happens when the physical system in question is taken to be the entire universe? Most physical systems one considers are proper subsets of the universe and thus admit observers who are outside the system and who can therefore measure the system and, as it were, collapse the state function. But when the universe as a whole is taken as the physical system in question, where is the observer to collapse the state function?[43] Everett's solution is to suppose that the state function does not collapse but rather splits into all different possible values that the state function could assume (mathematically this is very appealing—especially to quantum cosmologists—because it eliminates any break in dynamics resulting from state-function collapse). In effect, all possible quantum histories get lived out. Suppose, for instance, someone offers me a million dollars to play Quantum Russian Roulette (i.e., a quantum mechanical device is set up with six possibilities, each having probability one-sixth, and such that a bullet fires into my brain and kills me when exactly one of these possibilities occurs but leaves me unharmed otherwise). If I choose to play this game, then for every one quantum world in which I get a bullet to the head there are five in which I live happily ever after as a millionaire.

Next let us consider Lee Smolin's cosmological natural selection of self-reproducing black holes. Smolin's self-reproducing black holes constitute perhaps the most ambitious of the Z-factors we will consider. Smolin characterizes his project as explaining how the laws of physics have come to take the form they do, but in fact he is presenting a full-blown cosmogony in which Darwinian selection becomes the mechanism by which universes are generated and flourish. According to Smolin, quantum effects preclude singularities at which time stops. Consequently, time does not stop in a black hole but rather "bounces" in a new direction, producing a region of space-time inaccessible to ours except at the moment of its origination. Moreover, Smolin contends that during a "bounce" the laws of nature change their parameters but not their general form. Consequently, the formation of black holes follows an evolutionary algorithm in which parameters get continually tightened to maximize the production of black holes. Within Smolin's scheme the known universe is but one among innumerable black holes that have formed by this process and that in turn generate other black holes. Cosmological natural selection accounts not only for the generation of universes but also for their fine tuning and the possibility of such structures as life.

Finally, let us consider the possible worlds of David Lewis's extreme modal realist metaphysics. Lewis, unlike Guth, Everett, and Smolin, is not a scientist but a philosopher and in particular a metaphysician. For Lewis any logically possible world is as real as our world, which he calls the actual world. It is logically possible for a world to consist entirely of a giant tangerine. It is logically possible that the laws of physics might have been different, not only in their parameters but also in their basic form. It is logically possible that instead of turning to mathematics I might have become a rock and roll singer. For each of these logical possibilities Lewis contends that there are worlds as real as ours in which those possibilities are actualized. The only difference between those worlds and ours is that we happen to inhabit our world—that is what makes our world the actual world. Lewis's view is known as extreme modal realism. Modal realism asserts that logical possibilities are in some sense real (perhaps as abstractions in a mathematical space). *Extreme* modal realism emphasizes that logical possibilities are real in exactly the same way that the world we inhabit is real. Why does Lewis hold this view? According to him, possible worlds are indispensable for making sense of certain key philosophical problems, notably the analysis of counterfactual conditionals. What's more, he finds that all attempts to confer on possible worlds a status different from the actual world are incoherent (he refers to these disparagingly as *ersatz* possible worlds and finds them poor substitutes for his full-blown possible worlds).

I have provided only the briefest summary of the views of Alan Guth, Hugh Everett, Lee Smolin, and David Lewis. The problems these thinkers raise are important, and the solutions they propose need to be taken seriously. Moreover, except for David Lewis's possible worlds, which are purely metaphysical, the other three Z-factors considered make contact with empirical data. Lee Smolin even contends that his theory of cosmological natural selection has testable consequences—he even runs through several possible tests. The unifying theme in Smolin's tests is that varying the parameters for the laws of physics should tend to decrease the rate at which black holes are formed in the known universe. It is a consequence of Smolin's theory that for most universes generated by black holes, the parameters of the laws of physics should be optimally set to facilitate the formation of black holes. We ourselves are therefore highly likely to be in a universe where black hole formation is optimal. My own view is that our understanding of physics needs to proceed considerably further before we can establish convincingly that ours is a universe that optimally facilitates the formation of black holes. But even if this could be established now, it would not constitute independent evidence that a black hole is capable of generating a new universe. Smolin's theory, in positing that black holes generate universes, would explain why we are in a universe that optimally facilitates the formation of black holes. But it is not as though we would ever have independent evidence for Smolin's theory, say by looking inside a black hole and seeing whether there is a universe in it. Of all the objects in space (stars, planets, comets, etc.) black holes divulge the least amount of information about themselves.

The concept of *independent evidence* is important here and needs to be distinguished from *explanatory power*. Each of the four Z-factors considered here possesses explanatory power in the sense that each explains certain relevant data and thereby solves some problem of general interest and importance. These data are said to confirm or provide epistemic support for a Z-factor insofar as it adequately explains the relevant data and does not conflict with other recognized data. What's more, insofar as a Z-factor does not adequately explain the relevant data, it lacks explanatory power and is disconfirmed. In general, therefore, explanatory power entails testability in the weak sense that if a claim fails adequately to explain certain relevant data, it is to be rejected (thus failing the test).

Nevertheless, even though the four Z-factors considered here each possesses explanatory power, none of them possesses independent evidence for its existence. Independent evidence is by definition evidence that helps establish a claim apart from any appeal to the claim's explanatory power. The demand for independent evidence is neither frivolous nor tendentious. Instead, it is a necessary constraint on theory construction so that theory construction does not degenerate into total free-play of the mind.[44] Consider for instance the "gnome theory of friction." Suppose a physicist claims that the reason objects do not slide endlessly across surfaces is because tiny invisible gnomes inhabit all surfaces and push back on any objects pushed along the surfaces. What's more, the rougher a surface, the more gnomes inhabit it, and consequently the greater the resistance to an object moving across the surface. Suitably formulated, the gnome theory of friction can explain how objects move across surfaces just as accurately as current physical theory. So why do we not take the gnome theory of friction seriously? One reason (though not the only reason—the gnome theory has many more problems than described here) is the absence of independent evidence for gnomes.

Independent evidence and explanatory power need to work in tandem, and for one to outpace the other typically leads to difficulties. In spinning out their theories, conspiracy theorists place all their emphasis on explanatory power but ignore the demand for independent evidence. In enumerating countless low-level facts, crude inductivists place all their emphasis on independent evidence and thus miss the bold hypotheses and intuitive leaps that make for explanatory power and thus are capable of tying together their disparate facts. Independent evidence is the strict disciplinarian to explanatory power's carefree genius. Each is needed to balance the other. My favorite story illustrating the interplay between the two is due to John Leslie.[45] Suppose an arrow is fired at random into a forest and hits Mr. Brown. To explain such a chance occurrence it would suffice for the forest to be full of people. The forest being full of people therefore possesses explanatory power. Even so, this explanation remains but a speculative possibility until it is supported by independent evidence of people other than Mr. Brown in the forest.

The problem with the four Z-factors considered above is that none of them admits independent evidence. The only thing that confirms them is their ability to explain certain data or resolve certain problems. With regard to inflationary cosmology, we have no direct experience of hyper-rapid inflation nor have we observed any process that could reasonably be extrapolated to hyper-rapid inflation. With regard to the many-worlds interpretation of quantum mechanics, we always experience exactly one world and have no direct access to alternate parallel worlds. If there is any access at all to these worlds, it is indirect and circumstantial. Indeed, to claim that quantum interference signals the influence of parallel worlds is to impose a highly speculative interpretation on the data of quantum mechanics that is far from compelling.[46] With regard to black hole formation, there is no way for anybody on the outside to get inside a black hole, determine that there actually is a universe inside there, and then emerge intact to report as much. With regard to possible worlds, they are completely causally separate from each other—other possible worlds never were and never can be accessible to us, either directly or indirectly.

The absence of independent evidence for these Z-factors makes the problem of underdetermination especially acute for them. In general, when a hypothesis explains certain data, there are other hypotheses that also explain the data. In this way, data are said to underdetermine hypotheses. Nonetheless, it may be that one hypothesis explains the data better than the others so that it is possible to adjudicate among hypotheses simply on the basis of explanatory power. On the other hand, it may be that competing hypotheses exhibit identical explanatory power or that advocates of competing hypotheses claim that their preferred hypotheses exhibit the greater explanatory power. In either case, independent evidence will be required to adjudicate among the hypotheses. With the four Z-factors here considered no such independent evidence is forthcoming.

I want therefore next to examine these four Z-factors in relation to design to see whether design might be amenable to independent evidence in a way that the four Z-factors are not. As I defined it, a Z-factor is some entity, process, or stuff outside the known universe that helps explain certain data and thereby resolve some problem. Notably absent from the Z-factors described by Guth, Everett, Smolin, and Lewis is a designer. Their Z-factors are fully compatible with naturalism and thoroughly nonteleological. Now the interesting thing is that a designer, especially when fleshed out into a full-blown theistic deity, can be employed to resolve the very problems that the four Z-factors considered here were meant to resolve. The fine-tuning of the universe and the form of the laws of physics that are central to Guth's and Smolin's concerns can be attributed to a divine act of creation. Moreover, such a deity could collapse the state function of the universe and thereby resolve the measurement problem of quantum mechanics when this problem is applied to the universe taken as a whole. And finally, such a deity, by being suitably omniscient and thus possessing what philosophers of religion call "middle knowledge," could provide a semantics for counterfactual conditionals and resolve many of the other problems for which David Lewis thinks he requires possible worlds.[47]

Now I want to stress that I am not advocating these theistic alternatives to the four Z-factors considered above (I personally think there is something to the theistic fine-tuning arguments, but I am no fan of middle knowledge and have serious doubts about God's role as a state-function collapser). My point, rather, is this: Given that there are design-theoretic alternatives to the Z-factors considered here and given that such alternatives immediately raise the problem of underdetermination, the only way to resolve this problem is via independent evidence. So let me pose the question: Is there independent evidence that would allow us to distinguish the four Z-factors considered above from a design-theoretic alternative? We have already seen that there is no independent evidence that supports these four Z-factors. But could there be independent evidence that supports a design-theoretic alternative and in so doing also disconfirms these four Z-factors? I am going to argue that there is.

The four Z-factors considered here allow for unlimited probabilistic resources. Now the problem with unlimited probabilistic resources is that they allow us to explain absolutely everything by reference to chance—not just natural objects that actually did result by chance and not just natural objects that look designed, but also all artificial objects that are in fact designed. In effect, unlimited probabilistic resources collapse the distinction between apparent design and actual design and make it impossible to attribute anything with confidence to actual design. Was Arthur Rubinstein a great pianist or was it just that whenever he sat at the piano, he happened by chance to put his fingers on the right keys to produce beautiful music? It could happen by chance, and there is some possible world where everything is exactly as it is in this world except that the counterpart to Arthur Rubinstein cannot read music and happens to be incredibly lucky whenever he sits at the piano. Examples like this can be multiplied. There are possible worlds in which I cannot do arithmetic and yet sit down at my Macintosh computer and write probabilistic tracts about intelligent design. Perhaps Shakespeare was a genius. Perhaps Shakespeare was an imbecile who just by chance happened to string together a long sequence of apt phrases. Unlimited probabilistic resources ensure not only that we will never know, but also that we have no rational basis for preferring one to the other.

Given unlimited probabilistic resources, there is only one way to rebut this anti-inductive skepticism, and that is to admit that while unlimited probabilistic resources allow bizarre possibilities like this, these possibilities are nonetheless highly improbable in the little patch of reality that we inhabit. Unlimited probabilistic resources make bizarre possibilities unavoidable on a grand scale. The problem is how to mitigate the craziness entailed by them, and the only way to do this once such bizarre possibilities are conceded is to render them improbable on a local scale. Thus in the case of Arthur Rubinstein, there are worlds where someone named Arthur Rubinstein is a world famous pianist and does not know the first thing about music. But it is vastly more probable that in worlds where someone named Arthur Rubinstein is a world famous pianist, that person is a consummate musician. What's more, induction tells us that ours is such a world.

But can induction really tell us that? How do we know that we are not in one of those bizarre worlds where things happen by chance that we ordinarily attribute to design? Consider further the case of Arthur Rubinstein. Imagine it is January 1971 and you are at Orchestra Hall in Chicago listening to Arthur Rubinstein perform. As you listen to him perform Liszt's "Hungarian

Rhapsody," you think to yourself, "I know the man I'm listening to right now is a wonderful musician. But there's an outside possibility that he doesn't know the first thing about music and is just banging away at the piano haphazardly. The fact that Liszt's 'Hungarian Rhapsody' is pouring forth would thus merely be a happy accident. Now if I take seriously the existence of other worlds, then there is some counterpart to me pondering these very same thoughts, only this time listening to the performance of someone named Arthur Rubinstein who is a complete musical ignoramus. How, then, do I know that I'm not that counterpart?"[48]

Indeed, how do you know that you are not that counterpart? First off, let us be clear that the Turing Test is not going to come to the rescue here by operationalizing the two Rubinsteins and rendering them operationally indistinguishable. According to the Turing Test, if a computer can simulate human responses so that fellow humans cannot distinguish the computer's responses from an individual human's responses, then the computer passes the Turing Test and is adjudged intelligent.[49] This operationalizing of intelligence has its own problems, but even if we let them pass, success at passing the Turing Test is clearly not what is at stake in the Rubinstein example. The computer that passes the Turing Test presumably "knows" what it is doing (having been suitably programmed) whereas the Rubinstein who plays successful concerts by randomly positioning fingers on the keyboard does not have a clue. Think of it this way: Imagine a calculating machine whose construction guarantees that it performs arithmetic correctly and imagine another machine that operates purely by random processes. Suppose we pose the same arithmetic problems to both machines and out come identical answers. It would be inappropriate to assign arithmetic prowess to the random device, even though it is providing the right answers, because that is not its proper function—it is simply by chance happening upon the right answers. On the other hand, it is entirely appropriate to attribute arithmetic prowess to the other machine because it is constructed to perform arithmetic calculations accurately—that is its proper function. Likewise, with the real Arthur Rubinstein and his chance-performing counterpart, the real Arthur Rubinstein's proper function is, if you will, to perform music with skill and expression whereas the counterpart is just a lucky poseur. When Turing operationalized intelligence, he clearly meant intelligence to be a proper function of a suitably programmed computer and not merely a happy accident.[50]

How, then, do you know that you are listening to Arthur Rubinstein the musical genius and not Arthur Rubinstein the lucky poseur? To answer this question, let us ask a prior question: How did you recognize in the first place that the man called Rubinstein performing in Orchestra Hall was a consummate musician? Reputation, formal attire, and famous concert hall are certainly giveaways, but they are neither necessary nor sufficient. Even so, a necessary condition for recognizing Rubinstein's musical skill (design) is that he was following a prespecified concert program, and in this instance that he was playing Liszt's "Hungarian Rhapsody" note for note (or largely so—Rubinstein was not immune to mistakes). In other words, you recognized that Rubinstein's performance exhibited specified complexity. Moreover, the degree of specified complexity exhibited enabled you to assess just how improbable it was that someone named Rubinstein was playing the "Hungarian Rhapsody" with éclat but did not have a clue about music. Granted, you may have lacked the technical background to describe the performance in these terms, but the recognition of specified complexity was there nonetheless, and without that recognition there would have been no way to attribute Rubinstein's playing to design rather than chance.

Specified complexity is how we eliminate bizarre possibilities in which chance is made to account for things that we would ordinarily attribute to design. What's more, specified complexity is how we assess the improbability of those bizarre possibilities and therewith justify eliminating their chance occurrence. That being the case (and it certainly is the case for human artifacts), on what basis could we attribute chance to natural phenomena that exhibit specified complexity? Let us be clear that inflating probabilistic resources does not just diminish a universal probability bound and make it harder to attribute design—inflating probabilistic resources is not a matter of replacing one universal probability bound by another that is more stringent. Inflating probabilistic resources eliminates universal probability bounds entirely—the moment one posits unlimited probabilistic resources, anything of nonzero probability becomes certain (probabilistically this follows from the Strong Law of Large Numbers[51]). It seems, however, that in practical life we do allow for probability bounds to assess improbability and therewith specified complexity. A sentence or two verbatim repeated by another author can be enough to elicit the charge of plagiarism. It could happen by chance and given unlimited probabilistic resources there are patches of reality where it did happen by chance. But we do not buy it—at least not for our patch of reality. In practical life we tend not to be very conservative in setting probability bounds. They tend to be quite large, and certainly much larger than the universal probability bound of $10^{-150}$ that I have been advocating. For instance, in the Caputo case (see section 2.3) the probability bound employed by the New Jersey Supreme Court justices was not much smaller than $10^{-10}$.

The difficulty confronting unlimited probabilistic resources can now be put quite simply: There is no principled way to discriminate between using unlimited probabilistic resources to retain chance and using specified complexity to eliminate chance. You can have one or the other, but you cannot have both. And the fact is, we already use specified complexity to eliminate chance. Let me stress that there is no *principled* way to make the discrimination. It is, for instance, possible to invoke naturalism as a philosophical presupposition and use it to discriminate between using probabilistic resources to retain chance when designers unacceptable to naturalism are implicated (e.g., God) and using specified complexity to eliminate chance when designers acceptable to naturalism are implicated (e.g., Francis Crick's space aliens who seed the universe with life[52]). Thus for artifactual objects exhibiting specified complexity and for which an embodied intelligence could plausibly have been involved, we would attribute design; but for natural objects exhibiting specified complexity and for which no embodied intelligence could plausibly have been involved, we would invoke unlimited probabilistic resources and thus attribute chance (or perhaps simply plead ignorance). But this is entirely arbitrary. Indeed, the problem of unlimited probabilistic resources throws naturalism itself into question, and it does no good to invoke naturalism to resolve the problem.

It is important to understand that I am not arguing that the inflation of probabilistic resources entails anti-inductive skepticism. Indeed, my argument here is not anti-inductive but pro-specified complexity. I did offer an anti-inductive argument in chapter 6 of *The Design Inference*. My focus there was on the set of all logically possible worlds, and thus on worlds that instantiate every possible set of natural laws. In that case, inflating probabilistic resources entails inductive skepticism since there are far more worlds that agree with our world up to the present and go haywire afterward than there are worlds that continue to obey the regularities observed thus far. My argument here, however, allows that the worlds that inflate probabilistic resources obey laws of the same form as the laws of our universe. In that case, the vast majority of worlds in which Rubinstein delivers an exquisite performance are worlds in which Rubinstein is a skilled musician rather than a lucky poseur. But to convince ourselves for such worlds that Rubinstein is indeed a skilled musician rather than a lucky poseur requires specified complexity. Even with unlimited probabilistic resources, we need to distinguish design from nondesign, and specified complexity is how we do it. Consequently, there is no principled way to discriminate between using unlimited probabilistic resources to retain chance and using specified complexity to eliminate chance. And since we already use specified complexity to eliminate chance, invoking unlimited probabilistic resources to retain chance is not a defensible option. I am not arguing that inflating probabilistic resources destroys induction. I am arguing that inflating probabilistic resources does not destroy specified complexity. In particular, probabilistic resources from outside the known universe are irrelevant to assessing specified complexity.[53]

We are now in a position to see why a designer outside the known universe could in principle be supported by independent evidence whereas the Z-factors introduced by Guth, Everett, Smolin, and Lewis cannot. We already have experience of human and animal intelligences generating specified complexity. If we should ever discover evidence of extraterrestrial intelligence, a necessary feature of that evidence would be specified complexity. Thus, when we find evidence of specified complexity in nature for which no embodied, reified, or evolved intelligence could plausibly have been involved, it is a straightforward extrapolation to conclude that some unembodied intelligence must have been involved. Granted, this raises the question of how such an intelligence could coherently interact with the physical world (see section 6.5). But to deny this extrapolation merely because of a prior commitment to naturalism is not defensible. There is no principled way to distinguish between using specified complexity to eliminate chance in one instance and then in another invoking unlimited probabilistic resources to render chance plausible.

Design allows for the possibility of independent evidence whereas the Z-factors of Guth, Everett, Smolin, and Lewis do not. Specified complexity can be a point of contact between the known universe and an intelligence outside it—designers within the universe already generate specified complexity and a designer outside could potentially do the same. That is what allows for independent evidence to support unembodied designers. Provided nature supplies us with instances of specified complexity that cannot reasonably be attributed to any embodied intelligence (see chapter 5), the inference to an unembodied intelligence becomes compelling and any instances of specified complexity used to support that inference can rightly be regarded as independent evidence. By contrast, the Z-factors of Guth, Everett, Smolin, and Lewis provide no such palpable connection with the known universe. Indeed, what in our actual experience can straightforwardly be extrapolated to hyper-rapid expansion of space, quantum many worlds, cosmological natural selection, and causally inaccessible possible worlds? Is it, for instance, a straightforward extrapolation that takes us from biological natural selection of carbon-based life to cosmological natural selection of black holes? To be sure, there is an extrapolation here, but one where all meaningful analogies with actual experience break down.

Consequently, three crucial questions now face design: (1) Is specified complexity exhibited in any natural systems where no embodied intelligence could plausibly have been involved (see chapters 4 and 5)? (2) If so, does the design apparent in such systems match up meaningfully with known designs due to known embodied designers (see chapter 5)? (3) Does a theory of design that treats specified complexity as a reliable marker of intelligence possess sufficient explanatory power to render it interesting and fruitful for science (see chapter 6)? The aim of the succeeding chapters is to justify an affirmative answer to these three questions.

In closing this section I want to address a possible worry that might remain. I have argued that it does no good to look outside the known universe to increase one's probabilistic resources. But what about looking inside the known universe for additional probabilistic resources? Take, for instance, quantum computation. Peter Shor has described an algorithm for quantum computers that is capable of factoring numbers vastly larger than can be factored with conventional computers (thus threatening cryptographic schemes that depend on factorization constituting a hard computational problem).[54] David Deutsch therefore asks,

> When Shor's algorithm has factorized a number, using $10^{500}$ or so times the computational resources that can be seen to be present, where was the number factorized? There are only about $10^{80}$ atoms in the entire visible universe, an utterly minuscule number compared with $10^{500}$. So if the visible universe were the extent of physical reality, physical reality would not even remotely contain the resources required to factorize such a large number. Who did factorize it, then? How, and where, was the computation performed?[55]

In raising these questions, Deutsch is advocating a many-worlds interpretation of quantum mechanics. This interpretation is not mandated. Indeed, interpretations of quantum mechanics abound and all of them, insofar as they are coherent and empirically adequate, are empirically indistinguishable. As Anthony Sudbery remarks, "An interpretation of quantum mechanics is essentially an answer to the question 'What is the state vector?' Different interpretations cannot be distinguished on scientific grounds— they do not have different experimental consequences; if they did they would constitute different *theories*."[56] Yet if we resist the many-worlds interpretation of quantum mechanics and the unlimited probabilistic resources this interpretation provides, does not quantum mechanics, and quantum computation in particular, invite a huge number of probabilistic resources into our own known universe? I submit that it does not. True, quantum computation may alter the computational resources relevant to assessing the security of crypto-systems against brute force attacks that enlist the entire universe as a giant quantum computer. As a result, universal computation bounds will diverge from universal probability bounds—in the past they were largely identical because they were based on conventional computing whereas now they would diverge because of the increased computational resources due to quantum computing.

Even so, quantum computation provides no justification for altering the universal probability bound of $10^{-150}$. To see this, let us pose a related but different question from the one raised by Shor. Shor asked how large a number could be factored with quantum computers as opposed to conventional computers. He found that quantum computers vastly increased the size of the numbers that could be factored. But now let us ask how many numbers could be factored with quantum computers as opposed to conventional computers. To factor a given number on either a conventional or a quantum computer means entering it respectively as a specific sequence of bits or qubits, performing the relevant computation, and then identifying a specific output sequence as the answer. If we now ignore computation times, it follows that in terms of the sheer quantity of numbers that can be factored, quantum computation offers no advantage over conventional computation—specific numbers still have to be inputted and outputted. Input and output themselves take time, space, and material, and there are no more than $10^{150}$ specific numbers that computers, whether conventional or quantum, can ever input and output.

The lesson here is that specified complexity, precisely because it requires items of information to be specifically identified, provides no opening for quantum computation to exploit quantum parallelism or superposition and thereby generate specifications. We can imagine a quantum memory register of 1,000 qubits in a superposition of states representing every possible sequence of 0s and 1s of length 1,000. Nevertheless, this memory register is incapable of specifying even a single conventional bit string of length 1,000 until a measurement is taken and the superposition of states is projected onto an eigenstate.

Though quantum computation offers to dramatically boost computational power by allowing massively parallel computations, it does so by keeping computational states indeterminate until the very end of a computation. This indeterminateness of computational states takes the form of quantum superpositions, which are deliberately exploited in quantum computation to facilitate parallel computation. The problem with quantum superpositions, however, is that they are incapable of concretely realizing specifications. A quantum superposition is an indeterminate state. A specification is a determinate state. Measurement renders a quantum superposition determinate by producing an eigenstate, but once it does so we are no longer dealing with a quantum superposition. Because quantum computation thrives precisely where it exploits superpositions and avoids specificity, it offers no means for boosting the number of specifications that can be concretely realized in the known universe.[57]

Is there any place else to look for additional probabilistic resources inside the known universe? According to Robin Collins, quantum mechanics offers still one other loophole for inflating probabilistic resources and thereby undercutting specified complexity as a reliable indicator of design. Collins notes that the state function of a quantum mechanical system can take continuous values and thus assume infinitely many possible states. From this he draws the following conclusion: "This means that in Dembski's scheme one could only absolutely eliminate chance for events of zero probability!"[58] Presumably he thinks that because quantum systems can produce infinitely many possible events, this means that quantum systems also induce infinitely many probabilistic resources. And since infinitely many probabilistic resources coincide with a significance level of zero, my scheme could therefore only eliminate chance for events of probability zero. The problem here is that Collins fails to distinguish between the *range* of possible events that might occur and the *opportunities* for a given event to occur or be specified. A reference class of possibilities may well be infinite (as in the case of certain quantum mechanical systems). But the opportunities for sampling from such a reference class and thereby inducing information are always finite and extremely limited. Probabilistic resources always refer to the opportunities for sampling from a range of possible events. The range of possible events itself might well be infinite. But this has no bearing on the probabilistic resources associated with a given event in that range.

It appears, then, that we are back to our own known little universe, with its very limited number of probabilistic resources but therewith also its increased possibilities for detecting design. This is one instance where less is more, where having fewer probabilistic resources opens possibilities for knowledge and discovery that would otherwise be closed. Limited probabilistic resources enrich our knowledge of the world by enabling us to detect design where otherwise it would elude us. At the same time, limited probabilistic resources protect us from the unwarranted confidence in natural causes that unlimited probabilistic resources invariably seem to engender.

## 2.9   Design by Comparison[59]

In their review for *Philosophy of Science* of *The Design Inference*, Fitelson, Stephens, and Sober claim not only to show that my eliminative approach to detecting design is defective but also to offer an alternative likelihood approach that they regard as superior.[60] According to them, if design is to pass scientific muster, it must be able to generate predictions about observables.[61] By contrast, they see my approach as establishing the plausibility of design "merely by criticizing alternatives."[62] Sober's critique of my work on design did not end with this review. In his 1999 presidential address to the American Philosophical Association, Sober presented a paper titled "Testability."[63] In the first half of that paper he laid out what he regards as the proper approach for testing scientific hypotheses, namely, a likelihood approach in which hypotheses are confirmed to the degree that they render observations probable. In the second half of that paper he showed how the approach I develop for detecting design in *The Design Inference* diverges from this likelihood approach. Sober concluded that my approach to detecting design renders design untestable and therefore unscientific.

The likelihood approach that Sober and his colleagues advocate was familiar to me before I wrote *The Design Inference*. I found that approach to detecting design inadequate then and I still do. Sober's likelihood approach is a comparative approach to inferring design. In that approach all hypotheses are treated as chance hypotheses in the sense that they confer probabilities on states of affairs.[64] Thus in a competition between a design hypothesis and other hypotheses, one infers design by determining whether and the degree to which the design hypothesis confers greater probability than the others on a given state of affairs. Sober subscribes to a model of explanation known as *inference to the best explanation*, in which a "best explanation" always presupposes at least two competing explanations.[65] Inference to the best explanation eliminates hypotheses not by eliminating them individually but by setting them against each other and determining which comes out on top.

Why should eliminating a chance hypothesis always require additional chance hypotheses that compete with it? Certainly this is not a requirement for eliminating hypotheses generally. Consider the following hypothesis: "The moon is made of cheese." One does not need additional hypotheses (e.g., "The moon is a great ball of nylon.") to eliminate the moon-is-made-of-cheese hypothesis. There are plenty of hypotheses that we eliminate in isolation, and for which additional competing hypotheses do nothing to assist in eliminating them. Indeed, often with scientific problems we are fortunate if we can offer even a single hypothesis as a proposed solution (How many alternatives were there to Newtonian mechanics when Newton proposed it?). What's more, a proposed solution may be so poor and unacceptable that it can rightly be eliminated without proposing an alternative (e.g., the moon-is-made-of-cheese hypothesis). It is not a requirement of logic that eliminating a hypothesis means superseding it.

But is there something special about chance hypotheses? Sober advocates a likelihood approach to testing chance hypotheses according to which a hypothesis is confirmed to the degree that it confers increasing probability on a known state of affairs. Unlike Fisher's approach, the likelihood approach has no need for significance levels or small probabilities. What matters is the relative assignment of probabilities, not their absolute value. Also, unlike Fisher's approach (which is purely eliminative, eliminating a chance hypothesis without accepting another), the likelihood approach focuses on finding a hypothesis that confers maximum probability, thus making the elimination of hypotheses always a by-product of finding a better hypothesis. But there are problems with the likelihood approach, problems that severely limit its scope and prevent it from becoming the universal instrument for adjudicating among chance hypotheses that Sober intends. Indeed, I will argue that the likelihood approach is necessarily parasitic on Fisher's approach and that it can properly adjudicate only among hypotheses that Fisher's approach has thus far failed to eliminate.

One problem, though by no means fatal to the likelihood approach, is that there always exists a chance hypothesis that concentrates all the probability on the state of affairs in question. Within the likelihood approach, what tests a set of chance hypotheses is the probabilities they confer on a given state of affairs. Now it is always possible to stipulate a chance hypothesis that assigns probability 1 (i.e., the maximum probability allowable) to that state of affairs. To be sure, in most contexts such a hypothesis will be highly artificial and thus not up for consideration. But the point is that the likelihood approach depends on a context of inquiry that specifies only certain hypotheses and then adjudicates among them. The likelihood approach thus chooses the best among the hypotheses under consideration but is silent about how those hypotheses were put up for consideration in the first place. One therefore needs independent grounds for thinking that the hypotheses being considered are worthy of consideration and that adjudicating among them is valuable to our knowledge of the world. Bayesianism attempts to resolve this problem by assigning a prior probability distribution to the hypotheses. But this merely shifts the bump under the rug since those prior probabilities are typically impossible to justify. Fisher's approach, on the other hand, is not saddled with this problem of having to locate a hypothesis of interest within a broader class of hypotheses that compete with it. That is because Fisher's approach is able to eliminate chance hypotheses individually.

A much more serious problem is that even with independent grounds for thinking one has the right set of hypotheses, the likelihood approach can still lead to wholly unacceptable conclusions. Consider, for instance, the following experimental setup. There are two urns, one with five white balls and five black balls, the other with seven white balls and three black balls. One of these urns will be sampled with replacement a thousand times, but we do not know which. The chance hypothesis characterizing the first urn is that white balls should on average occur the same number of times as black balls, and the chance hypothesis characterizing the second urn is that white balls should on average outnumber black balls by a ratio of seven to three. Suppose now we are told that one of the urns was sampled and that all the balls ended up being white. The probability of this event by sampling from the first urn is roughly 1 in $10^{300}$ whereas the probability of this event by sampling from the second urn is roughly 1 in $10^{155}$.

The second probability is therefore almost 150 orders of magnitude greater than the first. Thus on the likelihood approach, the hypothesis that the urn had seven white balls is vastly better confirmed than the hypothesis that it only had five. But getting all white balls from the urn with seven white balls is a specified event of small probability, and on Fisher's approach to hypothesis testing should be eliminated as well (drawing with replacement from this urn 1000 times, we should expect on average around 300 black balls from this urn, and certainly not a complete absence of black balls). This comports with our best probabilistic intuitions: Given these two urns and a thousand white balls in a row, the only sensible conclusion is that *neither* urn was randomly sampled, and any superiority of the "urn two" hypothesis over the "urn one" hypothesis is utterly insignificant. To be forced to choose between these two hypotheses is like being forced to choose between the moon being made entirely of cheese or the moon being made entirely of nylon. Any superiority of the one hypothesis over the other drowns in a sea of inconsequentiality.

The likelihood principle, being an inherently comparative instrument, has nothing to say about the absolute value of the probability (or probability density) associated with a state of affairs, but only their relative magnitudes. Consequently, the vast improbability of either urn hypothesis in relation to the sample chosen (i.e., 1000 white balls) would on strict likelihood grounds be irrelevant to any doubts about either hypothesis. Indeed, I will argue that any such doubts could only arise by presupposing specified complexity as a general instrument for eliminating chance (more on this in section 2.10).

Having performed a likelihood analysis and having found that the hypothesis on which this analysis conferred maximal probability is dubious, one cannot on purely likelihood grounds add another hypothesis to the mix—there must be good independent reasons for expanding the range of chance hypotheses considered, for without some such constraint one can, as in the previous objection, simply add a hypothesis that confers a probability of one (i.e., the maximum allowable probability) on the state of affairs in question. Bayesians, to expand the range of hypotheses, resort to prior probabilities—in this case assigning prior probabilities that do not sum to 1 to the two urn hypotheses, thus conferring nonzero probability on some other hypothesis.

Expanding one's range of hypotheses in response to new evidence typically amounts to adding a "none-of-the-above hypothesis" to the mix. But this is illegitimate. A none-of-the-above hypothesis, which takes the hypotheses in a previous likelihood analysis and then merely asserts that none of these hypotheses accounts for a state of affairs in question, is not properly speaking even a hypothesis capable of conferring probabilities. Indeed, there is no way to assess the probability conferred on the selection of 1000 white balls by the (meta-)hypothesis that it was not an urn with five white and five black balls or an urn with seven white and three black balls.

It may help to recast the problem inherent in the urn example with an analogy. Suppose you are the admissions officer at a prestigious medical school. Lots of people want to get into your medical school; indeed, so many that even among qualified applicants you cannot admit them all. You feel bad about these qualified applicants who do not make it and wish them well. Nonetheless, you are committed to getting the best students possible, so among qualified applicants you choose only those at the very top. What's more, because your medical school is so prestigious, you are free to choose only from the top. There is, however, another type of student who applies to your medical school, one whose grades are poor, who shows no intellectual spark, and who lacks the requisite premedical training. These are the unqualified students, and you have no compunction about weeding them out immediately, nor do you care what their fate is in graduate school. In this analogy the unqualified students are the hypotheses weeded out by Fisher's approach to hypothesis testing whereas the qualified students are those sifted by the likelihood approach. If, perchance, only unqualified students apply one year (compare the two urns in our example, neither of which can plausibly account for the utter absence of black balls in 1,000 draws), the right thing to do would be to reject all of them rather than to admit the best of a bad lot.

Another problem with the likelihood approach is that there exist cases where one and only one statistical hypothesis is relevant and needs to be tested. For instance, suppose G is a compact topological group with group operation * and Haar probability measure **U** (**U** is invariant under the group operation—one can think of **U** as the uniform probability on G).[66] Suppose that (1) X is a random variable taking values in G but of unknown probability distribution; (2) Y is a random variable also taking values in G but with known probability distribution **U**; and (3) the processes responsible for X and Y are known to be causally independent (e.g., Y results from sampling a quantum mechanical device and X results from sampling some totally unrelated system). Then X and Y are stochastically independent. If we now define Z as $X*Y^{-1}$, there is only one hypothesis **H** relevant to Z, namely, whether Z has probability distribution **U**. This is because $X*Y^{-1}$ is the group product of X and the group inverse of Y, and the probability distribution induced by this product is always **U** provided that one of the group factors is distributed as **U** and that the two factors are stochastically independent (the mathematical justification is that convolving any probability measure with Haar measure yields Haar measure). In addressing whether Z has probability distribution **U**, **H** addresses whether X and Y are independently distributed. Given the experimental setup, this is the only hypothesis relevant to the statistical relation between X and Y. There are no other relevant statistical hypotheses.

I want next to examine the very idea of hypotheses conferring probability, an idea that is mathematically straightforward but that becomes problematic once the likelihood approach gets applied in practice. According to the likelihood approach, chance hypotheses confer probability on states of affairs, and hypotheses that confer maximum probability are preferred over others. But what exactly are these hypotheses that confer probability? In practice, the likelihood approach is too cavalier about the hypotheses it permits. Urn models as hypotheses are fine and well because they induce well-defined probability distributions. Models for the formation of functional protein assemblages might also induce well-defined probability distributions, though determining the probabilities here will be considerably more difficult (see section 5.10). But what about hypotheses like "Natural selection and random mutation together are the principal driving force behind biological evolution" or "God designed living organisms"? Within the likelihood approach, any claim can be turned into a chance hypothesis on the basis of which likelihood theorists then assign probabilities. Claims like these, however, do not induce well-defined probability distributions. And since most claims are like this (i.e., they fail to induce well-defined probability distributions), likelihood analyses regularly become exercises in rank subjectivism.

Consider, for instance, the following analysis taken from Sober's text *Philosophy of Biology*.[67] Sober considers the following state of affairs: E—"Living things are intricate and well-suited to the task of surviving and reproducing." He then considers three hypotheses to explain this state of affairs: $H_1$— "Living things are the product of intelligent design"; $H_2$—"Living things are the product of random physical processes"; $H_3$—"Living things are the product of random variation and natural selection." As Sober explains, prior to Darwin only $H_1$ and $H_2$ were live options, and E was more probable given $H_1$ than given $H_2$. Prior to Darwin, therefore, the design hypothesis was better supported than the chance hypothesis. But with Darwin's theory of random variation and natural selection, the playing field was expanded and E became more probable given $H_3$ than either $H_1$ or $H_2$.

Now my point is not to dispute whether in Darwin's day $H_3$ was a better explanation of E than either $H_1$ or $H_2$. My point, rather, is that Sober's appeal to probability theory to make $H_1$, $H_2$, and $H_3$ each confer a probability on E is misleading, lending an air or mathematical rigor to what really is just Sober's own subjective assessment of how plausible these hypotheses seem to him. Nowhere in this example do we find precise numbers attached to Sober's likelihoods. The most we see are inequalities of the form $P(E|H_1) \gg P(E|H_2)$, signifying that the probability of E given $H_1$ is much greater than the probability of E given $H_2$ (for the record, the "much greater" symbol "$\gg$" has no precise mathematical meaning). But what more does such an analysis do than simply assert that with respect to the intricacy and adaptedness of organisms, intelligent design is a much more convincing explanation to Sober than the hypothesis of pure chance? And since Sober presumably regards $P(E|H_3) \gg P(E|H_1)$, the Darwinian explanation is for him an even better explanation of the intricacy and adaptedness of organisms than intelligent design. The chance hypotheses on which Sober pins his account of scientific rationality and testability are not required to issue in well-defined probability distributions. Sober's probabilities are therefore probabilities in name only.

Or consider the supposed improvement that a likelihood analysis brings to one of my key examples for motivating the design inference, namely, the "Caputo case." Nicholas Caputo, a county clerk from New Jersey was charged with cheating because he gave the preferred ballot line to Democrats over Republicans 40 out of 41 times (the improbability here is that of flipping a fair coin 41 times and getting 40 heads). I have given my analysis of the

Caputo case in sections 2.3 and 2.7, but for now I want to focus on Sober's analysis. Here it is—in full detail:

> There is a straightforward reason for thinking that the observed outcomes favor Design over Chance. If Caputo had allowed his political allegiance to guide his arrangement of ballots, you would expect Democrats to be listed first on all or almost all of the ballots. However, if Caputo did the equivalent of tossing a fair coin, the outcome he obtained would be very surprising.[68]

Robin Collins takes the likelihood analysis no further:

> The actual ballot selection pattern was much more probable under the cheating hypothesis [i.e., the design hypothesis] than under the hypothesis that it occurred by chance. Thus, the ballot pattern confirms that Caputo cheated. Indeed, it is so much more probable under the one hypothesis than under the other that the confirmation turns out to be extremely strong, so strong that virtually any court would conclude cheating was involved.[69]

Such analyses do not go far enough. To see this, ask yourself how many times Caputo had to give the Democrats the top ballot line before it became evident that he was cheating. Two for the Democrats, one for the Republicans? Three for the Democrats, one for the Republicans? Four for the Democrats, one for the Republicans? Etc. On a likelihood analysis, any disparity favoring the Democrats provides positive evidence for Caputo cheating. But where is the cutoff? There is a point up to which giving his own Democratic party the top ballot line could be regarded as entirely innocent. There is a point after which it degenerates into obvious cheating and manipulation (residents of New Jersey recognized that Caputo had reached that point when they unofficially baptized him "the man with the golden arm"). Specified complexity—and not a likelihood analysis—determines the cutoff. Indeed, specified complexity determines when advantages apparently accruing from chance can no longer legitimately be attributed to chance.

So simple and straightforward do Sober and his co-authors regard their likelihood analysis that they mistakenly conclude: "Caputo was brought up on charges and the judges found against him."[70] Caputo was brought up on charges of fraud, but in fact the New Jersey Supreme Court justices did not find against him.[71] The probabilistic analysis that Sober and fellow likelihood theorists find so convincing is viewed with skepticism by the legal system, and for good reason. Within a likelihood approach, the probabilities conferred by design hypotheses are notoriously imprecise and readily lend themselves to miscarriages of justice.[72]

And this brings us to the final problem with the likelihood approach that I want to consider, namely, its treatment of design hypotheses as chance hypotheses. For Sober any hypothesis can be treated as a chance hypothesis in the sense that it confers probability on a state of affairs. As we have seen, there is a problem here because Sober's probabilities float free of well-defined probability distributions and thus become irretrievably subjective. But even if we bracket this problem, there is a problem treating design hypotheses as chance hypotheses, using design hypotheses to confer probability (now conceived in a loose, subjective sense) on states of affairs. To be sure, designing agents can do things that follow well-defined probability distributions. For instance, even though I act as a designing agent in writing this book, the distribution of letter frequencies in it follow a well-defined probability distribution in which the relative frequency of the letter *e* is approximately 13 percent, that of *t* approximately 9 percent, etc.—this is the distribution of letters for English texts.[73] Such probability distributions ride, as it were, epiphenomenally on design hypotheses.[74] Thus in this instance, the design hypothesis identifying me as author of this book confers a certain probability distribution on the letter frequencies of it. (But note, if these letter frequencies were substantially different, a design hypothesis might well be required to account for the difference. In 1939 Ernest Vincent Wright published a novel of over 50,000 words titled *Gadsby* that contained no occurrence of the letter *e*. Clearly, the absence of the letter *e* was designed.[75])

Sober, however, is much more interested in assessing probabilities that bear directly on a design hypothesis than in characterizing chance events that ride epiphenomenally on it. In the case of letter frequencies, the fact that letters in this book appear with certain relative frequencies reflects less about the design hypothesis that I am its author than about the (impersonal) spelling rules of English. Thus with respect to intelligent design in biology, Sober wants to know what sorts of biological systems should be expected from an intelligent designer having certain characteristics, and not what sorts of random epiphenomena might be associated with such a designer. What's more, Sober claims that if the design theorist cannot answer this question (i.e., cannot predict the sorts of biological systems that might be expected on a design hypothesis), then intelligent design is untestable and therefore unfruitful for science.

Yet to place this demand on design hypotheses is ill-conceived. We infer design regularly and reliably without knowing characteristics of the designer or being able to assess what the designer is likely to do. Sober himself admits as much in a footnote that deserves to be part of his main text: "To infer watchmaker from watch, you needn't know exactly what the watchmaker had in mind; indeed, you don't even have to know that the watch is a device for measuring time. Archaeologists sometimes unearth tools of unknown function, but still reasonably draw the inference that these things are, in fact, *tools*."[76]

Sober is wedded to a Humean inductive tradition in which all our knowledge of the world is an extrapolation from past experience.[77] Thus for design to be explanatory, it must fit our preconceptions, and if it does not, it must lack epistemic support. For Sober, to predict what a designer would do requires first looking to past experience and determining what designers in the past have actually done. A little thought, however, should convince us that any such requirement fundamentally misconstrues design. Sober's likelihood approach puts designers in the same boat as natural laws, locating their explanatory power in an extrapolation from past experience. To be sure, designers, like natural laws, can behave predictably (designers often institute *policies* that are dutifully obeyed). Yet unlike natural laws, which are universal and uniform, designers are also innovators. Innovation, the emergence to true novelty, eschews predictability. A likelihood analysis generates predictions about the future by conforming the present to the past and extrapolating from it. It therefore follows that design cannot be subsumed under a likelihood framework. Designers are inventors. We cannot predict what an inventor would do short of becoming that inventor.

But the problem goes deeper. Not only can Humean induction not tame the unpredictability inherent in design; it cannot account for how we recognize design in the first place. Sober, for instance, regards the design hypothesis for biology as fruitless and untestable because it fails to confer sufficient probability on biologically interesting propositions. But take a different example, say from archeology, in which a design hypothesis about certain aborigines confers a large probability on certain artifacts, say arrowheads. Such a design hypothesis would on Sober's account be testable and thus acceptable to science. But what sort of archeological background knowledge had to go into that design hypothesis for Sober's likelihood analysis to be successful? At the very least, we would have had to have past experience with arrowheads. But how did we recognize that the arrowheads in our past experience were designed? Did we see humans actually manufacture those arrowheads? If so, how did we recognize that these humans were acting deliberately as designing agents and not just randomly chipping away at random chunks of rock (carpentry and sculpting entail design; but whittling and chipping, though performed by intelligent agents, do not)? As is evident from this line of reasoning, the induction needed to recognize design can never get started.[78] Our ability to recognize design must therefore arise independently of induction and therefore independently of Sober's likelihood framework.

The direction of Sober's logic is from design hypothesis to designed object, with the design hypothesis generating predictions or expectations about the designed object. Yet in practice we start with objects that initially we may not know to be designed. Then by identifying general features of those objects that reliably signal design, we infer to a designing intelligence responsible for those objects. Still further downstream in the logic is an investigation into the specific design characteristics of those objects (e.g., How was the object constructed? How could it have been constructed? What is its function? What effect have natural causes had on the original design? Is the original design recoverable? How much has the original design been perturbed? How much perturbation can the object allow and still remain functional?—see section 6.1). But what are those general features of designed objects that set the design inference in motion and reliably signal design? The answer I am urging is *specification* and *complexity*. Indeed, the entire argument of this book centers on the claim that specified complexity is a reliable empirical marker of intelligent design.

## 2.10 Design by Elimination

The defects in Sober's likelihood approach are, in my view, so grave that it cannot provide an adequate account of how design hypotheses are inferred.[79] The question remains, however, whether specified complexity can provide an adequate account for how design hypotheses are inferred. The worry here centers on the move from specified complexity to design. Specified complexity is a statistical notion. Design, as generally understood, is a causal notion. How do the two connect? In chapter 1 I argued the connection as follows. First (section 1.6) I offered an inductive argument, showing that in all cases where we know the causal history and where specified complexity was involved, an intelligence was involved as well. The inductive generalization that follows is that all cases of specified complexity involve intelligence. Next (section 1.7) I argued that choice is the defining feature of intelligence and that specified complexity is how in fact we identify choice.

Although I regard these two arguments as utterly convincing, critics do not. The problem according to Sober is that specified complexity detects design purely by elimination, telling us nothing positive about how an intelligent designer might have produced an object we observe. Sober regards this as a defect. I regard it as a virtue. I will come back to why I regard it as a virtue, but for the moment let us consider this criticism on its own terms.

Take, for instance, a biological system, one that exhibits specified complexity, but for which we have no clue how an intelligent designer might have produced it. To employ specified complexity as a marker of design here seems to tell us nothing except that the object is designed. Indeed, when we examine the logic of detecting design via specified complexity, at first blush it looks purely eliminative. The "complexity" in "specified complexity" is a measure of improbability. Now probabilities are always assigned in relation to chance hypotheses. Thus, to establish specified complexity requires defeating a set of chance hypotheses. Specified complexity therefore seems at best to tell us what is not the case, not what is the case.

In response to this criticism, note first that even though specified complexity is established via an eliminative argument, it is not fair to say that it is established via a *purely* eliminative argument. If the argument were purely eliminative, one might be justified in saying that the move from specified complexity to a designing intelligence is an argument from ignorance (i.e., not X therefore Y). But unlike Fisher's approach to hypothesis testing, in which individual chance hypotheses get eliminated without reference to the entire set of relevant chance hypotheses that might explain a phenomenon, specified complexity presupposes that the entire set of relevant chance hypotheses has first been identified.[80] This takes considerable background knowledge. What's more, it takes considerable background knowledge to come up with the right pattern (i.e., specification) for eliminating all those chance hypotheses and thus for inferring design. Design inferences that infer design by identifying specified complexity are therefore not purely eliminative. They do not merely exclude, but they exclude from an exhaustive set in which design is all that remains once the inference has done its work (which is not to say that the set is logically exhaustive; rather, it is exhaustive with respect to the inquiry in question—that is all we can ever do in science). Design inferences, by identifying specified complexity, exclude everything that might in turn exclude design.

The question remains, however, What is the connection between design as a statistical notion (i.e., specified complexity) and design as a causal notion (i.e., the action of a designing intelligence)? Now it is true that simply knowing that an object is complex and specified tells us nothing positive about its causal history. To be sure, it tells us something negative about its causal history, namely, that the phenomenon in question was not produced by an undirected natural process rendering it probable. Even so, specified complexity by itself provides no causal details and thus, from a likelihood perspective, no explanation. Sober regards this as a defect of the concept. Yet it might equally well be regarded as a virtue for enabling us neatly to separate whether something is designed from how it was produced. Once specified complexity tells us that something is designed, there is nothing to stop us from inquiring into its production. A design inference therefore does not avoid the problem of how a designing intelligence might have produced an object. It simply makes it a separate question.

The claim that design inferences are purely eliminative is false, and the claim that they provide no (positive) causal story is true but hardly relevant—causal stories must always be assessed on a case-by-case basis independently of general statistical considerations. So where is the problem in connecting design as a statistical notion (i.e., specified complexity) to design as a causal notion (i.e., the action of a designing intelligence), especially given the widespread use of specified complexity in circumstantial evidence to identify the activity of intelligent agents, and also given the absence of counterexamples for generating specified complexity apart from intelligence?

In fact, the absence of counterexamples is very much under dispute. I have addressed and answered what I regard as the most important counterexamples and counterarguments in earlier sections of this chapter (see especially sections 2.5 and 2.8, which focused on whether specified complexity is well-defined and whether universal probability bounds can be sustained in the face of unlimited probabilistic resources). Even so, there is one line of objection I still want to take up. Sober and colleagues argue that specified complexity is unable to handle conjunctive, disjunctive, and mixed explananda.[81] Let us deal with these in order. Conjunctions are supposed to present a problem for specified complexity because a conjunction can exhibit specified complexity even though none of its conjuncts do individually. Thus, if specified complexity is taken as an indicator of design, this means that even though the conjunction gets attributed to design, each of the conjuncts get attributed to chance. Although this may seem counterintuitive, it is not clear why it should be regarded as a problem. Consider a Scrabble board with Scrabble pieces. Chance can explain the occurrence of any individual letter at any individual location on the board. Nevertheless, meaningful conjunctions of those letters arranged sequentially on the board are not attributable to chance. It is important to understand that chance is always a provisional designation, which can be overturned once closer examination reveals specified complexity. Thus attributing chance to the isolated positioning of a single Scrabble piece does not contradict attributing design to the joint positioning of multiple Scrabble pieces into a meaningful arrangement.

Disjunctions are a little trickier. Disjunctions are supposed to pose a problem in the case where some of the disjuncts exhibit specified complexity but the disjunction itself is no longer complex and therefore no longer exhibits specified complexity. Thus we would have a case where a disjunct signifies design, but the disjunction does not. How can this run into trouble? Certainly there is no problem in the case where one of the disjuncts is highly probable. Consider the disjunction, either the arrow lands in the target or outside. If the target is sufficiently small, the arrow landing in the target would constitute a case of specified complexity. But the disjunction itself is a tautology and the event associated with it can readily be attributed to chance.

How else might specified complexity run into trouble with disjunctions? Another possibility is that all the disjuncts are improbable. For instance, consider a lottery in which there is a one-to-one correspondence between players and winning possibilities. Suppose further that each player predicts he or she will win the lottery. Now form the disjunction of all these predictions. This disjunction is a tautology, logically equivalent to the claim that some one of the players will win the lottery (which is guaranteed since players are in one-to-one correspondence with winning possibilities). Clearly, as a tautology, this disjunction does not exhibit specified complexity and therefore does not signify design. But what about the crucial disjunct in this disjunction, namely, the prediction by the winning lottery player? As it turns out, this disjunct can never exhibit specified complexity either. This is because the number of disjuncts count as probabilistic resources. This number is the same as the number of lottery players and ensures that the prediction by the winning lottery player never attains the degree of complexity/improbability needed to exhibit specified complexity (see section 2.7). A lottery with N players has at least N probabilistic resources, and once these are factored in, the correct prediction by the winning lottery player is no longer improbable. In general, once all the relevant probabilistic resources connected with a disjunction are factored in, apparent difficulties associated with attributing a disjunct to design and the disjunction to chance disappear.

Finally, the case of mixed explananda is easily dispatched. Suppose we are given a conjunction of two conjuncts in which one exhibits specified complexity and the other does not. In that case one will be attributed to design and the other to chance. And what about the conjunction? The conjunction will be at least as improbable/complex as the first conjunct (the one that exhibits specified complexity). What's more, the pattern qua specification that delimits the first conjunct will necessarily delimit the conjunction as well (conjunctions always restrict the space of possibilities more than their conjuncts). Consequently the conjunction will itself exhibit specified complexity and be attributed to design. Note that this is completely unobjectionable. Specified complexity, in signaling design, merely says that an intelligent agent was involved. It does not require that intelligent agency account for every aspect of a thing in question. The physical medium that you, the reader, are now viewing as you read exhibits random variations due to chance, but the text that this physical medium conveys is due to intelligent agency.

Although death by counterexample would certainly be a legitimate way for specified complexity to fail as a reliable indicator of intelligence, Sober, Collins, and other critics suggest that there is still another way for it to fail. Namely, employing specified complexity to detect design fails as a rational reconstruction of how we detect design in common life. I have argued in this chapter that the reconstruction works, especially when Fisher's concept of statistical significance testing is supplemented with the notion of a probabilistic resource. What's more, I have argued in this chapter that the likelihood approach fails to give an adequate account of how we detect design. Still more problematic for these critics, however, is that the likelihood approach—even if it could be made to work—can make sense of design only by presupposing specified complexity.

To see this, take an event that is the product of intelligent design but for which we have not yet seen the relevant pattern that makes its design clear to us (take a Search for Extraterrestrial Intelligence example in which a long sequence of prime numbers, say, reaches us from outer space, but suppose we have not yet seen that it is a sequence of prime numbers). Without that pattern we will not be able to distinguish between the probability that this event takes the form it does given that it is the result of chance, and the probability that it takes the form it does given that it is the result of design. Consequently, we will not be able to infer design for this event. Only once we see the pattern will we, on a likelihood analysis, be able to see that the latter probability is greater than the former. But what are the right sorts of patterns that allow us to see that? Not all patterns signal design. What's more, the pattern needs to delimit an event of sufficient improbability (i.e., complexity) for otherwise the event can readily be referred to chance. We are back, then, to needing some account of complexity and specification. Thus a likelihood analysis that pits competing design and chance hypotheses against each other must itself presuppose the legitimacy of specified complexity as a reliable indicator of intelligence.

Nor is the likelihood approach salvageable. Lydia and Timothy McGrew, philosophers at Western Michigan University, think that likelihoods are ideally suited for detecting design in the natural sciences but that my Fisherian approach to specified complexity breaks down. Taking issue with both Sober and me, they argue that the presence of irreducible complexity in biological systems constitutes a state of affairs upon which the design hypothesis confers greater probability than the Darwinian hypothesis.[82] Irreducible complexity is biochemist Michael Behe's notion. According to Behe, a system is irreducibly complex if it is "composed of several well-matched, interacting parts that contribute to the basic function, wherein the removal of any one of the parts causes the system to effectively cease functioning."[83]

The McGrews are looking for some property of biological systems upon which the design hypothesis confers greater probability than its naturalistic competitors. This sounds reasonable until one considers such properties more carefully. For the McGrews specified complexity is disallowed because it is a statistical property that depends on Fisher's approach to hypothesis testing, and they regard this approach as not rationally justified (which I have argued it is once one introduces the notion of a probabilistic resource). What they apparently fail to realize, however, is that any property of biological systems upon which a design hypothesis confers greater probability than a naturalistic competitor must itself presuppose specified complexity.

Ultimately what enables irreducible complexity to signal design is that it is a special case of specified complexity. Behe admits as much in his public lectures whenever he points to my work in *The Design Inference* as providing the theoretical underpinnings for his own work on irreducible complexity. The connection between irreducible complexity and specified complexity is easily seen. The irreducibly complex systems Behe considers require numerous components specifically adapted to each other and each necessary for function. On any formal complexity-theoretic analysis, they are complex (see section 5.10). Moreover, in virtue of their function, these systems embody independently given patterns that can be identified without recourse to actual living systems. Hence these systems are also specified. Irreducible complexity is thus a special case of specified complexity.

But the problem goes even deeper. Name any property of biological systems that favors a design hypothesis over its naturalistic competitors, and you will find that what makes this property a reliable indicator of design is that it is a special case of specified complexity—if not, such systems could readily be referred to chance. William Paley's adaptation of means to ends,[84] Harold Morowitz's minimal complexity,[85] Marcel Schützenberger's functional complexity,[86] and Michael Behe's irreducible complexity all, insofar as they reliably signal design, have specified complexity at their base. Thus, even if a likelihood analysis could coherently assign probabilities conditional upon a design hypothesis (a claim I disputed in section 2.9), the success of such an analysis in detecting design would depend on a deeper probabilistic analysis that finds specified complexity at its base.

Consequently, if there is a way to detect design, specified complexity is it. But specified complexity does not just render design detectable; it also renders it testable (see section 6.9). Indeed, specified complexity tests design by precluding the very explanations that alone could preclude design. In fine, all the evidence to date suggests that specified complexity—and not a likelihood analysis—is the key to detecting design and turning intelligent design into a fully testable scientific research program.

# Notes

1. For a brief summary of Fisher's views on tests of significance and null hypotheses, see Ronald A. Fisher, *The Design of Experiments* (New York: Hafner, 1935), 13–17. For Fisher's approach contrasted with the Neyman-Pearson approach, see Bernard Lindgren, *Statistical Theory*, 3rd ed. (New York: Macmillan, 1976), 288–290.

2. Colin Howson and Peter Urbach, *Scientific Reasoning: The Bayesian Approach*, 2nd ed. (LaSalle, Ill.: Open Court, 1993), 178. For a similar criticism see Ian Hacking, *Logic of Statistical Inference* (Cambridge: Cambridge University Press, 1965), 81–83.

3. See A. G. Hamilton, *Logic for Mathematicians* (Cambridge: Cambridge University Press, 1978), 25.

4. See Heinz Bauer, *Probability Theory and Elements of Measure Theory*, trans. R. B. Burckel, 2nd English ed. (New York: Academic Press, 1981), 131–132.

5. William A. Dembski, *Design Inference: Eliminating Chance through Small Probabilities* (Cambridge: Cambridge University Press, 1998), ch. 6. See also section 1.5 of this book.

6. Note that the probabilistic resources here are not experimental trials (which would automatically be factored into any probabilities computed for an experiment), but the relevant number of experiments that might be performed and that might therefore result in a false positive.

7. The details of justifying Fisher's approach to hypothesis testing can be found in chapter 6 of *The Design Inference*.

8. For a full account of such probabilities, see my article "Uniform Probability," *Journal of Theoretical Probability* 3(4) (1990): 611–626. On the real line this measure is just Lebesgue measure.

9. Strictly speaking, $P(\cdot|H)$ can be represented as $f \cdot dU$ only if $P(\cdot|H)$ is absolutely continuous with respect to $U$, i.e., all the subsets of $\Omega$ that have zero probability with respect to $U$ must also have zero probability with respect to $P(\cdot|H)$ (this is just the Radon-Nikodym Theorem—see Bauer, *Probability Theory*, 86). Nonetheless, when $\Omega$ is finite, any probability measure whatsoever is absolutely continuous with respect to the uniform probability. What's more, in most cases where $\Omega$ is infinite, it is possible to approximate probability measures of the form $P(\cdot|H)$ arbitrarily closely by probability measures of the form $f \cdot dU$ (the latter probabilities can be made to "converge weakly" to the former—see Patrick Billingsley, *Convergence of Probability Measures*, 2nd ed. [New York: Wiley, 1999],

ch. 1). Physicists and engineers bypass questions of absolute continuity and weak convergence by thinking of f as a generalized function—see for instance Paul Dirac, *The Principles of Quantum Mechanics*, 4th ed. (Oxford: Oxford University Press, 1958), 48 or Michael Reed and Barry Simon, *Methods of Modern Mathematical Physics I: Functional Analysis*, revised and enlarged (New York: Academic Press, 1980), 148.

10. William Feller, *An Introduction to Probability Theory and Its Applications*, vol. 1, 3rd ed. (New York: Wiley, 1968), 167–169.

11. The statistics text in question is David Freedman, Robert Pisani, and Roger Purves, *Statistics* (New York: Norton, 1978), 426–427. Fisher's original account can be found in Ronald A. Fisher, *Experiments in Plant Hybridisation* (Edinburgh: Oliver and Boyd, 1965), 53. For a more recent reevaluation of Mendel's data, which still concludes that "the segregations are in general closer to Mendel's expectations than chance would dictate," see A. W. F. Edwards, "Are Mendel's Results Really Too Close?" *Biological Review* 61 (1986): 295–312.

12. See Bauer, *Probability Theory*, 44. Indicator functions are also known as characteristic functions.

13. For the *New York Times* account quoted here, see the 23 July 1985 issue, page B1. For a statistics textbook that analyzes the Caputo case, see David S. Moore and George P. McCabe, *Introduction to the Practice of Statistics*, 2nd ed. (New York: W. H. Freeman, 1993), 376–377.

14. For finite probability spaces, the canonical measure is the counting measure. Consequently, probabity density functions on such spaces are always relative to the counting measure and thus simply evaluate to the probability of individual points in the probability space.

15. Gregory J. Chaitin, "On the Length of Programs for Computing Finite Binary Sequences," *Journal of the Association for Computing Machinery* 13 (1966): 547–569; Andrei Kolmogorov, "Three Approaches to the Quantitative Definition of Information," *Problemy Peredachi Informatsii* (in translation) 1(1) (1965): 3–11; Ray J. Solomonoff, "A Formal Theory of Inductive Inference, Part I," *Information and Control* 7 (1964): 1–22 and Ray J. Solomonoff, "A Formal Theory of Inductive Inference, Part II," *Information and Control* 7 (1964): 224–254.

16. See Hartley Rogers Jr., *Theory of Recursive Functions and Effective Computability* (1967; reprinted Cambridge, Mass.: MIT Press, 1987).

17. For an overview of the algorithmic information theoretic approach to randomness, see Peter Smith, *Explaining Chaos* (Cambridge: Cambridge University Press, 1998), ch. 9.

18. See C. H. Woo, "Laws and Boundary Conditions," in *Complexity, Entropy and the Physics of Information*, ed. W. H. Zurek (Reading, Mass.: Addison-Wesley, 1990), 132–133, where Woo offers some preliminary remarks about the connection between thermodynamic and algorithmic entropy.

19. Bauer, *Probability Theory*, 139.

20. Howson and Urbach, *Scientific Reasoning*, 181–183.

21. See Earl A. Coddington and Norman Levinson, *Theory of Ordinary Differential Equations* (1955; reprinted Malabar, Florida: Krieger, 1984), ch. 1, which deals entirely with the existence and uniqueness of solutions to ordinary differential equations.

22. G. H. Hardy and E. M. Wright, *An Introduction to the Theory of Numbers*, 5th ed. (Oxford: Clarendon Press, 1979), 128.

23. Cf. the conventionality of language. No string of symbols or sounds is intrinsically a word, though any such string could be a word in some language. But whether it is a word depends on the actual usage of that string within a linguistic community.

24. John R. Searle, *The Construction of Social Reality* (New York: Free Press, 1995), 8–12.

25. See Dembski, *The Design Inference*, ch. 5.

26. One possibility worth considering is that we do not know enough because there is not objectively enough "out there" that could be known to partition $\Omega$ with detachable rejection regions of size $\alpha$ (provided $\alpha$ is small enough). I am indebted to Bruce Gordon for pointing out this possibility to me.

27. It seems to me that these considerations effectively answer Willard Quine's arguments for the indeterminacy of translation. See Willard Quine, *Word and Object* (Cambridge, Mass.: MIT Press, 1960), 72–79.

28. Dembski, *The Design Inference*, sec. 5.3.

29. See Dembski, *The Design Inference*, ch. 4. If the subject S is a computational system, then $\varphi$ is a computational complexity measure (e.g., the degree of compressibility of a specification as measured within algorithmic information theory—see section 2.4). I like to think of $\varphi$ as measuring S's disposition to output a given specification. Given the rejection region qua specification R, we imagine S without any knowledge of E or R outputting patterns in $\Omega$ that correspond to events of probability no more than the probability of R. S's strategy is to output as many patterns as possible, hoping that R is among them. Thus to output R, S might have to output other patterns as well. It is therefore not as though S has one, and only one, opportunity to hit the right pattern. The crucial thing is that the target pattern R be among those patterns outputted. Thus we imagine S outputting a list of patterns: $R_0, R_1, R_2, R_3, \ldots$ Moreover, the order is determined by S's relative disposition to output patterns (for $i < j$, S is more disposed to output $R_i$ than $R_j$—for simplicity let us assume that ties are not a problem). The first place that R appears on this list is then the complexity of R: $\varphi(R) = n$ for $R = R_n$ and $R \neq R_i$ for $i < n$. This is, of course, intuitive. The formal treatment is given in the text.

30. Robin Collins, "An Evaluation of William A. Dembski's *The Design Inference*: A Review Essay," *Christian Scholar's Review* 30(3) (2001): 329–341. See also my response in that same issue—William A. Dembski, "Detecting Design by Eliminating Chance: A Response to Robin Collins," *Christian Scholar's Review* 30(3) (2001): 343–357.

31. See Dembski, *The Design Inference*, 207–209.

32. See Kenneth W. Dam and Herbert S. Lin, eds., *Cryptography's Role in Securing the Information Society* (Washington, D.C.: National Academy Press, 1996), 380, n. 17. In examining cryptography's role in securing the information society, Dam and Lin propose a universal probability bound of $10^{-95}$. Note that quantum computers cannot circumvent universal probability bounds—see Dembski, *The Design Inference*, 210, n. 16. I address quantum computation as well in section 2.8.

33. Stuart Kauffman, *Investigations* (New York: Oxford University Press, 2000). Al-

though Kauffman does not explicitly mention the phrase "specified complexity," his emphasis throughout this book is on the complexity of biological systems, and the type of complexity he is concerned to explain is in fact specified complexity.

34. Ibid., 144.

35. Ibid.

36. Ibid., 138.

37. Ibid., 137–138, 144, 162, 167.

38. See Michael Hallett, *Cantorian Set Theory and Limitation of Size* (Oxford: Oxford University Press, 1984), 55–56.

39. Peter Rüst refers to such numbers as "transastronomical." See Peter Rüst, "How Has Life and Its Diversity Been Produced?" *Perspectives on Science and Christian Faith* 44(2) (1992): 80. Emile Borel referred to the reciprocal of such numbers as "probabilities which are negligible on the supercosmic scale." See Emile Borel, *Probabilities and Life*, trans. M. Baudin (New York: Dover, 1962), 28–30.

40. See Kauffman, *Investigations*, 144, where he switches indiscriminately between referring to "the known universe" and simply "the universe."

41. Dembski, *The Design Inference*, sec. 6.6.

42. See respectively Alan Guth, *The Inflationary Universe: The Quest for a New Theory of Cosmic Origins* (Reading, Mass.: Addison-Wesley, 1997); Hugh Everett III, " 'Relative State' Formulation of Quantum Mechanics," *Reviews of Modern Physics* 29 (1957): 454–462; Lee Smolin, *The Life of the Cosmos* (Oxford: Oxford University Press, 1997); and David Lewis, *On the Plurality of Worlds* (Oxford: Basil Blackwell, 1986).

43. Strictly speaking an observer is not necessary. All that is necessary for quantum measurement is that to each eigenstate for a subsystem there correspond a unique relative state for the remainder of the whole system. If the subsystem is the whole universe, however, then there is no remainder and nothing (apparently) to do the measuring. Everett's solution is to deny that state functions collapse to eigenstates and assert instead that all possible eigenstates are realized. Simon Saunders thinks that sense can be made of Everett's solution without postulating many worlds. See Simon Saunders, "Decoherence, Relative States, and Evolutionary Adaptation," *Foundations of Physics* 23 (1993): 1553–1595.

44. The need for independent evidence to confirm a scientific theory has frequently been noted in connection with intelligent design. Philip Kitcher, for instance, citing Leibniz, describes the need for "independent criteria of design" before design can be taken seriously in science (*Abusing Science: The Case against Creationism* [Cambridge, Mass.: MIT Press, 1982], 138). The present book is an attempt to answer Kitcher's challenge in the case of intelligent design. Nonetheless, it is a challenge that all scientific theories must at some point face, the Z-factors considered here being a case in point.

45. John Leslie, *Universes* (London: Routledge, 1989), 10, 12.

46. David Deutsch would reject my claim that the many-worlds interpretation lacks independent evidence. Describing the double-slit experiment in *The Fabric of Reality: The Science of Parallel Universes—and Its Implications* (New York: Penguin, 1997), Deutsch writes, "A real, tangible photon behaves differently according to what paths are open,

elsewhere in the apparatus, for something to travel along and eventually intercept the tangible photon. Something does travel along those paths, and to refuse to call it 'real' is merely to play with words. 'The possible' cannot interact with the real: non-existent entities cannot deflect real ones from their paths. If a photon is deflected, it must have been deflected by something, and I have called that thing a 'shadow photon' " (49).

For Deutsch shadow photons reside in universes different from our own and yet causally interact with our universe by, for instance, deflecting photons. In fact, to read Deutsch one would think that the many-worlds, or as he calls it the "multiverse," interpretation of quantum mechanics is the only one that is coherent and experimentally supported. As he writes, "I have merely described some physical phenomena and drawn inescapable conclusions. . . . Quantum theory describes a multiverse" (50). Or, "The quantum theory of parallel universes is not the problem, it is the solution. It is not some troublesome, optional interpretation emerging from arcane theoretical considerations. It is the explanation—the only one that is tenable—of a remarkable and counter-intuitive reality" (51).

But in fact, one can interpret the double-slit experiment and other quantum mechanical results without multiple worlds and do so coherently—i.e., without internal contradiction and without contradicting any empirical data. And there are plenty such interpretations. The uniting feature of these different interpretations is that they are empirically equivalent—if not, there would be multiple quantum theories. As it is, there is only one quantum theory and many interpretations. See Anthony Sudbery, *Quantum Mechanics and the Particles of Nature* (Cambridge: Cambridge University Press, 1984), 212–225.

Deutsch sees the deflection of photons in a double-slit experiment as sure evidence of parallel universes interacting with our own. Deutsch's very reference to "deflected photons" is a throwback to metaphors of classical physics that have no proper place in quantum mechanics. To invoke them as independent evidence of the many-worlds interpretation of quantum mechanics is to confuse what needs to be explained with what adjudicates among competing explanations or interpretations. The behavior of photons passing through two slits and exhibiting an interference pattern on a screen needs to be explained, but that behavior does not single out the many-worlds interpretation as, to quote Deutsch, "the only one that is tenable." Deutsch's uncompromising advocacy of the many-worlds interpretation of quantum mechanics is as dogmatic as it is unfounded.

47. For a sampling of theistic solutions to such problems consult the essays in William Lane Craig and J. P. Moreland, eds., *Naturalism: A Critical Analysis* (London: Routledge, 2000) and Michael J. Murray, ed., *Reason for the Hope Within* (Grand Rapids, Mich.: Eerdmans, 1999).

48. Note that I am not wedded to any particular metaphysical position about counterparts. My actual argument in the text treats counterparts as separate individuals and thus not as a single transworld individual. But for my argument to work it is enough that separate persons with similar cognitive faculties and background beliefs exist in separate worlds and be listening to separate Rubinsteins, the one real and the other fake. Transworld identity is therefore not required nor is a theory of counterparts. For David Lewis's theory of counterpart relations and his critique of transworld identity in modal metaphys-

ics see Lewis, *On the Plurality of Worlds*, 9–13 and 210–220 respectively. For Alvin Plantinga's indexical account of transworld identity and his critique of Lewis's counterpart theory see Alvin Plantinga, *The Nature of Necessity* (Oxford: Clarendon Press, 1974), 88–101 and 102–120 respectively.

49. Alan Turing, "Computing Machinery and Intelligence," *Mind* 59 (1950): 434–460.

50. For more on proper function see Alvin Plantinga, *Warrant and Proper Function* (Oxford: Oxford University Press, 1993).

51. For the Strong Law of Large Numbers see Bauer, *Probability Theory*, 172.

52. See Francis Crick and Leslie E. Orgel, "Directed Panspermia," *Icarus* 19 (1973): 341–346.

53. I am grateful to Rob Koons for pressing me to clarify this point.

54. Peter Shor, "Algorithms for Quantum Computation: Discrete Logarithms and Factoring," *Proceedings of the 35th Annual Symposium on Foundations of Computer Science* (1994): 124–134.

55. Deutsch, *Fabric of Reality*, 217.

56. Sudbery, *Quantum Mechanics and the Particles of Nature*, 212.

57. For an overview of quantum computation see Colin P. Williams and Scott H. Clearwater, *Explorations in Quantum Computing* (New York: Springer-Verlag, 1998). See also Anthony J. G. Hey, ed., *Feynman and Computation: Exploring the Limits of Computers* (Reading, Mass.: Perseus, 1999).

58. Collins, "An Evaluation of William A. Dembski's *The Design Inference*," 336, n. 7.

59. In this section I respond to what has come to be called the likelihood approach to hypothesis testing—see Richard M. Royall, *Statistical Evidence: A Likelihood Paradigm* (London: Chapman & Hall, 1997). The likelihood approach is inherently comparative, requiring several hypotheses, pitting them against each other, and thereby determining which comes out on top. The main criticisms I raise against the likelihood approach also hold against the Bayesian and Neyman-Pearson approaches to hypothesis testing. The Bayesian approach has all the same elements as the likelihood approach and therefore all the same problems, but in addition raises the problem of setting prior probabilities (see Royall, *Statistical Evidence*, 172–173). The Neyman-Pearson approach, though focusing not on the probability of individual events but rather on the probability of critical regions, nonetheless encounters the same difficulty as the likelihood approach in making hypothesis testing depend on the relative value of competing probabilities rather than on a direct evaluation of the size of a single probability. In addressing the likelihood approach in this section, I therefore see myself as addressing comparative approaches to hypothesis testing generally. Note that advocates of the likelihood approach view it as a way of assessing the statistical evidence or degree of confirmation for hypotheses and thus would feel uncomfortable characterizing their approach as hypothesis testing in the stricter sense of merely providing a decision procedure for choosing one hypothesis over another. I intend hypothesis testing in the broader sense of any statistical methodology for making sense of hypotheses.

60. Branden Fitelson, Christopher Stephens, and Elliott Sober, "How Not to Detect

Design—Critical Notice: William A. Dembski, *The Design Inference*," *Philosophy of Science* 66 (1999): 472–488.

61. Ibid., 487.

62. Ibid.

63. Elliott Sober, "Testability," *Proceedings and Addresses of the American Philosophical Association* 73(2) (1999): 47–76.

64. In Sober's account, probabilities are conferred on observations. Because states of affairs constitute a broader category than observations, in the interest of generality I prefer to characterize the likelihood approach in terms of probabilities conferred on states of affairs.

65. Elliott Sober, *Philosophy of Biology* (Boulder, Colo.: Westview, 1993), 30–36.

66. Donald L. Cohn, *Measure Theory* (Boston: Birkhäuser, 1980), ch. 9.

67. Sober, *Philosophy of Biology*, 33.

68. Fitelson et al., "How Not to Detect Design," 475.

69. Collins, "An Evaluation of William A. Dembski's *The Design Inference*," 337.

70. Fitelson et al., "How Not to Detect Design," 474.

71. According to the *New York Times* (23 July 1985, B1): "The court suggested—but did not order—changes in the way Mr. Caputo conducts the drawings to stem 'further loss of public confidence in the integrity of the electoral process.' . . . Justice Robert L. Clifford, while concurring with the 6-to-0 ruling, said the guidelines should have been ordered instead of suggested." The court did not conclude that cheating was involved, but merely suggested safeguards so that future drawings would be truly random.

72. See Laurence Tribe's analysis of the Dreyfus affair in "Trial by Mathematics: Precision and Ritual in the Legal Process," *Harvard Law Review* 84 (1971): 1329–1393.

73. See Simon Singh, *The Code Book: The Evolution of Secrecy from Mary Queen of Scots to Quantum Cryptography* (New York: Doubleday, 1999), 19.

74. This epiphenomenal riding of chance on design is well-knomn. For instance, actuaries, marketing analysts, and criminologists all investigate probability distributions arising from the actions of intelligent agents (e.g., murder rates). I make the same point in *The Design Inference* (46–47). Fitelson et al.'s failure to recognize this point, however, is no criticism of my project: "Dembski treats the hypothesis of independent origination as a Chance hypothesis and the plagiarism hypothesis as an instance of Design. Yet, both describe the matching papers as issuing from intelligent agency, as Dembski points out (47). Dembski says that context influences how a hypothesis gets classified (46). How context induces the classification that Dembski suggests remains a mystery." ("How Not to Detect Design," 476) There is no mystery here. Context tells us when the activity of an intelligent agent has a well-defined probability distribution attached to it.

75. Ernest Vincent Wright, *Gadsby* (Los Angeles: Wetzel, 1939).

76. Sober, "Testability," 73, n. 20.

77. Hume himself rejected induction as sufficient for knowledge and regarded past experience as the source of a non-reflective habituation of belief.

78. Thomas Reid argued as much over 200 years ago: "No man ever saw wisdom, and if he does not [infer wisdom] from the marks of it, he can form no conclusions respecting

anything of his fellow creatures. . . . But says Hume, unless you know it by experience, you know nothing of it. If this is the case, I never could know it at all. Hence it appears that whoever maintains that there is no force in the [general rule that from marks of intelligence and wisdom in effects a wise and intelligent cause may be inferred], denies the existence of any intelligent being but himself." See Thomas Reid, *Lectures on Natural Theology*, eds. E. Duncan and W. R. Eakin (1780; reprinted Washington, D.C.: University Press of America, 1981), 56.

79. Fitelson et al. ("How Not to Detect Design," 475) write, "We do not claim that likelihood is the whole story [in evaluating Chance and Design], but surely it is relevant." In fact, a likelihood analysis is all they offer. What's more, such an analysis comes into play only after all the interesting statistical work has already been done.

80. Fitelson et al. ("How Not to Detect Design," 479) regard this as an impossible task: "We doubt that there is any general inferential procedure that can do what Dembski thinks the [criterion of specified complexity] accomplishes." They regard it as "enormously ambitious" to sweep the field clear of chance in order to infer design. Nonetheless, we do this all the time. This is not to say that we eliminate every logically possible chance hypothesis. Rather, we eliminate the ones relevant to a given inquiry. The chance hypotheses relevant to a combination lock, for instance, do not include a chance hypothesis that concentrates all the probability on the actual combination. Now it can happen that we may not know enough to determine all the relevant chance hypotheses. Alternatively, we might think we know the relevant chance hypotheses, but later discover that we missed a crucial one. In the one case a design inference could not even get going; in the other, it would be mistaken. But these are the risks of empirical inquiry, which of its nature is fallible. Worse by far is to impose as an a priori requirement that all gaps in our knowledge must ultimately be filled by non-intelligent causes.

81. Ibid., 486.

82. Lydia McGrew, "Likely Machines: A Response to Elliott Sober's 'Testability'," typescript, presented at conference titled *Design and Its Critics* (Mequon, Wis.: Concordia University, 22–24 June 2000).

83. Michael Behe, *Darwin's Black Box* (New York: Free Press, 1996), 39.

84. See the watchmaker argument in William Paley, *Natural Theology: Or Evidences of the Existence and Attributes of the Deity Collected from the Appearances of Nature* (1802; reprinted Boston: Gould and Lincoln, 1852), ch. 1.

85. Harold J. Morowitz, *Beginnings of Cellular Life: Metabolism Recapitulates Biogenesis* (New Haven, Conn.: Yale University Press, 1992), 59–68.

86. Interview with Marcel Schützenberger, "The Miracles of Darwinism," *Origins and Design* 17(2) (1996): 11.

# CHAPTER THREE

~

# Specified Complexity as Information

## 3.1 Information

Even though the technical literature on information theory is vast and different types of information abound, the basic idea behind information is straightforward and easily stated. Robert Stalnaker puts it this way: "To learn something, to acquire information, is to rule out possibilities. To understand the information conveyed in a communication is to know what possibilities would be excluded by its truth."[1] If I tell you that it is either going to rain or not rain tomorrow, I have not told you anything you did not already know. Rain-or-not-rain exhausts all possibilities, so telling you that it is either going to rain or not rain is uninformative. You already knew the range of possibilities. Consequently, the only way to convey information is by restricting that range of possibilities. Thus, if I tell you it will rain tomorrow, I do indeed communicate information because I have excluded the possibility of not-rain.

Information always presupposes a range of possibilities, and conveying information means ruling out some of those possibilities. It follows that information can be quantified. Indeed, the more possibilities that get ruled out, the more information gets conveyed. Fred Dretske elaborates: "Information theory identifies the amount of information associated with, or generated by, the occurrence of an event (or the realization of a state of affairs) with the reduction in uncertainty, the elimination of possibilities, represented by that event or state of affairs."[2] Even so, to measure information it is not enough simply to count the number of possibilities that were eliminated and present that number as the relevant measure of information. The problem is that a simple enumeration of eliminated possibilities tells us nothing about how those possibilities were individuated.

Consider, for instance, the following individuation of poker hands:

RF    A royal flush.
~RF   Everything else.

To learn that something other than a royal flush was dealt (i.e., possibility ~RF) is clearly to acquire less information than to learn that a royal flush was dealt (i.e., possibility RF). A royal flush is highly specific. We have acquired a lot of information when we learn that a royal flush was dealt. On the other hand, we have acquired hardly any information when we learn that something other than a royal flush was dealt. Most poker hands are not royal flushes, and we expect not to be dealt them. Nevertheless, if our measure of information is simply an enumeration of eliminated possibilities, the same numerical value must be assigned in both instances since in each instance a single possibility is eliminated.

It follows that how we measure information needs to be independent of whatever procedure we use to individuate the possibilities under consideration. The way to do this is not simply to count possibilities but to assign probabilities to those possibilities. For a thoroughly shuffled deck of cards, the probability of being dealt a royal flush (i.e., possibility RF) is approximately .000002 whereas the probability of being dealt anything other than a royal flush (i.e., possibility ~RF) is approximately .999998.

Probabilities by themselves, however, are not information measures. Although probabilities distinguish possibilities by the amount of information they contain, probabilities are an inconvenient way to measure information. There are two reasons for this. First, the scaling and directionality of the numbers assigned by probabilities need to be recalibrated. We are clearly acquiring more information when we learn someone was dealt a royal flush than when we learn someone was not dealt a royal flush. And yet the probability of being dealt a royal flush (i.e., .000002) is minuscule compared to the probability of being dealt something other than a royal flush (i.e., .999998). Smaller probabilities signify more information, not less.

The second reason probabilities are inconvenient for measuring information is that they are multiplicative rather than additive. If we learn that Alice was dealt a royal flush playing poker at Caesar's Palace and that Bob was dealt a royal flush playing poker at the Mirage, the probability that both Alice and Bob were dealt royal flushes is the product of the individual probabilities. On the other hand, it is convenient for information to be measured additively so that the measure of information assigned to Alice and Bob jointly being dealt royal flushes equals the measure of information assigned to Alice being dealt a royal flush plus the measure of information assigned to Bob being dealt a royal flush. Now there is a straightforward mathematical way to transform probabilities that circumvents both these difficulties, and that is to apply a negative logarithm to the probabilities. Applying a negative logarithm assigns more information to less probability and, because the logarithm of a product is the sum of the logarithms, transforms multiplicative probability measures into additive information measures.

Moreover, in deference to communication theorists, it is customary to use the logarithm to the base 2. The rationale for this choice of logarithmic base is as follows: The most convenient way for communication theorists to measure information is in bits. Any message sent across a communication channel can be viewed as a string of 0s and 1s. For instance, the ASCII code[3] uses strings of eight 0s and 1s to represent the characters on a typewriter, with whole words and sentences in turn represented as longer strings that encompass such character strings. Similarly, all communication may be reduced to the transmission of sequences of 0s and 1s. Given this reduction, the obvious way for communication theorists to measure information is in number of bits transmitted across a communication channel. And since the negative logarithm to the base 2 of a probability corresponds to the average number of bits needed to identify an event of that probability, the logarithm to the base 2 is the canonical logarithm for communication theorists. Thus, we define the measure of information in an event of probability p as $-\log_2 p$.[4]

To see that this information measure is additive, return to the example of Alice being dealt a royal flush playing poker at Caesar's Palace and that Bob being dealt a royal flush playing poker at the Mirage. Let us call the first event A and the second B. Since randomly dealt poker hands are probabilistically independent, the probability of A and B taken jointly equals the product of the probabilities of A and B taken individually. Symbolically, $P(A\&B) = P(A) \times P(B)$.[5] Given our logarithmic definition of information, we therefore define the amount of information in an arbitrary event E as $I(E) =_{def} -\log_2 P(E)$. It then follows that $P(A\&B) = P(A) \times P(B)$ if and only if $I(A\&B) = I(A) + I(B)$. Since in the example of Alice and Bob $P(A) = P(B) = .000002$, $I(A) = I(B) = 19$, and $I(A\&B) = I(A) + I(B) = 19 + 19 = 38$. Thus, the amount of information inherent in Alice and Bob jointly obtaining royal flushes is 38 bits.

Since lots of events are probabilistically independent, information measures exhibit lots of additivity. But since lots of events are also correlated, information measures exhibit lots of nonadditivity as well. In the case of Alice and Bob, Alice being dealt a royal flush is probabilistically independent of Bob being dealt a royal flush, and so the amount of information in Alice and Bob both being dealt royal flushes equals the sum of the individual amounts of information. But consider next a different example. Alice and Bob together toss a coin five times. Alice observes the first four tosses but is distracted, and so misses the fifth toss. On the other hand, Bob misses the first toss, but observes the last four tosses. Let us say the actual sequence of tosses is 11001 (1 = heads, 0 = tails). Thus Alice observes 1100* and Bob observes *1001 (asterisks denote missed coin tosses). Let A denote the first observation, B the second. It follows that the amount of information in A&B is the amount of information in the complete sequence 11001, namely, 5 bits. On the other hand, the amount of information in A alone is the amount of information in the incomplete sequence 1100*, namely 4 bits. Similarly, the amount of information in B alone is the amount of information in the incomplete sequence *1001, also 4 bits. This time information does not add up: $5 = I(A\&B) \neq I(A) + I(B) = 4 + 4 = 8$.

Here A and B are correlated. Alice knows all but the last bit of information in the complete sequence 11001. Thus when Bob gives her the incomplete sequence *1001, all Alice really learns is the last bit in this sequence. Similarly, Bob knows all but the first bit of information in the complete sequence 11001. Thus when Alice gives him the incomplete sequence 1100*, all Bob really learns is the first bit in this sequence. What appears to be four bits of information actually ends up being only one bit of information once Alice and Bob factor in their prior information. We need therefore to introduce the idea of conditional information. $I(B|A)$ denotes the conditional information of B given A and signifies the amount of information in Bob's observation once Alice's observation is taken into account. This, as we just saw, is 1 bit. It follows that $5 = I(A\&B) = I(A) + I(B|A) = 4 + 1$.

$I(B|A)$, like $I(A\&B)$, $I(A)$, and $I(B)$, can be represented as the negative logarithm to the base 2 of a probability, only this time the probability under the logarithm is a conditional as opposed to an unconditional probability. By definition $I(B|A) =_{def} -\log_2 P(B|A)$, where $P(B|A)$ is the conditional probability of B given A. Whereas the unconditional probability $P(B)$ is the probability assigned to B apart from any additional assumptions, the conditional probability $P(B|A)$ is the probability assigned to B under the assumption that A obtains. For instance, the unconditional probability of rolling a die and obtaining a six is 1/6. The conditional probability of rolling a die and obtaining a six *given that* we know that an even number was thrown (i.e., either a two or four or six) is 1/3. Now since $P(B|A)$ is by definition the quotient $P(A\&B)/P(A)$, and since the logarithm of a quotient is the difference of the logarithms, it follows that $\log_2 P(B|A) = \log_2 P(A\&B) - \log_2 P(A)$, and so $-\log_2 P(B|A) = -\log_2 P(A\&B) + \log_2 P(A)$, which is just $I(B|A) = I(A\&B) - I(A)$. This last equation is equivalent to

$$(*) \qquad\qquad I(A\&B) = I(A) + I(B|A).$$

Since the information measure $I$ is always nonnegative, this formula implies that $I(A\&B) \geq I(A)$ for all A and B. Formula (*) holds with full generality, reducing to $I(A\&B) = I(A) + I(B)$ when A and B are probabilistically independent (in which case $P(B|A) = P(B)$ and thus $I(B|A) = I(B)$).

Formula (*) asserts that the information in both A and B jointly is the information in A plus the information in B that is not in A. Its point, therefore, is to spell out how much additional information B contributes to A. As such, this formula places tight constraints on the generation of new information. Does, for instance, a computer program (call it A) by outputting some data (call the data B) generate new information? Computer programs are fully deterministic, and so B is fully determined by A. It follows that $P(B|A) = 1$, and thus $I(B|A) = 0$ (the logarithm of 1 is always 0). From formula (*) it therefore follows that $I(A\&B) = I(A)$, and therefore that the amount of information in A and B jointly is no more than the amount of information in A by itself. This is an instance of what Peter Medawar calls the Law of Conservation of Information.[6]

For an example in the same spirit consider that there is no more information in two copies of Shakespeare's *Hamlet* than in a single copy. This is of course patently obvious, and any formal account of information had better agree. To see that our formal account does indeed agree, let A denote the printing of the first copy of *Hamlet*, and B the printing of the second copy. Once A is given, B is entirely determined. Indeed, the correlation between A and B is perfect. Probabilistically this is expressed by saying the conditional probability of B given A is 1, namely, $P(B|A) = 1$. In information-theoretic terms this is to say that $I(B|A) = 0$. As a result, $I(B|A)$ drops out of formula (*), and so $I(A\&B) = I(A)$. Our information-theoretic formalism therefore agrees with our intuition that two copies of *Hamlet* contain no more information than a single copy.

## 3.2 Syntactic, Statistical, and Algorithmic Information

The account of information presented in the last section is quite general—indeed so general that it may not be immediately evident how it matches up with what information theorists typically mean by information. Typically when information theorists think of information, they think of either the Shannon or the Chaitin-Kolmogorov-Solomonoff theory of information (the latter also being referred to as "algorithmic information theory"). Both these approaches to information are special cases of the general framework just outlined.

Shannon's theory of information is a syntactic theory. It accounts for the transmission of character strings across a communication channel where the characters derive from a fixed alphabet. The alphabet is assumed to have at least two distinct characters. In case there are only two characters, they are typically represented by "0" and "1," and the strings derived from these characters are referred to as "bit strings" ("bit" for "binary digit"). Note that alphabets with only a single character can also transmit information, but the mode of transmission is very cumbersome. With only one character in the alphabet, information is transmitted by counting the number of times that alphabetic character is transmitted (i.e., if the one alphabetic character in question is "x," then all communications have the form "x" or "xx" or "xxx" or "xxxx" . . .). To convey information a communication channel must allow a multiplicity of distinct possible signals any one of which might be sent.

In the Shannon theory, the reference class of possibilities from which information gets generated is the sum total of character strings from the relevant alphabet. To convey information within the Shannon theory is therefore to identify a character string within that reference class of possibilities and send it across the communication channel. In this way a possibility is identified, other possibilities are ruled out, and information is generated (in the sense defined in the last section).

Since the reference class of possible character strings is typically huge, identifying a single string will vastly reduce the reference class of possibilities and therefore generate a huge amount of information. Theoretical interest in Shannon's theory lies in quantifying the information in such character strings, treating the statistical properties of such strings when they are sent across a noisy communication channel (noise, typically, is represented by a stochastic process that disrupts the strings in statistically well-defined ways), preserving the strings despite the presence of noise (i.e., the theory of error-correcting codes), and transforming the strings into other strings to maintain their security (i.e., cryptography).

Though Shannon's theory starts out as a syntactic theory (deriving its reference class of possibilities from character strings based on a fixed alphabet), it quickly becomes a statistical theory. Characters from the alphabet will often have different probabilities of occurrence (cf. the letters from our ordinary alphabet, which occur with widely varying frequencies—in English the letter *e* occurs roughly 12 percent of the time, the letter *q* less than 1 percent of the time; what's more, *u* follows *q* with probability one). These probabilities in turn determine how much information any given string can convey.

It is easily proven mathematically that character strings will on average convey maximal information if and only if all the letters in the alphabet are equally probable and stochastically independent (i.e., all letters have the same probability and the probability of a given letter is unaffected by the occurrence of letters elsewhere in the string). Given n distinct alphabetic characters $a_1, \ldots a_n$ with probabilities respectively $p_1, \ldots p_n$, the average information per character in a string is given by the entropy $\mathbf{H}$:

$$\mathbf{H}(a_1, \ldots a_n) =_{\text{def}} \Sigma_i -p_i \log_2 p_i = \Sigma_i p_i \mathbf{I}(a_i).$$

Here the summation over i goes from 1 to n, and $\mathbf{I}(a_i) = -\log_2 p_i$ is the information in any given alphabetical character, with $\mathbf{I}$ being the information measure defined in section 3.1. $\mathbf{H}$ is maximal when all the $p_i$s are identical (i.e., each $p_i = 1/n$).[7] As an aside, this information-theoretic entropy measure is mathematically identical to the Maxwell-Boltzmann-Gibbs entropy from statistical mechanics provided the alphabet $a_1, \ldots a_n$ is reinterpreted as a partition of phase space and the probabilities $p_1, \ldots p_n$ are reinterpreted as the probabilities of particles being in those corresponding partition elements.[8]

Like the Shannon theory, the Chaitin-Kolmogorov-Solomonoff theory of information is a syntactic theory with a strong statistical component. Where it differs from the Shannon theory is in adding a computational component. Also known as algorithmic information theory, this theory attempts to characterize what makes a bit string random. Since probability theory is incapable of distinguishing bit strings of identical length (if we think of bit strings as sequences of coin tosses, then any sequence of n flips has probability 1 in $2^n$), something else is required. According to algorithmic information theory, a bit string is random to the degree that it is incompressible (see section 2.4). It is a combinatorial fact that the vast majority of sequences of 0s and 1s have as their shortest description just the sequence itself. Thus most sequences are random in the sense of being algorithmically incompressible.

How does algorithmic information theory connect to the account of information given in the last section? First note that the limitation to sequences of 0s and 1s is not intrinsic to algorithmic information theory. Indeed, just like the Shannon theory, algorithmic information theory applies to character strings based on arbitrary alphabets. Algorithmic information theory therefore treats the same reference class of possibilities as the Shannon theory. Yet unlike the Shannon theory, which focuses on how character strings traverse communication channels, algorithmic information theory focuses on whether character strings are compressible into shorter strings, interpreting the shorter strings as computer programs within some prespecified programming environment. What then constitutes information—that is, the identification of possibilities and ruling out of others—within algorithmic information theory? In this instance information consists not in identifying individual strings but in identifying collections of strings that exhibit a given degree of randomness (i.e., incompressibility). Collections of highly nonrandom (i.e., highly compressible) strings constitute the information of principal interest in algorithmic information theory.

In closing this section I want to remark on some work of Murray Gell-Mann that attempts to combine Shannon's statistical theory of information with the Chaitin-Kolmogorov-Solomonoff algorithmic theory of information into a comprehensive theory of complexity and information for science. Gell-Mann starts with the observation that the complexity that interests us in practice is not pure randomness but patterned regularities that remain once the effects of randomness have been factored out. Gell-Mann therefore defines "effective complexity" as the complexity inherent in these patterned regularities. Moreover, he defines "total information" as the effective complexity together with the complexity inherent in the effects of randomness that were factored out. He then characterizes effective complexity mathematically in terms of an algorithmic information measure that measures the extent to which patterned regularities can be compressed into a minimal representation (he calls such representations "schemata"). Moreover, he characterizes the residual effects of randomness mathematically in terms of a Shannon information measure that measures the extent to which random deviations depart from the patterned regularities in question. Total information thus becomes the sum of an algorithmic information measure and a Shannon information measure.[9]

Gell-Mann's theory of effective complexity attempts to account for how complex adaptive systems like us make sense out of a world that exhibits regularities as well as random deviations from those regularities. Though richly suggestive, applying Gell-Mann's mathematical formalism in practice is largely intractable since it requires taking conceptual schemata of patterned regularities appropriate to some inquiry, mapping them onto a computational data structure, and then seeing how such data structures can be reduced in size while faithfully preserving the conceptual structures that map from conceptual to computational space. Thus far Gell-Mann's theory has resisted detailed applications to real-world problems.

Why then do I consider it here? There are two reasons. First, while Gell-Mann's theory is well-suited for describing how regularities of nature that are continually subjected to random perturbations match our conceptual schemata, it is not capable of handling contingencies in nature that are unaccounted (and perhaps unaccountable) by any regularities but that happen all the same to match our conceptual schemata. It is this latter possibility that complex specified information addresses (see sections 3.5 and 3.9).

The second reason for taking up Gell-Mann's theory is related to the first. According to philosopher David Roche, design theorists like me are all mixed up about information theory and complexity.[10] Thus Roche argues that the Darwinian mechanism is well able to account for biological complexity once we are clear about the type of complexity that is actually at stake in biology. The problem, according to Roche, is that design theorists are using the wrong notion of complexity. What is the right notion? Roche claims Gell-Mann's concept of effective complexity is the right one for biology. Thus he writes, "[Once] we interpret 'information' to mean [effective] complexity, then we are simply left with answering the familiar question of how the Darwinian process could give rise to such complex organs as the vertebrate eye; a question already thoroughly dealt with by many biologists (e.g., Dawkins 1986)."[11] In fact, it is very much under dispute whether biologists have adequately demonstrated the power of the Darwinian process to account for biological complexity.

Assimilating biological complexity to Gell-Mann's notion of effective complexity guarantees that biological complexity must be understood in terms of some regularity or other (the regularity of choice these days being natural selection). Gell-Mann's effective complexity thus effectively precludes design. Moreover, it does so prejudicially by ruling out all but regularities from the definition of complexity. On the other hand, complex specified information as I develop it in this chapter allows for an unprejudiced examination of the role of regularity, chance, and actual design in the emergence of biological complexity. For Gell-Mann's "total information" to be truly total it must decompose into a chance component (Shannon information), a regularity or necessitarian component (algorithmic information), and a design component (complex specified information).

## 3.3   Information in Context

Information presupposes a reference class of possibilities, and for information to be generated requires that some of those possibilities be identified and others excluded. This was the upshot of section 3.1. It was illustrated in section 3.2 for both Shannon and algorithmic information. An obvious question now arises: What determines the reference class of possibilities from which information gets generated? A given possibility constitutes information only in relation to other possibilities that were excluded. Information is thus fundamentally a relational notion. But what are those other possibilities in relation to which a given possibility becomes information?

Consider the following example from Ivar Ekeland's *The Broken Dice*.[12] Ekeland describes how the kings of Norway and Sweden back in the Middle Ages decided to cast a pair of dice to determine ownership of a settlement on the Island of Hising, a settlement that alternately had belonged to both countries. The highest totaling sum was to determine the winner. The king of Sweden went first and rolled double sixes. It would therefore seem that the king of Norway could at best tie the king of Sweden, though the more likely outcome was that the Hising settlement would end up in the hands of Sweden. With six faces on a die and faces numbered one through six, the sum of any pair of faces from a pair of dice could total no less than two and no more than twelve. The reference class of possible outcomes for the pair of dice could therefore be represented by the set {2, 3, 4, 5, 6, 7, 8, 9, 10, 11, 12}. What's more, the king of Sweden had just rolled the optimal possibility in this set, namely, 12.

What happened next was therefore remarkable: "Thereupon Olaf, king of Norway, cast the dice, and one six showed on one of them, but the other split in two, so that six and one turned up; and so he took possession of the settlement."[13] Since in this game of dice higher sums trump lower sums, thirteen ( = 6 + 6 + 1) trumps twelve ( = 6 + 6), and so the king of Norway was declared the winner. Typically, any game with a pair of dice reckons with at most a pair of faces on any throw. Given this constraint, the reference class of possible sums for a pair of dice faces will be {2, 3, 4, 5, 6, 7, 8, 9, 10, 11, 12}. Yet given the possibility of a die splitting in two and showing two faces, the reference class of possible sums would have to be expanded to include at least {2, 3, 4, 5, 6, 7, 8, 9, 10, 11, 12, 13} and possibly even more.

Extraordinary possibilities, precisely because they are extraordinary, are often omitted from the reference class of possibilities presupposed in an information-theoretic analysis. Such omissions do not in most instances impair an information-theoretic analysis. But they point up the importance of including all relevant possibilities in an information-theoretic analysis and exercising caution in just what possibilities we regard as extraordinary. The king of Sweden was confident that the relevant reference class of possibilities comprised {2, 3, 4, 5, 6, 7, 8, 9, 10, 11, 12}. Given the king of Norway's roll of the dice, he should have reckoned with a reference class that included at least {2, 3, 4, 5, 6, 7, 8, 9, 10, 11, 12, 13}.

In general, we are safer erring on the side of abundance in assigning possibilities to a reference class. If we err on the side of meagerness, we are likely to omit possibilities that might actually arise and thereby undercut our information-theoretic analysis (as the king of Sweden did, though through no fault of his own). A reference class of possibilities that is richer than we are likely to need is easily handled by focusing on the relevant subclass of possibilities that we regard as "live" or "realistic." Mathematically this means concentrating the probability measure $P$ (the one used to measure information via the corresponding information measure $I$) on a suitable subset from that reference class of possibilities. Consider, for instance, a pair of loaded dice guaranteed to land seven. The relevant reference class of possibilities for the sum of the faces of this pair of dice is thus the singleton set {7}. In this case the occurrence of a seven has probability one and information zero ($P(\{7\}) = 1$ and $I(\{7\}) = -\log_2 P(\{7\}) = 0$). Even so, it is safer to embed this reference class in the standard reference class of possibilities for the sum of two dice, namely, {2, 3, 4, 5, 6, 7, 8, 9, 10, 11, 12}, assigning $P(\{7\}) = 1$ and $P(\{2, 3, 4, 5, 6, 8, 9, 10, 11, 12\}) = 0$.

The question remains, What determines the reference class of possibilities from which information gets generated? It is important to understand that a reference class of possibilities never forces itself on us. Rather, it is we, human inquirers, who must identify the reference class of possibilities appropriate to our inquiries. This identification of a reference class of possibilities will depend on our background knowledge, assumptions about world, values, local circumstances, and interests—in short, our context of inquiry. Our context of inquiry determines what possibilities we regard as plausible. In turn, plausibility determines what possibilities we take seriously enough to include in our reference class of possibilities.

The king of Sweden regarded a pair of dice as capable of showing no more than two faces. Given his context of inquiry, the only plausible possibilities for the sum of the faces of a pair of dice would be {2, 3, 4, 5, 6, 7, 8, 9, 10, 11, 12}. This was the reference class of possibilities with which the king of Sweden reckoned. The king of Norway, on the other hand, King Olaf Haraldsson, a saint in the Catholic Church and one supposedly endued with miraculous powers, would not be bound to this set of possibilities. Dice splitting in two with faces totaling more than twelve would be entirely plausible within the mystical world of medieval saints. Thus King Olaf's reference class of possibilities would include at least {2, 3, 4, 5, 6, 7, 8, 9, 10, 11, 12, 13}.

To sum up, context determines what possibilities we regard as plausible (or, if you will, not unduly extraordinary), and plausibility determines what possibilities we include in our reference class of possibilities. Shift the context, and our reference class of possibilities will shift accordingly. Within a Newtonian context, freely moving objects proceed rectilinearly because spacetime is Euclidean. Within a relativistic context, freely moving objects proceed curvilinearly because spacetime is curved. This is not to say that all contexts are created equal. Einsteinian relativity corrects serious deficiencies in Newtonian mechanics and thus provides a more adequate reference class of possible motions for actual objects moving in the actual world. Nonetheless, given a Newtonian context of inquiry, it is entirely appropriate to omit curvilinear paths for freely moving objects since they find no place in Newton's theory.

A context of inquiry can be problematic in the sense that its aims, methods, and presuppositions may be faulty (cf. Newtonian mechanics with its faulty assumptions about absolute space and time). Even so, we can still talk about a reference class of possibilities being appropriate or inappropriate to that context. For the king of Sweden rolling a pair of dice and taking their sum, the reference class {2, 3, 4, 5, 6, 7, 8, 9, 10, 11, 12} was entirely appropriate. To be sure, the king of Sweden did not reckon with a broader reference class being required because his gaming partner the king of Norway was a Christian saint endued with miraculous powers. On the other hand, the king of Sweden would have been seriously mistaken to omit one or more of the numbers in the reference class that he actually chose. Indeed, any such reference class would have been inappropriate to the king of Sweden's context of inquiry. In identifying a reference class of possibilities for an information-theoretic analysis, we therefore need to be clear about our context of inquiry and we need to assess the appropriateness of the reference class to that context. What's more, we want to err on the side of abundance and include as many possibilities as might plausibly obtain within that context.

In being as inclusive as possible with a reference class of possibilities, we can do no better than include all the possibilities that we can for now conceive. What we can for now conceive, however, is not fixed, and what we can conceive down the road may greatly exceed what we can for now conceive. This is one of Stuart Kauffman's main points in *Investigations*. The phase space (i.e., reference class of possibilities) for the biosphere is, as he puts it, "not finitely prestatable."[14] Indeed, the emergence of new and undreamt of possibilities is one of the things that makes biology so exciting.

Kauffman's insight is therefore vitally important to scientific inquiry in that it keeps us ever open to new possibilities and therewith to an expanded vision of the world. At the same time, however, Kauffman's insight provides no warrant for doubting or undermining information already derived from a finitely prestated reference class of possibilities.

The problem with a finitely prestated reference class of possibilities is that down the line we may have to add new possibilities to the reference class. Nonetheless, possibilities identified and therefore information derived from an old reference class remain valid with the addition of new possibilities. For instance, the king of Sweden's roll of double sixes continued to make perfect sense in light of the king of Norway's miraculous roll of double sixes plus a one. What's more, the amount of information associated with a possibility from a finitely prestated reference class of possibilities can only increase as new possibilities are added. A possibility realized from a finitely prestated reference class of possibilities rules out not only the remaining possibilities in the old reference class but also the new ones that were added. This observation is important because later in this chapter we will define specified complexity as a form of information that reliably signals design provided the amount of information associated with a possibility attains a certain level. Adding new possibilities to a reference class of possibilities does nothing to diminish the amount of information associated with already realized possibilities.

## 3.4 Conceptual and Physical Information

To generate information is to rule out possibilities. But who or what rules out those possibilities? In practice, there are two sources of information: intelligent agency and physical processes. This is not to say that these sources of information are mutually exclusive—human beings, for instance, are both intelligent agents and physical systems. Nor is this to say that these sources of information exhaust all logically possible sources of information—it is conceivable that there could be nonphysical random processes that generate information.

Although physical processes that are not also intelligent agents can generate information, there is a sense in which information, whatever its source, is irreducibly conceptual and thus presupposes intelligent agency. This is because the very reference class of possibilities that sets the backdrop for the generation of information must invariably be delineated by an intelligent agent (see section 3.3). Thus information, whatever else we might want to say about it, can never be entirely mind-independent or concept-free.

Nevertheless, once an intelligent agent identifies a reference class of possibilities according to some relevant context of inquiry, it is a separate question whether information generated from that reference class results from an intelligent agent or a physical process. An intelligent agent may explicitly identify a pattern within the reference class of possibilities and thereby generate information. Alternatively, a physical process can produce an event, represented as a possibility within the reference class of possibilities, and thereby generate information. Let us refer to the former type of information as *agent-induced* or *conceptual information* and to the latter as *event-induced* or *physical information* (see figure 3.1).

Think of the two types of information this way. The reference class of possibilities $\Omega$ represents a collection of possible events identified by an intelligent agent S. The event E, denoted by an "x," is an outcome that actually occurred. The target T, denoted by the squiggly shaded area, is a pattern identified by the intelligent agent S. Since $\Omega$ represents possible events, both E and T denote events. The difference is in how E and T were actualized. In the case of E, a physical process caused E to happen. In the case of T, an intelligent agent S explicitly identified a pattern within the reference class of possibilities (irrespective of whether T actually happened).

To illustrate these two types of information, take as a reference class of possibilities all bit strings of length 100, and suppose each string represents a sequence of 100 tosses with a fair coin so that 1 corresponds to heads and 0 to tails. There are two ways of identifying a sequence of 100 coin tosses from this reference class of possibilities and thereby generating information. One is for an intelligent agent simply as a cognitive act to identify a bit string without doing any coin tossing (e.g., an intelligent agent S identifies the possibility 100-heads-in-a-row). The other is for a fair coin to be flipped 100 times and generate a bit string (e.g., 1100101 . . .). In the former case, an intelligent agent identifies a pattern or target within the reference class of possibilities (thereby generating conceptual information). In the latter case, a physical process produces an event represented within the reference class of possibilities (thereby generating physical information).

Besides illustrating the two types of information, this example also illustrates why conceptual or agent-induced information is not reducible to physical or event-induced information. Even if one assumes that intelligent agents are at base purely physical systems, this does not mean that conceptual information automatically collapses into physical information. For instance, in the coin-tossing example the reference class of possibilities consists of $2^{100}$ bit strings of length 100, which represents all possible coin tosses of that length. This reference class, however, does not represent the

# Reference Class of Possibilities Ω

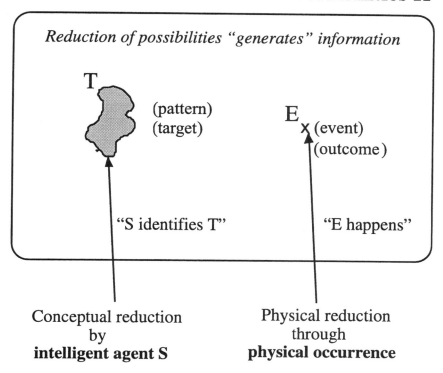

*Reduction of possibilities "generates" information*

T (pattern) (target)

E x (event) (outcome)

"S identifies T"    "E happens"

Conceptual reduction
by
**intelligent agent S**

Physical reduction
through
**physical occurrence**

***Conceptual Information:*** *Intelligent agent S identifies a pattern and thereby conceptually reduces the reference class of possibilities.*

***Physical Information:*** *Event E occurs and thereby physically reduces the reference class of possibilities.*

Figure 3.1.    Two Types of Information.

underlying physical events that on a reductionist account of mentality would induce an intelligent agent to identify a sequence of 100 coin tosses and thereby generate information (in the case of human agents, the physical events would be neural events). The reference class here represents coin tossing and not neurophysiology.

Getting a reference class of possible events to provide a complete and adequate representation of internal cognitive states remains for now an intractable problem. Indeed, providing an empirically adequate representation of such states is the great unsolved problem of cognitive science and one some cognitive scientists doubt will ever admit resolution. Thus, even if within a physicalist ontology intelligent agency is ultimately reducible to event-causation, as a practical matter we cannot dispense with the twin categories of conceptual and physical information.

## 3.5 Complex Specified Information

The information measure described in section 3.1 is a complexity measure.[15] Complexity measures arise whenever we assign numbers to degrees of complication. A reference class of possibilities will often admit varying degrees of complication, ranging from extremely simple to extremely complicated. Complexity measures assign nonnegative numbers to these possibilities so that 0 corresponds to the most simple and ∞ to the most complicated. For instance, computational complexity is always measured in terms of either time (i.e., number of computational steps) or space (i.e., size of memory, usually measured in bits or bytes) or some combination of the two. The more difficult a computational problem, the more time and space are required to run the algorithm that solves the problem, and correspondingly the bigger the complexity.

For information measures, degree of complication is measured in bits. Given an event A of probability $P(A)$, $I(A) = -\log_2 P(A)$ measures the average number of bits required to specify an event with that probability. We therefore speak of the "complexity of information" and say that the complexity of information increases as $I(A)$ increases (or, correspondingly, as $P(A)$ decreases). We also speak of "simple" and "complex" information according to whether $I(A)$ signifies few or many bits of information.

This information-theoretic account of complexity is entirely consistent with the account of complexity given in sections 1.3 and 1.5. Likewise, the account of specification given in sections 1.4 and 2.5 carries over to information. It follows that information can be complex, specified, or both. Information that is both complex and specified will be called *complex specified information*, or CSI for short (see figure 3.2).

Think of complex specified information this way. An intelligent agent S identifies a reference class of possibilities $\Omega$ representing a collection of possible events. The event E, denoted by an "x," is an outcome that occurred via some physical process. The target T, denoted by the squiggly shaded area, is

# Reference Class of Possibilities $\Omega$

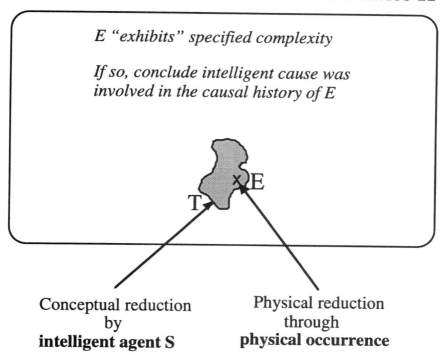

*E "exhibits" specified complexity*

*If so, conclude intelligent cause was involved in the causal history of E*

Conceptual reduction
by
**intelligent agent S**

Physical reduction
through
**physical occurrence**

*Complex Specified Information: The coincidence of conceptual and physical information where the conceptual information is both identifiable independently of the physical information and also complex.*

**Figure 3.2. Complex Specified Information.**

a pattern identified by the intelligent agent S without recourse to the event E. Since $\Omega$ represents possible events, both T and E denote events. The ordered pair (T,E) now constitutes *specified information* provided that the event E is included in the event T and provided that T can be identified independently of E (i.e., is detachable from E—see section 2.5). Moreover, if T also has high complexity (or correspondingly small probability—see sections 1.5 and 2.8), then (T,E) constitutes *complex specified information* or CSI. It follows that for (T,E) to constitute complex specified information is logically equivalent to E satisfying the complexity-specification criterion of section 1.3. Provided E satisfies this criterion, we also say that E exhibits *specified complexity*. As we saw in chapter 1 and then justified in chapter 2, this means that an intelligent cause was involved in E's causal history.

Complex specified information is a souped up form of information. To be sure, complex specified information or CSI is consistent with the basic idea behind information, which is the reduction of possibilities from a reference class of possibilities. But whereas the traditional understanding of information is unary, conceiving of information as a single reduction of possibilities, complex specified information is a binary form of information. Complex specified information, and specified information more generally, depends on a dual reduction of possibilities, namely a conceptual reduction (i.e., conceptual information) combined with a physical reduction (i.e., physical information). Moreover, these dual reductions must be coordinated so that the conceptual information subsumes the physical information.

To get from specified information to complex specified information requires that both the conceptual and physical information that constitute the specified information be complex. This joint complexity requirement admits a simplification: Since the conceptual component of specified information always subsumes the physical, it is enough simply to require that the conceptual component be complex, since that forces the physical component to be complex as well. More precisely, for specified information (T,E), since T subsumes E (i.e., the occurrence of E entails the occurrence of T), the probability of E cannot exceed that of T. This in turn means that the information of T (as measured in bits) is no more than the information of E (also measured in bits). Thus, as long as T is complex (i.e., requires many bits to represent it), so is E.

To illustrate specified information, consider the roll of a single die. The reference class of possibilities $\Omega$ is the set {1, 2, 3, 4, 5, 6}. The event E, denoted in figure 3.2 by an "x," will then be the outcome that occurred by rolling the die, say a six. Probabilists distinguish between outcomes or elementary events and events generally. To roll a six with a single die is an outcome or elementary event. On the other hand, to roll an even number with a single die is an event that includes (or subsumes) the outcome of rolling a six, but also includes rolling a four or two. Let us call this event of rolling an even number T. Then E = {6} and T = {2, 4, 6}. Events (other than the null or impossible event) include at least one elementary event or outcome (like E), but may include more (like T).

The ordered pair (T,E) constitutes specified information. To see this it is enough to see that T subsumes E and that T can be identified independently of E. T clearly subsumes E since E is an elementary event that is part of T. What's more, T can be identified independently of E since the even integers are defined mathematically without reference to random rolls of the die (this is easily formalized in terms of detachability—see section 2.5). But note, even though (T,E) constitutes specified information, it does not constitute complex specified information. Since the tacit assumption here is that all rolls of the die are equiprobable, the probability of T is 1/2, which implies that the information of T is 1 bit: $P(T) = 1/2$ and $I(T) = -\log_2 P(T) = 1$.

To illustrate not just specified information but complex specified information, consider instead all bit strings of length 1000. Suppose each string represents a sequence of 1000 tosses with a fair coin such that 1 corresponds to heads and 0 to tails. This collection of bit strings constitutes the reference class of possibilities $\Omega$. Now as we have seen, there are two ways of identifying a sequence of 1000 coin tosses from this reference class of possibilities (and thereby generating information). One is for an intelligent agent simply by a cognitive act to identify a bit string without doing any coin tossing. Thus an intelligent agent S might identify the following target T:

```
11011101111101111110111111111101111111111111011111111
11111111101111111111111111111011111111111111111111111101
11111111111111111111111111110111111111111111111111111111
11111011111111111111111111111111111111111111101111111111
11111111111111111111111111111101111111111111111111111111
11111111111111110111111111111111111111111111111111111111
11111111111011111111111111111111111111111111111111111111
11111111111011111111111111111111111111111111111111111111
11111111111111111011111111111111111111111111111111111111
11111111111111111111111111111110111111111111111111111111
11111111111111111111111111111111111111111111111110111111
11111111111111111111111111111111111111111111111111111111
11111111111011111111111111111111111111111111111111111111
11111111111111111111111111111111111111110111111111111111
11111111111111111111111111111111111111111111111111111111
```

```
1111111111101111111111111111111111111111111111111111111111
1111111111111111111111111111111111111111111111111011111111
1111111111111111111111111111111111111111111111111111111111
1111111111.
```

T is the prime numbers from 2 to 89 along with some filler at the end. As in the movie *Contact* (see section 1.3), prime numbers are represented in unary notation with a given prime number corresponding to adjacent 1s and consecutive prime numbers separated by 0s.

Instead of an intelligent agent identifying this possible sequence of coin tosses, it is also possible (logically though not realistically) for a fair coin to be flipped 1000 times and deliver as the event E this same sequence of prime numbers. T and E are then identical events, with T constituting conceptual information and E physical information. In this case the ordered pair (T,E) constitutes not just specified information but complex specified information. Formally justifying this is straightforward: Since T and E are identical, T clearly subsumes E; as a sequence of prime numbers, T is readily seen to be detachable from E and therefore constitutes a specification (see section 2.5); finally, by having probability 1 in $2^{1000}$ or approximately 1 in $10^{300}$, T has probability less than the universal probability bound of 1 in $10^{150}$ and is therefore complex (see section 1.5). Alternatively, since a universal probability bound of 1 in $10^{150}$ corresponds to a universal complexity bound of 500 bits of information, (T,E) constitutes CSI because T subsumes E, T is detachable from E, and T measures at least 500 bits of information.

CSI is what all the fuss over information has been about in recent years, not just in biology but in science generally. It is CSI that for Manfred Eigen constitutes the great mystery of life's origin, and one he hopes eventually to unravel in terms of algorithms and natural laws.[16] It is CSI that Michael Behe has uncovered with his irreducibly complex biochemical machines (see chapter 5).[17] It is CSI that for cosmologists underlies the fine-tuning of the universe, and which the various anthropic principles attempt to understand.[18] It is CSI that David Bohm's quantum potentials are extracting when they scour the microworld for what Bohm calls "active information."[19] It is CSI that enables Maxwell's demon to outsmart a thermodynamic system tending toward thermal equilibrium (see section 3.10).[20] It is CSI that for Roy Frieden unifies the whole of physics.[21] It is CSI on which David Chalmers hopes to base a comprehensive theory of human consciousness.[22] It is CSI that within the Chaitin-Kolmogorov-Solomonoff theory of algorithmic information identifies the highly compressible, nonrandom strings of digits (see section 2.4).[23]

CSI makes clear the connection between design and information theory: To infer design by means of the complexity-specification criterion (see section 1.3) is equivalent to detecting complex specified information. All the elements in the complexity-specification criterion that lead us to infer design find their counterpart in the detection of complex specified information. For an event to satisfy the complexity-specification criterion, it must first of all be contingent. But contingency, as we have seen, is the chief characteristic of information (recall Robert Stalnaker's quote in section 3.1). What's more, for a contingent event to be complex and specified is precisely what it means for that event in conjunction with a specification to constitute complex specified information (CSI). It follows that the complexity-specification criterion attributes design if and only if it detects CSI.

## 3.6   Semantic Information

I want next to relate CSI to semantic information. Within the conventional understanding of information, semantic information is one of the four main aspects of information. These are *mereology*, *statistics*, *syntactics*, and *semantics*. The order here is significant, with one building on the next. Mereology is the most basic aspect of information. "Mereology" derives from the Greek word for part (i.e., *meros*). As we saw at the beginning of this chapter, the fundamental idea behind information is the identification of one possibility to the exclusion of others within a reference class of possibilities. Mereology in this connection refers to the individuation of possibilities in a reference class so that they form identifiable parts, thereby making the generation of information explicit.

Once possibilities within a reference class have been individuated (i.e., once mereology is in place), we will want to measure the amount of information associated with a given possibility. This requires a measure of information that is independent of the procedure used to individuate the possibilities in a reference class. Otherwise the same possibility can be assigned different amounts of information depending on how the other possibilities in the reference class are individuated (thus making the information measure ill-defined). As we saw in section 3.1, the way to achieve a well-defined information measure is not simply to count possibilities but to assign probabilities to those possibilities. For convenience we then recalibrated the resulting probability measure $P$ with a logarithmic transformation. This yielded our preferred means of measuring information, namely, the information measure $I$ where $I =_{\text{def}} -\log_2 P$. $I$ constitutes a statistical measure of information and thus supplements mereology with statistics.

Once mereology and statistics are in place, we can focus on reference classes of possibilities that comprise symbol strings from a fixed alphabet. This is traditionally where most of the action in information theory has been because historically information theory developed as a branch of communication theory, and communication theory is concerned with transmitting messages (represented as symbol strings) across communication channels. A reference class of possibilities that comprises symbol strings thus naturally evokes not just mereology and statistics, but also syntax. Symbol strings tend to follow certain rules. Moreover, these rules tend to have statistical properties. In English, for instance, the letter *e* occurs roughly 12 percent of the time; *u* invariably follows *q*; *x* hardly ever begins a word; etc. Traditional information theory combines mereology, statistics, and syntactics.

The jumps from mereology to statistics and then from statistics to syntactics are mathematically and logically straightforward. The final jump from syntactics to semantics is not. It is here that the mathematical theory of information has uniformly failed to make progress. Indeed, the jump from syntactics to semantics constitutes one of the most hotly disputed areas within the philosophy of language. Is there an independent realm of meaning that legitimately attaches to syntactic structures? Is meaning a purely conventional social construction that supervenes on syntactic structures? Is there only syntax and no meaning?[24]

This is not the place to explore these questions. Nonetheless, it is important to realize that the semantic aspect of information is, at least in the popular conception, the one that is most significant. The mereological, statistical, and syntactic aspects of information are fine and well, but most people are interested in information for its semantic content. People find information important because it tells them important things about their lives and the world—because it tells them whether it is going to rain tomorrow, whether their favorite stock is going to go up, and whether their car can survive a road trip they have planned. The meaning inherent in information, and not the precise linguistic structure by which the information is conveyed, is what is important to most people.

Thus, even though mereology, statistics, syntactics, and semantics figure into the traditional view of information, pride of place has traditionally gone to semantics, with the other three aspects of information viewed as ancillary to semantics. At the same time, semantic information has not submitted to the same mathematical and logical analysis that the other three aspects of information have. What's more, only the semantic aspect of information has been traditionally associated with intelligent agency. The connection between CSI and semantic information therefore requires some clarification:

Unlike semantic information, CSI submits to a mathematical and logical analysis; yet like semantic information, CSI is associated with intelligent agency. To define CSI requires only the mereological and statistical aspects of information. No syntax and no theory of meaning is required. For the ordered pair (T,E) to constitute complex specified information, an intelligent agent need only be able to identify the pattern T independently of E. How the intelligent agent identifies the pattern is irrelevant. In particular, the intelligent agent need not assign a meaning to the pattern.

Is this a weakness of CSI? Not at all. Counterintuitive as it may seem, semantics, far from helping to detect design, can actually hinder its detection. Consider that the Smithsonian Institution devotes a room to obviously designed artifacts for which no one has a clue what those artifacts do.[25] The meaning of those artifacts is lost. This loss of semantic information, however, does not prevent those artifacts from exhibiting complex specified information and thereby reliably signaling design. Semantic information and complex specified information are distinct categories of information. Indeed, to require that semantic information be made explicit before one can infer design is artificially to restrict the design inference. CSI depends on a coincidence of agent-induced and event-induced information. It does not depend on an agent assigning semantic content to that information. That is not to say that semantic content is necessarily lacking from CSI. But it is not required.

Neither CSI nor semantic information presupposes the other. This in my view is a tremendous asset of CSI, for it allows one to detect design without necessarily determining the function, purpose, or meaning of a thing that is designed (which is not to say that function, purpose, or meaning may not be useful in identifying a specification, but they are not mandated). Mereology and statistics, not syntactics or semantics, are the rock-bottom foundational aspects of information. Indeed, a scientifically fecund study of information can proceed solely on the basis of mereology and statistics (which is not to say that syntactics and semantics do not enrich the study of information). The sufficiency of mereology and statistics as a foundation for information and the dependence of CSI solely on these foundational aspects means that CSI bypasses many an impasse that semantic information has had to confront. CSI is robust and resolves many of the difficulties traditionally associated with information.

## 3.7 Biological Information

In *Steps Towards Life* Manfred Eigen characterizes the central problem of origins-of-life research as follows: "Our task is to find an algorithm, a natural law that leads to the origin of information."[26] Eigen is only half right. To determine how life began, it is indeed necessary to understand the origin of information. Neither algorithms nor natural laws, however, can produce the sort of information required for the origin of life. The great myth of contemporary evolutionary biology is that the information needed to explain complex biological structures can be purchased without intelligence. My aim throughout this book is to dispel that myth.

Manfred Eigen, Bernd-Olaf Küppers, and their circle identify the origin of information as the central problem of biology.[27] But what sort of information are they talking about? Clearly, if they are talking about a purely statistical sort of information, then information cannot be said to constitute a deep problem for biology or science generally. In that case information can readily be gotten on the cheap without recourse to intelligence. Just flip a coin 1000 times and you will witness an incredibly improbable (and hence highly complex) event. The information content of that sequence computes to 1000 bits of information. Chance can generate huge amounts of statistical information. Consequently Eigen and his colleagues must have something else in mind besides information simpliciter when they describe the origin of information as the central problem of biology.

I submit that what they have in mind is specified complexity, or what equivalently we have been calling in this chapter complex specified information or CSI. Certainly the complexity of biological information is not at issue. Living things are complex in the sense required by any purely statistical account of information. Nor is specification, or as it is also called biological specificity, at issue. For instance, historian of biology Horace Freeland Judson attributes the twentieth-century revolution in biology to "the development of the concept of biological specificity."[28]

Biological specification always refers to function. An organism is a functional system comprising many functional subsystems. In virtue of their function, these systems embody patterns that are objectively given and can be identified independently of the systems that embody them. Hence these systems are specified in the sense required by the complexity-specification criterion (see sections 1.3 and 2.5). The specification of organisms can be cashed out in any number of ways. Arno Wouters cashes it out globally in terms of the *viability* of whole organisms.[29] Michael Behe cashes it out in terms of the *minimal function* of biochemical systems.[30] Darwinist Richard Dawkins cashes out biological specification in terms of the *reproduction* of genes. Thus, in *The Blind Watchmaker* Dawkins writes, "Complicated things have some quality, specifiable in advance, that is highly unlikely to have been acquired by random chance alone. In the case of living things, the quality that is specified in advance is . . . the ability to propagate genes in reproduction."[31]

The central problem of biology is therefore not simply the origin of information but the origin of complex specified information. Paul Davies emphasized this point in his recent book *The Fifth Miracle* where he summarizes the current state of origin-of-life research: "Living organisms are mysterious not for their complexity *per se*, but for their tightly specified complexity."[32] The problem of specified complexity has dogged origin-of-life research now for decades. Leslie Orgel recognized the problem in the early 1970s: "Living organisms are distinguished by their specified complexity. Crystals such as granite fail to qualify as living because they lack complexity; mixtures of random polymers fail to qualify because they lack specificity."[33]

Where, then, does complex specified information or CSI come from, and where is it incapable of coming from? According to Manfred Eigen, CSI comes from algorithms and natural laws. As he puts it, "Our task is to find an algorithm, a natural law that leads to the origin of [complex specified] information."[34] The only question for Eigen is which algorithms and natural laws explain the origin of CSI. The logically prior question of whether algorithms and natural laws are even in principle capable of explaining the origin of CSI is one he ignores. And yet it is a question that undermines the entire project of naturalistic origins-of-life research. Algorithms and natural laws are in principle incapable of explaining the origin of CSI. To be sure, algorithms and natural laws can explain the flow of CSI. Indeed, algorithms and natural laws are ideally suited for transmitting already existing CSI. As we shall see next, what they cannot do is explain its origin.[35]

## 3.8   The Origin of Complex Specified Information

Manfred Eigen's search for algorithms and natural laws to account for biological information is properly subsumed under the more general search for a naturalistic account of complex specified information. Such an account would have to identify natural causes capable of generating complex specified information. Now, as we saw in chapter 1, natural causes are characterized by necessity, chance, or a combination of the two. Moreover, within science necessity is usually conceived in terms of deterministic natural laws (cf. Newton's law of gravitational attraction), chance is usually conceived in terms of randomness or "pure chance" (cf. the radioactive emission of an alpha particle), and the combination of chance and necessity is conceived in terms of nondeterministic natural laws (cf. natural selection acting on random variation).

These three ways of characterizing natural causes are represented mathematically by nonstochastic functions (representing deterministic natural laws and therefore necessity), random sampling from a probability distribution (representing randomness or pure chance), and stochastic processes (representing nondeterministic natural laws and therefore the combination of chance and necessity). Of these, stochastic processes constitute the most general mathematical formalism (by zeroing out the stochastic element one recovers a nonstochastic function and therefore necessity; by focusing purely on the stochastic element one recovers random sampling from a probability distribution and therefore pure chance).

In this section I will present an in-principle mathematical argument for why natural causes are incapable of generating complex specified information. I will show that neither nonstochastic functions nor random sampling from a probability distribution nor stochastic processes can do better than transmit already existing complex specified information. It follows that any claim that natural causes can generate complex specified information not only cannot be justified but cannot even be situated within an information-theoretic framework where it could be justified.

Justifying the claim that natural causes cannot generate complex specified information is technically demanding. Before justifying this claim mathematically, let me therefore try to spell out in plain English why natural causes are not up to the task of generating CSI. Using natural causes to explain CSI commits a category mistake. It is like using plumbing supplies to explain oil painting—the one is irrelevant to the other and to conflate the two only leads to confusion. The problem with using natural causes to explain CSI is essentially this. Natural causes are properly represented by nondeterministic functions (stochastic processes). Just what these are in precise mathematical terms is not important. The important thing is that functions map one set of items to another set of items and in doing so map a given item to one and only one other item. Thus for a natural cause to "generate" CSI would mean for a function to map some item to an item that exhibits CSI. But that means the complexity and specification in the item that got mapped onto gets pushed back to the item that got mapped. In other words, natural causes just push the CSI problem from the effect back to the cause, which now in turn needs to be explained. It is like explaining a pencil in terms of a pencil-making machine. Explaining the pencil-making machine is as difficult as explaining the pencil. In fact, the problem typically gets worse as one backtracks CSI.

Stephen Meyer makes this point beautifully for DNA.[36] Suppose some natural cause is able to account for the sequence specificity of DNA (i.e., the CSI in DNA). The four nucleotide bases are attached to a sugar-phosphate backbone and thus cannot influence each other via bonding affinities. In other words, there is complete freedom in the sequencing possibilities of the nucleotide bases. In fact, as Michael Polanyi observed in the 1960s, this must be the case if DNA is going to be optimally useful as an information bearing molecule.[37] Indeed, any limitation on sequencing possibilities of the nucleotide bases would hamper its information carrying capacity. But that means that any natural cause that brings about CSI in DNA must admit at least as much freedom as is in the DNA sequencing possibilities (if not, DNA sequencing possibilities would be constrained by physico-chemical laws, which we know they are not). Consequently, any CSI in DNA tracks back via natural causes to CSI in the antecedent circumstances responsible for the sequencing of DNA. To claim that natural causes have "generated" CSI is therefore totally misleading—natural causes have merely shuffled around preexisting CSI.

Let us now turn to the mathematical justification for why natural causes cannot generate CSI. We begin with deterministic natural laws. Within mathematics, such laws are represented as functions, that is, relations between two sets which to every member in one set (called the domain) associates one, and only one, member in the other set (called the range). Typically we say that the function *maps* an element in the domain to its associated element in the range. Functions are fully deterministic: given an element in the domain, a function maps it to a unique element in the range. The algorithms of computer science are functions in which the domain comprises input data and the range output data. But functions also meet us in the less mathematical aspects of our lives. There is a function that maps every U.S. citizen to his or her Social Security number. This is a function because everyone's Social Security number is unique (at least for those citizens currently living). There are also functions that map each of us uniquely to our fathers and mothers (each of us has only one father and only one mother). On the other hand, the relation connecting parents to their children is nonfunctional: a given father or mother may have more than one offspring.

When deterministic natural laws are represented as functions, the domain comprises initial and boundary conditions, and the range comprises physical states at subsequent times t. Let us now try to imagine what it would mean for a deterministic natural law, when represented as a function, to generate complex specified information. Suppose therefore that we had some CSI $j$ and a function (representing a deterministic natural law) $f$ that, à la Manfred

152 ⌣ Chapter 3

Eigen, led to the origin of $j$. That would mean some element in the domain of $f$, call it $i$, when acted on by $f$, yielded the output $j$. Mathematicians represent this relationship by writing $f(i) = j$. But this functional relationship hardly explains the origin of $j$. One problem has been solved by creating another, for now the origin of $i$ must be explained. Worse yet, the newly created problem is no easier than the one we started with. Functional relationships at best preserve what information is already there, or else degrade it—they never add to it. Thus, however much information resides in $j$ will be contained in any $i$ that via the function $f$ maps onto $j$. What's more, if $j$ is specified, then the inverse image under the function $f$ will also be specified (the inverse image of $j$ under $f$ are all the elements in its domain that $f$ associates with $j$). In particular, since $i$ maps onto $j$ via $f$, $i$ is in this inverse image. In short, if $j$ constitutes complex specified information and $f$ is a function that maps $i$ onto $j$, then $i$ constitutes specified information at least as complex as $j$.

Thus, instead of explaining the origin of CSI, algorithms and natural laws shift the problem elsewhere—in fact, to a place where the origin of CSI will be at least as difficult to explain as before. Formula (*), which we derived in section 3.1, bears this out. According to this formula, for all items of information A and B

(*) $$I(A\&B) = I(A) + I(B|A).$$

Since $i$ fully determines $j$ with respect to $f$, $I(j|i) = 0$. Thus, applying formula (*) to $i$ and $j$ yields $I(i\&j) = I(i)$. It follows that $j$ contains no information that was not already contained in $i$.

The argument just given was perhaps a bit too fast. What's more, its connection to the formulation of complex specified information given in section 3.5 may not be immediately evident, since in that chapter we characterized CSI as comprising ordered pairs of physical and conceptual information, whereas here we seem to be sidestepping this feature of CSI, referring to CSI with single letters like $i$ and $j$. Let us therefore back up and reframe the preceding argument in terms of the formal apparatus of section 3.5. (The next two paragraphs contain more mathematics than most readers may care to endure. Nonetheless, they are necessary to connect the present discussion to the account of CSI given in section 3.5. Readers who are willing to take my word for it can skip these paragraphs.)

The argument can then be reframed as follows: We are given CSI $j = (T_1, E_1)$ based on a reference class of possibilities $\Omega_1$. There is a function $f$ that maps another reference class $\Omega_0$ (possibly identical with $\Omega_1$) into $\Omega_1$. $E_1$ is an actual event constituting physical information and is subsumed within a target $T_1$, which constitutes not just conceptual information but also a specification in virtue of its capacity to be identified independently of $E_1$ (i.e., its detachability from $E_1$). Now, to account for the origin of CSI $j$ via the function $f$ would mean that some element in the domain of $f$ (the domain being $\Omega_0$), which when acted on by $f$, yielded the output $j$. Call this element in $\Omega_0$ $i$. Clearly $i$ will have to include an event $E_0$ that under $f$ maps onto $E_1$, that is, $f(E_0) = E_1$. So far $E_0$ is just a generic item of information from $\Omega_0$, leaving it for the moment undetermined whether functions need to input CSI to output CSI. In fact they do. To see that functions need to input CSI to output CSI, consider that for any function $f$ that maps $\Omega_0$ to $\Omega_1$, $f^{-1}$ defines a function from the subsets of $\Omega_1$ to the subsets of $\Omega_0$ such that if E is an arbitrary subset of $\Omega_1$, then $f^{-1}(E) =_{\text{def}} \{x \in \Omega_0 \mid f(x) \in E\}$ is the subset of $\Omega_0$ comprising all the elements that $f$ maps into E ($f^{-1}(E)$ is known alternately as the *preimage, counterimage,* or *inverse image* of E).[38] The function $f^{-1}$ is a *homomorphism of boolean algebras* from the powerset of $\Omega_1$ to the powerset of $\Omega_0$. This means that $f^{-1}$ is a well-defined function from the subsets of $\Omega_1$ back to the subsets of $\Omega_0$ that preserves the set-theoretic structure of both reference classes $\Omega_0$ and $\Omega_1$.

Consequently, not only does $f$ map $E_0$ to $E_1$ (i.e., $f(E_0) = E_1$), but $f^{-1}$ maps the target $T_1$ back to a subset of $\Omega_0$, which we can call $T_0$. What's more, because $f^{-1}$ is a homomorphism of boolean algebras, $T_0$ subsumes $E_0$. So too, because $f^{-1}$ is a homomorphism of boolean algebras, a probability measure $\mathbf{P}$ on $\Omega_0$ induces a probability $\mathbf{P}°f^{-1}$ on $\Omega_1$ ($\mathbf{P}°f^{-1}$ is the functional composition of $\mathbf{P}$ and $f^{-1}$—i.e., take an event in the reference class $\Omega_1$ and apply $f^{-1}$ to it; then take what you get and apply $\mathbf{P}$ to it).[39] Thus, if $\mathbf{P}$ characterizes the probability of $E_0$ occurring and $f$ characterizes the physical process that led from $E_0$ to $E_1$, then $\mathbf{P}°f^{-1}$ characterizes the probability of $E_1$ occurring and $\mathbf{P}(E_0) \leq \mathbf{P}°f^{-1}(E_1)$ since $f(E_0) = E_1$ and thus $E_0 \subset f^{-1}(E_1)$. Moreover, since $T_0$ is by definition the event in $\Omega_0$ gotten by mapping the target $T_1$ via $f^{-1}$, it follows that $T_0 = f^{-1}(T_1)$, and thus $\mathbf{P}(T_0) = \mathbf{P}°f^{-1}(T_1)$. Finally, since $f$ is a well-defined function and since $T_1$ can be identified independently of $E_1$, it follows that $T_0 = f^{-1}(T_1)$ can be identified independently of $E_0$ (to be identified $T_0$ requires only $f^{-1}$ and $T_1$). Formalizing this in terms of the definition of detachability in section 2.5 is straightforward, with $f$ merely needing to be composed with the rejection function on $\Omega_1$: if g is the rejection function on $\Omega_1$ that induces the rejection region $T_1$ that is detachable from $E_1$, then g°$f$, the composition of g and $f$, is the rejection function on $\Omega_0$ that induces the rejection region $T_0$ that is detachable from $E_0$). Consequently, if $f$ generates CSI $j = (T_1, E_1)$, then the information $i = (T_0, E_0)$ that $f$ maps to $j$ is itself CSI with the degree of complexity in both being identical (because

$P(T_0) = P \circ f^{-1}(T_1)$ and therefore $I(T_0) = I(f^{-1}(T_1))$). Bottom line: for functions to generate CSI they must employ preexisting CSI.

It follows that functions (and therefore deterministic natural laws represented by functions) do not explain the origin of complex specified information but only make the information problem worse. Suppose, for instance, you look at the *Statistical Abstract of the United States* and find that the average income of a U.S. citizen is so-much-and-so-much. How did this item of information originate? Clearly, the census bureau had to contact all the U.S. citizens, record their individual incomes, add the incomes together, and divide by the number of U.S. citizens. To take an average is to apply a function—given the input data (all the individual U.S. incomes), the output data are uniquely determined. But more significantly, to take an average is also to compress data. The information inherent in the record of all individual incomes far exceeds the information inherent in their average. Taking an average is a statistical technique for compressing data. In an information age, information inundates us. To assist the information seeker, information providers will therefore often compress information.

There is one subtlety we need now to consider. I have just argued that when a function acts to yield information, what the function acts upon has at least as much information as what the function yields. This argument, however, treats functions as mere conduits of information, and does not take seriously the possibility that functions might add information. I gave the example of taking an average whereby data are compressed and information is lost. But consider the function that maps library call numbers to their corresponding books. Clearly, there is less information in the call numbers than in the books. Here we have a function that is adding information. Moreover, it is adding information because the information is embedded in the function itself.

Although this observation seems to undermine my previous argument (i.e., that the output of a function can contain no more information than its input), in fact it leaves the argument virtually unchanged. The point is that instead of the function $f$ now merely serving as a conduit, mapping information $i$ to information $j$, the information in $f$ must now itself be taken into account. The way to do this is to employ the universal composition function $U$, which to an ordered information-function pair $(i, f)$ assigns the information obtained by applying $f$ to $i$—in this case $j$. Thus $U(i, f) = f(i) = j$. Unlike $f$, which may well incorporate information, $U$, the universal composition function, incorporates no information of its own, but is merely a conduit for information. By simply taking ordered pairs and treating the second element as a function applied to the first, $U$ introduces no information of its own. Note that in the case of algorithms $U$ is a universal Turing machine (i.e., an algorithm capable of running all other algorithms).[40]

The form of the original argument is therefore unchanged: the information $j$ arises by applying $U$ (cf. $f$ in the original argument) to the information $(i,f)$ (cf. $i$ in the original argument). Just as in computer science the distinction between data and programs is not hard and fast, so the distinction between functions and information is not hard and fast. We can therefore treat the ordered pair $(i,f)$ as information which via the universal composition function maps to the information $j$. Clearly, the information inherent in $(i, f)$ is no less than that in $j$. Formula (*) confirms this as well. This argument, by employing the universal composition function, is perfectly general. In particular, it answers the attempt by complexity-theorists to account for the origin of CSI in terms of self-organizing dynamical systems. Once we examine the precise informational antecedents to $j$, the illusion that we can generate CSI for free disappears. Like a bulge under a rug, the information problem can be shifted around, but it does not go away.

What mathematicians call functions and what scientists call deterministic natural laws cannot explain the origin of CSI. Because the processes that such functions or laws describe are deterministic, these processes cannot yield contingency, and without contingency there can be no information. The problem with deterministic laws is that they invariably yield only a single live possibility. Take a computer algorithm that performs addition. Let us say the algorithm has a correctness proof, so that it performs its additions correctly. Given the input data $2 + 2$, can the algorithm output anything other than 4? Computer algorithms are wholly deterministic. They allow for no contingency, and thus can generate no information. At best, therefore, laws can shift information around or lose it, as when data get compressed. What laws cannot do is produce contingency; and without contingency they cannot generate information, to say nothing of complex specified information.[41]

If not by means of laws, how then does contingency—and hence information—arise? Two, and only two, answers are possible here. Either the contingency is a blind, purposeless contingency—which is chance (whether pure chance or chance constrained by necessity); or it is a guided, purposeful contingency—which is intelligent causation. We shall return to intelligent causation in due course, but for now let us examine whether chance is capable of generating CSI. First, let us be clear that pure chance, entirely unsupplemented and left to its own devices, is incapable of generating CSI. Chance can generate complex unspecified information, and chance can generate noncomplex specified information. What chance cannot generate is information that is both complex and specified.

To see this, consider a typist at a keyboard. By randomly typing a long sequence of letters, the typist will generate complex unspecified information: the precise sequence of letters typed will constitute a highly improbable unspecified event, yielding complex unspecified information (recall that high probability corresponds to low complexity whereas low probability—i.e., high improbability—corresponds to high complexity). Alternatively, the typist, even if typing randomly, might by chance type the short sequence of letters *t-h-e*, thereby generating noncomplex specified information: typing *t-h-e* constitutes a specified high-probability event, instancing noncomplex specified information. What random typing cannot do is produce an extended meaningful text, thereby generating information that is both complex and specified.

Why can this not happen by chance? According to the complexity-specification criterion of chapter 1, once the improbabilities (i.e., complexities) become too vast and the specifications too tight, chance is eliminated and design is implicated. Just where the probabilistic cutoff is can be debated, but that there is a probabilistic cutoff beyond which chance becomes an unacceptable explanation is clear. The universe will experience heat death before random typing at a keyboard produces a Shakespearean sonnet. The French mathematician Emile Borel proposed 1 in $10^{50}$ as a universal probability bound below which chance could definitely be precluded—that is, any specified event as improbable as this could not be attributed to chance.[42] Borel based his universal probability bound on cosmological considerations, looking to the opportunities for repeating and observing events throughout cosmic history. Borel's 1 in $10^{50}$ probability bound translates to 166 bits of information.

In sections 1.5 and 2.8 I justify a more stringent universal probability bound of 1 in $10^{150}$ based on the number of elementary particles in the observable universe, the duration of the observable universe until heat death or collapse, and the Planck time.[43] A probability bound of 1 in $10^{150}$ translates to 500 bits of information. Accordingly, specified information of complexity greater than 500 bits cannot reasonably be attributed to chance. This 500-bit ceiling on the amount of specified information attributable to chance constitutes a *universal complexity bound* for CSI. If we now define CSI as any specified information whose complexity exceeds 500 bits of information, it follows immediately that chance cannot generate CSI. Throughout the sequel we take the "C" in "CSI" to denote at least 500 bits of information.

Biologists by and large do not dispute that pure chance, in the sense of random sampling from a probability distribution, cannot generate CSI. Most biologists reject pure chance as an adequate explanation of CSI. Besides flying in the face of every canon of statistical reasoning, pure chance is scientifically unsatisfying as an explanation of CSI. To explain CSI in terms of pure chance is no more instructive than pleading ignorance or proclaiming CSI a mystery. It is one thing to explain the occurrence of heads on a single coin toss by appealing to chance. It is quite another, as Küppers points out, to take the view that "the specific sequence of the nucleotides in the DNA molecule of the first organism came about by a purely random process in the early history of the earth."[44] CSI cries out for explanation, and pure chance will not do it. Or as Richard Dawkins puts it, "We can accept a certain amount of luck in our explanations, but not too much."[45]

We can allow our scientific theorizing only so much luck—after that science degenerates into handwaving and mystery. A universal probability bound of $10^{-150}$, or a corresponding universal complexity bound of 500 bits of information, sets a conservative limit on the amount of luck we can allow ourselves in our scientific theorizing. Such a limitation on luck is crucial to the integrity of science. If we allow ourselves too many "wildcard" bits of information—either by giving ourselves too many lucky guesses or nature too many lucky occurrences—we can explain everything by reference to chance. This is as bad as explaining everything by reference to design. A precondition for any mode of explanation to be fruitful for science is that it not explain everything. Neither design as developed in this book nor chance and necessity are cover-all modes of explanation.

Thus far we have established the following: (1) Chance (as represented by random sampling from a probability distribution) generates contingency, but not complex specified information; (2) Deterministic natural laws (as represented by functions) generate neither contingency, nor information, much less complex specified information; and (3) Functions at best transmit already present information or else lose it. The next order of business is therefore to show that no combination of chance and deterministic natural laws (i.e., nondeterministic natural laws) is going to generate complex specified information either.

The theoretical justification for why nondeterministic natural laws cannot generate CSI is virtually the same as the theoretical justification given earlier in this section for why deterministic natural laws cannot generate CSI. But instead of considering a deterministic function $f(i)$ in one variable, we need to consider a nondeterministic function $f(i,\omega)$ in two variables where the first variable signifies the object on which the function acts and the second signifies the randomizing component (i.e., the chance variable). We then define the universal composition function $U$ that inputs the object-chance-function ordered triple $(i,\omega,f)$ and outputs $f(i,\omega) = j$, that is, $U(i,\omega,f) = f(i,\omega) = j$. As in the deterministic case, the universal composition function $U$ incorporates no information of its own, but is merely a conduit for information. The formalism just described for characterizing nondeterministic natural laws is perfectly general. In mathematics $f$ is known as a *stochastic process*.[46] Stochastic processes can model everything from the neo-Darwinian mutation-selection mechanism to the probabilistic algorithms of computer science (e.g., genetic algorithms).

Now suppose we have some CSI $j$ and a nondeterministic function $f$ (i.e., a stochastic process) that, à la Manfred Eigen, leads to the origin of $j$. The origin of $j$ can then be broken into two stages. In the first stage, a chance outcome $\omega$ occurs. Once $\omega$ occurs and is fixed, the function $f$ becomes deterministic, that is, $f$ becomes a function in one variable: $f(\cdot,\omega) = f_\omega(\cdot)$, $\omega$ now being treated as a fixed parameter of the function $f$. This is the standard probabilistic move for transforming stochastic processes into *random functions*, which, once the random element $\omega$ is fixed, become what are called *sample paths* (stochastic processes and random functions are mathematically equivalent).[47] In the second stage, the parameterized deterministic function $f_\omega(\cdot)$ (i.e., sample path) gets applied to some element in its domain, call it $i$, yielding the item of interest, the CSI $j$.

From this two-stage analysis it becomes clear that no CSI is generated in the production of $j$. The first stage involves only chance, and therefore, as was shown earlier in this section, cannot generate CSI. The second stage involves no chance, but only a deterministic function, and therefore, as was shown in section 3.1, cannot generate CSI either. Thus, at no point in the transition from $\omega$ to $f_\omega(\cdot)$ to $f_\omega(i) = j$ is CSI generated. Whatever CSI is inherent in $j$ was already inherent in the nondeterministic function $f$ together with the nonrandom element in the domain of $f$, namely, $i$. This argument holds for Darwin's mutation-selection mechanism, for genetic algorithms, and indeed for any other chance-law combination. Just as chance or necessity left to themselves individually cannot purchase CSI, so their joint action cannot purchase CSI either.[48]

This argument that chance and necessity together cannot generate CSI holds with perfect generality. $f(i,\omega) = j$ is a stochastic process. Stochastic processes provide the most general mathematical means for modeling the joint action of chance and necessity.[49] In fact, by zeroing out the randomizing component $\omega$, stochastic processes can also model deterministic natural laws and therefore necessity. Moreover, by zeroing out the nonrandom component $i$, stochastic processes can also model pure chance. Stochastic processes are capable of modeling chance, necessity, or any combination of the two. Since chance, necessity, and their combination characterize natural causes, it now follows that *natural causes are incapable of generating CSI.*

## 3.9  The Law of Conservation of Information

In section 3.1 we saw that fully deterministic processes like computer programs satisfy a conservation law in which the amount of information outputted by a deterministic process never exceeds the amount inputted. Peter Medawar referred to this law as the Law of Conservation of Information (abbreviated LCI).[50] This law follows from the following formula, which has played such a key role throughout this chapter:

$$(*) \qquad\qquad I(A\&B) = I(A) + I(B|A).$$

If A fully determines B, then B is certain given A and therefore $P(B|A) = 1$. This in turn means that $I(B|A) = -\log_2 P(B|A) = 0$ (the logarithm of 1 is always 0). It therefore follows that the last term in formula (*) drops out and that $I(A\&B) = I(A)$. Thus when A fully determines B, the amount of information in A and B jointly is identical with the amount of information in A by itself. For a purely deterministic process leading from A to B, Medawar's Law of Conservation of Information can therefore be formulated as follows:

$$(\text{LCIdet}) \qquad\qquad I(A\&B) = I(A).$$

This last formula describes a deterministic version of the Law of Conservation of Information and applies to information generally and not just to CSI. Moreover, since chance can generate information generally (though not CSI), it follows that (LCIdet) has no analogue for chance processes so long as no restriction is placed on the information to which this formula is applied. Nevertheless, when restricted to CSI, this deterministic version of the Law of Conservation of Information admits a powerful extension. In the last section we saw that: (1) natural causes are characterized by chance, necessity, or a combination of the two; (2) such causal processes can be represented mathematically by stochastic processes; and (3) in outputting CSI, stochastic processes need to input preexisting CSI whose complexity is at least that which was outputted. The broad conclusion of the last section was therefore that *natural causes are incapable of generating CSI.*

Within the formalism developed thus far, we can interpret this last claim as follows: If a natural cause produces some event $E_2$ that exhibits specified complexity, then for any antecedent event $E_1$ that is causally upstream from

$E_2$ and that under the operation of natural causes is sufficient to produce $E_2$, $E_1$ likewise exhibits specified complexity. More precisely, if a natural cause produces some event $E_2$ that exhibits specified complexity, then $E_2$ is the second component of some instance of CSI, call it $B = (T_2,E_2)$. Furthermore, for any antecedent event $E_1$ that under the operation of natural causes is sufficient to produce $E_2$, $E_1$ is the second component of some other instance of CSI, call it $A = (T_1,E_1)$. And finally, $T_1$ is such that the amount of information in $T_1$ and $T_2$ together is essentially identical to the amount of information in $T_1$ by itself, where any difference in the two is less than the universal complexity bound (which we abbreviate UCB and which throughout this book we take to be 500 bits of information—see section 3.5). We can abbreviate this relation between these two quantities of information by saying that they are equal modulo the UCB—i.e., $I(T_1\&T_2) = I(T_1)$ mod UCB. The word "modulo" here refers to the wiggle room within which $I(A)$ can differ from $I(A\&B)$. To say that these two quantities of information are equal modulo UCB is to say that they are essentially the same except for a difference no greater than the UCB.

The general form of the Law of Conservation of Information may now be stated as follows:

**Law of Conservation of Information.** Given an item of CSI, call it $B = (T_2,E_2)$, for which $E_2$ arose by natural causes, any event $E_1$ causally upstream from $E_2$ that under the operation of natural causes is sufficient to produce $E_2$ belongs to an item of CSI, call it $A = (T_1,E_1)$, such that

(LCIcsi)                    $I(A\&B) = I(A)$ mod UCB

where by definition the quantity of information in an item of specified information is the quantity of information in the conceptual component (i.e., $I(A) =_{def} I(T_1)$ and $I(A\&B) =_{def} I(T_1\&T_2)$).

Some elaboration is in order. For simplicity we can assume that A and B are both defined in relation to the same reference class of possibilities $\Omega$ (if A is defined in relation to $\Omega_1$ and B in relation to $\Omega_2$, we can let $\Omega$ be the Cartesian product of $\Omega_1$ and $\Omega_2$ and then embed A and B canonically in $\Omega$). For $A = (T_1,E_1)$ and $B = (T_2,E_2)$, we define A&B, the conjunction of these two items of specified information, as $(T_1\&T_2,E_1\&E_2)$. Moreover, we define the quantity of information associated with a generic item of specified information as the quantity of information in the first member of the ordered pair (i.e., the quantity of information in the conceptual as opposed to the physical component of the pair). Thus for $A = (T_1,E_1)$, $B = (T_2,E_2)$, and A&B

$= (T_1 \& T_2, E_1 \& E_2)$, we have $I(A) = I(T_1)$, $I(B) = I(T_2)$, and $I(A\&B) = I(T_1 \& T_2)$. This way of defining the quantity of information associated with an item of specified information is clearly the way to go since what turns specified information into complex specified information is that the quantity of information in the conceptual component is large. The quantity of information in the physical component will then be large as a matter of course (see section 3.5—intuitively the idea here is that hitting a target does not happen by chance so long as the target is small enough).

Since conjoining items of information can only increase the number of possibilities ruled out from the reference class of possibilities and therefore can only increase the amount of information, it follows in (LCIcsi) that $I(A\&B)$ is always greater than or equal to $I(A)$. Consequently, (LCIcsi) is equivalent to $I(A\&B) \leq I(A) + UCB$. In sections 1.5 and 2.8 I showed that for a universal probability bound of $10^{-150}$ any specified event of probability that small or smaller cannot reasonably be attributed to chance. Now it turns out that $-\log_2 10^{-150}$ is just less than 500 (see section 3.5). A probability bound of 1 in $10^{150}$ therefore translates to 500 bits of information. Accordingly, specified information of complexity greater than 500 bits cannot reasonably be attributed to chance. This 500-bit ceiling on the amount of specified information potentially attributable to chance constitutes a universal complexity bound for CSI. Because small amounts of specified information can be produced by chance, this 500-bit tolerance factor needs to be included in the Law of Conservation of Information. Moreover, because this number is contingent on the probabilistic resources of the known universe, I prefer to write $I(A\&B) = I(A)$ mod UCB rather than $I(A\&B) = I(A)$ mod 500. The number 500, conceived as a universal complexity bound, is in principle revisable whereas the Law of Conservation of Information, when stated in relation to the appropriate universal complexity bound, is not (following as it does strictly on mathematical grounds). For now, however, we are justified treating UCB as equal to 500.

How is the Law of Conservation of Information a *conservation* law? Ordinarily, when something is conserved, some quantity characterizing that thing remains unchanged. This is certainly the case with conservation of energy. On the other hand, this is not so clear with conservation of information. In our ordinary experience information can actually increase under the operation of natural causes—for instance, random coin tossing generates information (though not CSI). At the same time, ordinary experience also tells us that complex specified information originates from intelligence. Moreover, it tells us that when CSI is given over to natural causes it either remains unchanged (in which case information is conserved) or disintegrates (in which case information diminishes). For instance, the best thing that can happen to a book on a library shelf is that it remains as it was when originally published and thus preserves the CSI inherent in its text. Over time, however, what usually happens is that a book gets old, pages fall apart, and the information on the pages disintegrates. The Law of Conservation of Information is therefore more like a thermodynamic law governing entropy than a conservation law governing energy, with the focus on degradation rather than conservation. Nonetheless, given that the statement of the law involves an equality between a given instance of CSI and a causally prior one and given that this reference to the law already has some currency (cf. Peter Medawar's use of the term), it seems appropriate to refer to this law as a conservation law. The crucial point of the Law of Conservation of Information is that natural causes can at best preserve CSI (augmenting it by no more than the UCB), may degrade it, but cannot generate it.

It is important to be clear just what is being denied when the Law of Conservation of Information claims that natural causes cannot generate CSI. To deny that natural causes can generate CSI is not the same as denying that natural causes can produce events that exhibit CSI. As has been stressed repeatedly, natural causes are ideally suited as conduits for CSI. It is in this sense, then, that natural causes can be said to "produce CSI." But natural causes never produce things *de novo* or *ex nihilo*. Whenever natural causes produce things, they do so by reworking other things. Thus to the question How did natural causes produce X? it is never enough to assert that natural causes simply did it. Rather, one must point to some antecedent Y that is causally sufficient to account for X. This is the case regardless whether the natural cause that produced X operated deterministically or nondeterministically. Gravity, operating deterministically, is sufficient to account for the falling of a metal ball near the earth's surface. Radioactivity, operating nondeterministically, is sufficient to account for the decay of uranium into lead and helium.

The sufficiency of natural causes to produce an effect is crucial to naturalistic explanations since without it the door is open to intelligent agency. Naturalistic explanations by definition exclude appeals to intelligent agency. Thus to say that a natural cause produced X, and then to point to some antecedent Y and say that Y under the operation of natural causes only contributed to the production of X, is not enough for a bona fide naturalistic explanation. The reason it is not enough is because it leaves the door open to intelligent agency acting in tandem with natural causes. Such a mixture of intelligent and natural causes is clearly not what is being denied when the Law of Conservation of Information states that CSI cannot be generated by natural causes. The Law of Conservation of Information is not saying that natural causes in tandem with intelligent causes cannot generate CSI but that natural causes apart from intelligent causes cannot generate CSI. Thus to attribute X to natural causes is a call for explanation in terms of some antecedent circumstance Y upon which natural causes—and only natural causes—operate and suffice to produce X. The Law of Conservation of Information says that if X exhibits CSI, then so does Y. It follows that natural causes do not and indeed cannot generate CSI but merely shuffle it around.

LCI has profound implications for science. Among its immediate corollaries are: (1) The CSI in a closed system of natural causes remains constant or decreases. (2) CSI cannot be generated spontaneously, originate endogenously, or organize itself (as these terms are used in origins-of-life research). (3) The CSI in a closed system of natural causes either has been in the system forever or was at some point added exogenously (implying that the system, though now closed, was not always closed). (4) In particular, any closed system of natural causes that is also of finite duration received whatever CSI it contains before it became a closed system.

The first corollary can be understood in terms of data storage and retrieval. Data can constitute a form of CSI. Ideally data would stay unaltered over time. Nonetheless, entropy being the corrupting force that it is, data tend to degrade and need constantly to be restored. Over time magnetic tapes deteriorate, pages yellow, print fades, and books disintegrate. Information may be eternal, but the physical media that house information are subject to natural causes and are thoroughly ephemeral. The first corollary acknowledges this fact.

The second and third corollaries assert that CSI cannot be explained in terms other than itself. CSI cannot be reduced to self-organizational properties of matter, for these would just be natural causes, and LCI rejects natural causes as adequate to generate CSI. Given an instance of CSI, these corollaries allow but two possibilities: either the CSI was always present or it was inserted. Proponents of intelligent design differ about which of these two possibilities obtains (see section 6.6). This debate is not new.[51] The German teleo-mechanists and the British natural theologians engaged in much the same debate, with the Germans arguing that teleology was intrinsic to the world, the British arguing that it was extrinsic.[52] However this debate gets resolved, CSI is an empirically detectable entity that is not reducible to natural causes.

The fourth and final corollary shows that scientific explanation is not identical with reductive explanation. This corollary is especially relevant to science. Richard Dawkins, Daniel Dennett, and many scientists and philosophers are convinced that proper scientific explanations must be reductive, moving from the complex to the simple.[53] The Law of Conservation of Information, however, shows that CSI cannot be explained reductively. To explain an instance of CSI requires either a direct appeal to an intelligent agent who via a cognitive act originated the CSI in question, or locating an antecedent instance of CSI that contains at least as much CSI as we started with. A pencil-making machine is more complicated than the pencils it makes. A clock factory is more complicated than the clocks it produces. What's more, tracing back the causal chains from pencil to pencil-making machine or clock to clock factory all in the end terminate in an intelligence. Intelligent causes generate CSI whereas natural causes transmit preexisting CSI (and usually imperfectly).[54]

Thus to explain CSI in terms of natural causes is to fill one hole by digging another. With CSI the information problem never goes away short of locating the intelligence that originated the CSI. We have known this since elementary school. The telephone game, where one person whispers information to the next person who whispers it to the next, etc., illustrates how information degrades over time. The players of this game are links in a chain. With each transmission of information from one link to the next, there is the potential for losing information. Ideally each person would repeat exactly the information given by the preceding person in the chain and thus preserve the information given at the start of the chain. In general, however, that does not happen. In fact, the fun of the telephone game is to see how information degrades as it passes from the first to the last person in the chain.

The telephone game has more serious analogues. Consider the textual transmission of ancient manuscripts. A textual critic's task is to recover as much of the original text of an ancient manuscript as possible. Almost always the original text is unavailable. Instead the textual critic confronts multiple variant texts, each with a long genealogy tracing back to the original text. Fifty generations of copies may separate a given manuscript from the original text. The original text is copied in the first generation, then that copy is itself copied, then that second copy is in turn copied, and so on fifty times before we get to the manuscript in our possession. We assume that most of the copyists were trying to preserve the text faithfully. Even so, they were bound to introduce errors now and then. Worse yet are the naughty copyists who use copying as a pretext for inserting their own pet ideas into a text. The textual critic must therefore identify errors introduced by careless copying as well as errors stemming from a copyist's personal agenda. This can be enormously difficult. Even so, there is always a fixed reference point: Because the copyist presupposes an original text as the source from which all the variant manuscripts ultimately derive, the original text constitutes the initial CSI on which the transmission of the text depends.[55]

In both the telephone game and the transmission of texts, an intelligent cause rather than a natural cause not only generates the initial CSI but also transmits it. Is that a problem here given that the Law of Conservation of Information applies, strictly speaking, only to the transmission of CSI by natural causes? Although this law is concerned solely with placing limits on natural and not intelligent causes, it still applies. Intelligent causes can mimic natural causes, and that is what they are doing here. In both the telephone game and in the transmission of ancient texts, the persons transmitting information are supposed to repeat what they have been given. Repetition is an automatic process for which natural causes are ideally suited and for which intelligent causes are not required.

We can see this more clearly by considering successive photocopies of a black and white photograph. Natural causes govern a photocopy machine's operation. Take, therefore, a black and white photograph, photocopy it, then photocopy that copy, and keep doing this fifty times. The successive photocopies will show increasing degradation of the original photograph (i.e., the initial CSI). Depending on the quality of the photocopy machine and the resolution of the original photograph, the original photograph may be unrecognizable by the fiftieth photocopy. If the two previous examples were merely suggestive of the Law of Conservation of Information, the photocopy example illustrates this law exactly.

The most interesting application of the Law of Conservation of Information is the reproduction of organisms. Since reproduction proceeds by natural causes, there is no question that the law applies. In reproduction one organism transmits its CSI to the next generation. For most evolutionary biologists, however, this is not the end of the story. Most evolutionists would argue that the Darwinian mechanism of natural selection and random variation introduces novel CSI into an organism, supplementing the CSI of the parent(s) with CSI from the environment. Accordingly, the genetic contribution from the parent(s) and the Darwinian contribution through natural selection and random variation together constitute the CSI of an organism.

Because I will take up this claim in the next chapter, I will not dwell on it here. Nevertheless, it is important to understand a feature of CSI that will count decisively against generating CSI from the environment via natural selection and random variation or any other naturalistic mechanism for that matter. The crucial feature of CSI is that it is *holistic*. Although "holism" and "holistic" have become buzzwords, in reference to CSI these terms have a well-defined meaning. To say that CSI is holistic means that individual items of information (be they simple, complex, specified, or even complex specified) cannot simply be added together and thereby form a new item of complex specified information. CSI is not the aggregate of its constituent items of information. What this means for biology is that a gradual, incremental approach to generating CSI will never work.

CSI holism is a case of the whole being greater than the sum of its parts. CSI requires not only having the right collection of parts, but also having the parts in proper relation. Consider, for instance, the aggregate {A, IS, IT, LIKE, WEASEL, METHINKS}. All the items of information here are specified (they represent known words in the English language). Of these, ME-THINKS is the most complex, having 8 letters. For sequences of capital letters and spaces (27 possibilities at each position), the complexity of ME-THINKS is bounded by $-\log_2 1/27^8 = 38$ bits of information. Now contrast the aggregate {A, IS, IT, LIKE, WEASEL, METHINKS} with the sentence METHINKS IT IS LIKE A WEASEL. This sentence not only includes all the items of information that appear in the aggregate, but also arranges them in a grammatical sequence with semantic content. Unlike the aggregate, for which only the individual words are specified, here the entire sentence is specified (it is a line from *Hamlet*). Moreover, because the sentence is a sequence of 28 letters and spaces, its complexity is bounded by $-\log_2 1/27^{28} = 133$ bits of information and far exceeds the complexity of any item in the set or for that matter the sum of the complexities of all items in the set. Thus we see that CSI does not emerge by merely aggregating component parts. Only if a specification for the whole is given can parts be suitably arranged to form CSI. CSI is therefore a top-down, not a bottom-up concept.

## 3.10  A Fourth Law of Thermodynamics?

In *Logic and Information* Keith Devlin reflects on the thermodynamic significance of information:

> Perhaps *information* should be regarded as (or maybe is) a basic property of the universe, alongside matter and energy (and being ultimately interconvertible with them). In such a theory (or suggestion for a theory, to be more precise), information would be an intrinsic measure of the structure and order in parts or all of the universe, being closely related to entropy (and in some sense its inverse).[56]

I want in the concluding section of this chapter to focus on the last point Devlin raises, namely, whether information appropriately conceived can be regarded as inverse to entropy and whether a law governing information might correspondingly parallel the second law of thermodynamics, which governs entropy. Given the previous exposition it will come as no shock that my answer to both questions is yes, with the appropriate form of information being complex specified information and the parallel law being the Law of Conservation of Information. Indeed, I want to claim that the elusive fourth law of thermodynamics about which there has been sporadic and inconclusive speculation in the scientific literature is properly identified with the Law of Conservation of Information.

The first reference to a "fourth law of thermodynamics" with which I am familiar stems from the economist Nicholas Georgescu-Roegen and dates back to the early 1970s. His economic formulation of the fourth law is essentially a corollary of the second law of thermodynamics. Just as the second law limits the amount of usable energy available in a system to do work, Georgescu-Roegen's fourth law limits the amount of usable materials available in a system to manufacture products. This formulation of the law is sometimes stated in terms of pollution controls and the impossibility of perfect recycling.[57]

Outside economics the fourth law has yet to receive a precise and cogent formulation. In the natural sciences the first reference to it with which I am familiar stems from Victor Weisskopf and dates back to an informal article of his on the nature of science that appeared in 1977.[58] In that article he refers to a "fourth law of thermodynamics" as an inverse to the second law of thermodynamics. Weisskopf's fourth law generates "organized complexity" by exploiting temperature gradients between interacting thermodynamic systems. Weisskopf's remarks in this article about the fourth law are inchoate and show little appreciation for the distinction between heat being shifted around subject to the second law and mass-energy being configured into informational structures consistent with but in no way required by the second law. George Stavropoulos, in commenting on Weisskopf's article, points out this distinction.[59] Michael Polanyi had noted this distinction ten years earlier.[60] In his reply to Stavropoulos, Weisskopf maintained that the formation of organized complexity is consistent with the second law; nonetheless, he also admitted that the second law is insufficient to account for organized complexity. In particular, Weisskopf conceded that "we understand pitifully little about the processes of life."[61]

With the rise of complexity theory, chaos, and self-organization, the fourth law nowadays typically refers to the generation of complex or ordered structures via the flow of energy from an energy source to an energy sink.[62] Self-organizational theorists hope to find in the fourth law an answer to how complex systems—especially biological systems—organize themselves and evolve. Per Bak and Stuart Kauffman are two of the better known theorists working in this area.[63] So far the fourth law as developed by complexity theorists provides at best a qualitative description of the emergence of complexity, turning the fact that nature exhibits order and complexity into a principle that—for some as yet undiscovered reasons—it must do so. What all the formulations of the fourth law have yet to answer is how the order and complexity inherent in biological systems is generated. In other words, none of the formulations of the fourth law to date propose a naturalistic mechanism for generating order and complexity.

Consider for instance Stuart Kauffman's four candidate laws for the fourth law of thermodynamics:[64]

Law 1. Communities of autonomous agents will evolve to the dynamical "edge of chaos" within and between members of the community, thereby simultaneously achieving an optimal coarse graining of each agent's world that maximizes the capacity of each agent to discriminate and act without trembling hands.

Law 2. A coassembling community of agents, on a short timescale with respect to coevolution, will assemble to a self-organized critical state with some maximum number of species per community. In the vicinity of that maximum, a power law distribution of avalanches of local extinction events will occur. As the maximum is approached the net rate of entry of new species slows, then halts.

Law 3. On a coevolutionary timescale, coevolving autonomous agents as a community attain a self-organized critical state by tuning landscape structure (ways of making a living) and coupling between landscapes, yielding a global power law distribution of extinction and speciation events and a power law distribution of species lifetimes.

Law 4. Autonomous agents will evolve such that causally local communities are on a generalized "subcritical-supracritical boundary" exhibiting a generalized self-organized critical average for the sustained expansion of the adjacent possible of the effective phase space of the community.

The precise meaning of these candidate laws is not as important as what these laws presuppose and what they purport to explain. Newton's law of gravity, for instance, presupposes space, time, and massive bodies that exert a force of attraction on each other. Moreover, it explains the orbiting of the planets around the sun. What do Kauffman's laws presuppose? In each case they presuppose autonomous agents, which Kauffman defines as follows: "A molecular autonomous agent is a self-reproducing molecular system able to carry out one or more thermodynamic work cycles."[65] Kauffman's laws also presuppose that such agents evolve into communities that operate as nonlinear dynamical systems. Moreover, Kauffman's laws purport to explain the increase in complexity of biological systems over the course of evolution.

There is, to be sure, more to these laws, and I am giving them only the briefest exposition. But what should be immediately evident from Kauffman's statement of these laws is that they are very different in form and function from the traditional three laws of thermodynamics. The traditional three laws of thermodynamics are each proscriptive generalizations, that is, they each make an assertion about what cannot happen to a physical system. The first law states that in an isolated system total energy neither increases nor decreases. The second law states that the entropy of an isolated system cannot decrease. The third law states that it is not possible to reduce the temperature of an isolated system to absolute zero in a finite number of operations.

Kauffman's candidate laws are nothing like this. Instead they provide qualitative descriptions of the emergence of complexity in nature, yet without proposing a mechanism that is causally sufficient to account for the emergence of that complexity. What's more, their warrant, even as qualitatively descriptive laws, is suspect since contingencies in nature seem not to require autonomous agents to evolve into anything even as minimally complex as the simplest cell. Kauffman sees it as a virtual certainty that nature will produce complex living systems.[66] Other theorists like Francis Crick and Fred Hoyle see such systems as incredibly improbable.[67] It is not clear how Kauffman's candidate laws connect with thermodynamics except in the loose sense that the autonomous agents presupposed in Kauffman's laws perform work, replicate, and do other tasks characterized by the traditional laws of thermodynamics. But as descriptions of the self-organization and evolution of autonomous agents, Kauffman's candidate laws are far from being well-confirmed.

In place of Kauffman's candidates for the fourth law of thermodynamics, which attempt to provide a positive account for the emergence of complexity, I want to propose a fourth law that in the spirit of the traditional three laws of thermodynamics imposes a limitation on the emergence of complexity, namely, the Law of Conservation of Information. As we saw in the last section, this law states that the complex specified information in an isolated system of natural causes does not increase (modulo the universal complexity bound). So formulated, the fourth law has the right form and spirit as the other laws. What remains to be established is its physical significance, especially in relation to thermodynamics. I submit there is a deep connection, and that so formulated the fourth law allows for a definitive resolution of the Maxwell demon paradox whereas until now this paradox has only admitted partial resolutions.

John Pierce describes the Maxwell demon paradox as follows:

> Maxwell's demon inhabits a divided box and operates a small door connecting the two chambers of the box. When he sees a fast molecule heading toward the door from the far side, he opens the door and lets it into his side. When he sees a slow molecule heading toward the door from his side he lets it through. He keeps the slow molecules from entering his side and the fast molecules from leaving his side. Soon, the gas in his side is made up of fast molecules. It is hot, while the gas on the other side is made up of slow molecules and it is cool. Maxwell's demon makes heat flow from the cool chamber to the hot chamber.[68]

Since the work performed by the demon in moving the door can be made arbitrarily small and the work extracted from moving the fast molecules to one side of the box and the slow ones to the other can be used to do substantial work (as in driving a piston), Maxwell's demon is supposed to provide a counterexample to the claim that the entropy of an isolated system (in this case the box) always increases.

Is Maxwell's demon thumbing his nose at the second law or is the demon on closer inspection in fact not violating the second law? A number of physicists and information theorists have argued that the demon in fact fails to violate the second law. According to John Pierce, "Since the demon's environment is at thermal equilibrium, the only light present is the random electromagnetic radiation corresponding to thermal noise, and this is so chaotic that the demon can't use it to see what sort of molecules are coming toward the door."[69] Pierce is picking up on Leo Szilard's argument in his 1929 paper titled "On the Decrease of Entropy in a Thermodynamic System by the Intervention of Intelligent Beings."[70] According to Szilard, in the very act of observation or measurement to determine which way and how fast the molecules are approaching the door, the demon must perform enough work to counterbalance the decrease in entropy caused by sorting the molecules. In this way a violation of the second law is prevented. More recently Charles Bennett has argued that Maxwell's demon must possess a memory and erase information to sort the molecules in the box. But, as Bennett puts it, "forgetting results, or discarding information, is thermodynamically costly."[71] In Bennett's resolution of the Maxwell demon paradox, the thermodynamic cost of resetting the measurement apparatus restores the validity of the second law.

All these resolutions of the Maxwell demon paradox preserve the second law by presupposing that the demon is a physically embodied information gathering and using system (or "IGUS," as Wojciech Zurek abbreviates it).[72] By being physically embodied and by physically manipulating information, an IGUS incurs thermodynamic costs and thereby offsets any apparent decrease in entropy. The problem with assuming that Maxwell's demon is an IGUS, however, is that this assumption can be dropped without contradicting the laws of thermodynamics. Moreover, when it is dropped, the paradox raised by Maxwell's demon reemerges with full force, challenging the second law. Yair Guttmann puts his finger on the problem:

> Suppose, next, we have a second container [i.e., a box] that is an exact replica of the first [i.e., the box with Maxwell's demon as in John Pierce's description of it]. This time, though, there is no demon near the door. There is only a random device that opens the door at irregular intervals. Suppose that it so happens that the random device opened the door exactly when the demon did. In such a case, too, the same conclusion will follow. A heat flow will be created from [one side of the box] to [the other]. Notice that the second case is completely consistent with the laws of physics. . . . It proves that, in very special circumstances, the "fast" molecules of a colder body might be transferred to a warmer one, thereby contradicting the laws of thermodynamics [specifically, the second law].[73]

But could such a device be truly random? Clearly, such a device would be exhibiting specified complexity since the overwhelming majority of successive openings and closings of the door separating the two sides of the box would have no effect on sorting the fast and slow molecules into different compartments. Since specified complexity eliminates chance, such a device could not be truly random. It may seem, then, that the resolution of the Maxwell demon paradox has again eluded us, but in fact a definitive resolution is now in sight. The issue raised by the Maxwell demon paradox was never whether a physically embodied information gathering and using system (i.e., an IGUS) could circumvent the second law, but whether information involving no energetic cost could do as much. The important thing to realize is that this question has real physical significance and does not merely raise an interesting though highly speculative philosophical thought experiment.

How can information that requires no expenditure of energy have physical significance? While it is true that the transmission of information is frequently mediated by transfers of energy across a communication channel, the transmission of information is properly understood in terms of the corre-lations between what happens at the two ends of a communication chan-nel—and thus without reference to any intervening physical process. Absent such an intervening physical process, the information relationships can still obtain. As Fred Dretske explains,

> It may seem as though the transmission of information . . . is a process that depends on the causal inter-relatedness of source and receiver. The way one gets a message from s to r is by initiating a sequence of events at s that culminates in a correspond-ing sequence at r. In abstract terms, the message is borne from s [to] r by a causal process which determines what happens at r in terms of what happens at s. The flow of information may, and in most familiar instances obviously does, depend on underlying causal processes. Nevertheless, the information relationships between s and r must be distinguished from the system of causal relationships existing be-tween these points.[74]

The resolution of the Maxwell demon paradox is therefore not to look for expenditures of energy by the demon to ensure that the second law is pre-served, but to note that any demon capable of overriding the second law must employ complex specified information to do so. And since according to the Law of Conservation of Information complex specified information cannot be generated by natural causes, this means that the complex specified information a demon employs to override the second law was either already there in the system or inputted from outside or generated by the demon in virtue of its being an intelligent agent and not merely a natural object opera-ting exclusively by natural causes.

Does Maxwell's demon therefore violate the second law after all? To think of the demon as violating the second law is the wrong way to look at the demon's activity. The second law is concerned with localized concentrations of energy that can be exploited to do work. What I am calling the fourth law, the Law of Conservation of Information, is concerned with arrangements or configurations of energy that exhibit complex specified information. Where the interplay of these laws becomes interesting is for energy configurations that are equivalent from the vantage of the second law but differ as to whether they exhibit complex specified information. A magnetic diskette recording random bits versus one recording, say, the text of this book are thermodynamically equivalent from the vantage of the second law. Yet from the vantage of the fourth law they are radically different. Two Scrabble boards with Scrabble pieces covering identical squares are thermodynami-cally equivalent from the vantage of the second law. Yet from the vantage of the fourth law they can be radically different, one displaying a random arrangement of letters, the other meaningful words and therefore CSI.

Although Maxwell's demon does not violate the second law, the demon does show that the second law is subject to the Law of Conservation of Information, or what I am proposing also to call the fourth law. For an open system (i.e., open to outside energy), entropy can readily decrease and the second law does not even apply. But for a closed system (i.e., closed to outside energy), for entropy to decrease means that the system has taken advantage of and utilized complex specified information (this was certainly the case with Maxwell's demon). CSI, whose source is ultimately intelligence, can override the second law. It is not fair, however, to call this overriding of the second law a violation of it. The second law is often stated nonstatistically as the claim that in a closed system operating by natural causes entropy is guaranteed to remain the same or increase. But the second law is properly a statistical law stating that in a closed system operating by natural causes entropy is overwhelmingly likely to remain the same or increase. The fourth law, as I am defining it, accounts for the highly unlikely exceptions.

## Notes

1. Robert Stalnaker, *Inquiry* (Cambridge, Mass.: MIT Press, 1984), 85.

2. Fred Dretske, *Knowledge and the Flow of Information* (Cambridge, Mass.: MIT Press, 1981), 4.

3. ASCII = American Standard Code for Information Interchange.

4. Claude Shannon and Warren Weaver, *The Mathematical Theory of Communication* (Urbana, Ill.: University of Illinois Press, 1949), 32.

5. Note that within a set-theoretic context we interpret the conjunction A&B as the intersection A ∩ B.

6. Peter Medawar, *The Limits of Science* (New York: Harper & Row, 1984), 78–82.

7. Fazlollah M. Reza, *An Introduction to Information Theory* (1961; reprinted New York: Dover, 1994), 83.

8. See Hubert Yockey, *Information Theory and Molecular Biology* (Cambridge: Cambridge University Press, 1992), 66–67, but note the errors in formulas 2.27 and 2.28.

9. Gell-Mann takes up these ideas about complexity and information at a popular level in his book *The Quark and the Jaguar: Adventures in the Simple and the Complex* (New York: Freeman, 1994), chs. 3–5. For a more technical treatment, see Murray Gell-Mann and Seth Lloyd, "Information Measures, Effective Complexity, and Total Information," *Complexity* 2(1) (1996): 44–52.

10. See David Roche, "A Bit Confused: Creationism and Information Theory," *Skeptical Inquirer* 25(2) (March/April 2001): 40–42. Roche writes: "Shannon information and [effective] complexity are quite distinct concepts. . . . A common mistake of those at-

tempting to use information theory to debunk Darwinian evolution is to confuse the two concepts. Dembski's 'complex specified information' is the most prominent example" (41). This criticism is remarkable. Gell-Mann's theory of effective complexity, as I am calling it, is a recent addition to the information-theoretic literature and has hardly become mainstream. Moreover, it places restrictions on the concepts of complexity and information that for purposes like the assessment of biological complexity are certainly dispensable and, nay, even prejudicial and unwarranted. What Roche calls "a common mistake" is in fact simply a refusal on the part of design theorists to employ a self-defeating conception of complexity and information that effectively prevents intelligent design from being confirmed by evidence.

11. Ibid. The reference here is to Richard Dawkins's *The Blind Watchmaker*.

12. Ivar Ekeland, *The Broken Dice* (Chicago: University of Chicago Press, 1993).

13. Ibid., 3.

14. Stuart Kauffman, *Investigations* (New York: Oxford University Press, 2000), 227.

15. See William Dembski, *The Design Inference* (Cambridge: Cambridge University Press, 1998), ch. 4.

16. Manfred Eigen, *Steps Towards Life: A Perspective on Evolution*, trans. P. Woolley (Oxford: Oxford University Press, 1992), 12.

17. Michael Behe, *Darwin's Black Box* (New York: Free Press, 1996). See also section 5.10 of this book.

18. John Barrow and Frank Tipler, *The Anthropic Cosmological Principle* (Oxford: Oxford University Press, 1986).

19. David Bohm, *The Undivided Universe: An Ontological Interpretation of Quantum Theory* (London: Routledge, 1993), 35–38.

20. Rolf Landauer, "Information is Physical," *Physics Today* (May 1991): 26.

21. Roy Frieden, *Physics from Fisher Information: A Unification* (Cambridge: Cambridge University Press, 1998).

22. David J. Chalmers, *The Conscious Mind: In Search of a Fundamental Theory* (New York: Oxford University Press, 1996), ch. 8.

23. The nonrandom strings form a very small (i.e., highly improbable and therefore highly complex) set within the space of all possible strings, most of which are random in the sense of being noncompressible. The nonrandom strings are also specified (compressibility provides the specification).

24. See Stephen Schiffer, *Remnants of Meaning* (Cambridge, Mass.: MIT Press, 1987).

25. Del Ratzsch, "Design, Chance, and Theistic Evolution," in *Mere Creation*, ed. W. A. Dembski (Downers Grove, Ill.: InterVarsity, 1998), 294.

26. Manfred Eigen, *Steps Towards Life*, 12.

27. See Bernd-Olaf Küppers, *Information and the Origin of Life* (Cambridge, Mass.: MIT Press, 1990).

28. Horace Freeland Judson, *The Eighth Day of Creation: Makers of the Revolution in Biology* (New York: Touchstone, 1979), 12.

29. Arno Wouters, "Viability Explanation," *Biology and Philosophy* 10 (1995): 435–457.

30. Behe, *Darwin's Black Box*, 45–46.

31. Richard Dawkins, *The Blind Watchmaker: Why the Evidence of Evolution Reveals a Universe without Design* (New York: Norton, 1987), 9.

32. Paul Davies, *The Fifth Miracle* (New York: Simon & Schuster, 1999), 112.

33. Leslie Orgel, *The Origins of Life* (New York: Wiley, 1973), 189.

34. Eigen, *Steps Towards Life*, 12.

35. See Dretske, *Knowledge and the Flow of Information* and Susantha Goonatilake, *The Evolution of Information: Lineages in Gene, Culture and Artefact* (London: Pinter Publishers, 1991). Both these books are typical of naturalistic accounts of information—they focus exclusively on the flow of information but ignore its origin. Naturalistic accounts of the flow of information are fine and well, but do nothing to account for the origin of complex specified information.

36. Stephen C. Meyer, "DNA by Design: An Inference to the Best Explanation for the Origin of Biological Information," *Rhetoric & Public Affairs* 1(4) (1998): 519–556.

37. Michael Polanyi, "Life Transcending Physics and Chemistry," *Chemical and Engineering News* (21 August 1967): 54–66; Michael Polanyi, "Life's Irreducible Structure," *Science* 113 (1968): 1308–1312.

38. James R. Munkres, *Topology: A First Course* (Englewood Cliffs, N.J.: Prentice-Hall, 1975), 17.

39. Heinz Bauer, *Probability Theory and Elements of Measure Theory*, trans. R. B. Burckel, 2nd English ed. (London: Academic Press, 1981), 90–92.

40. Klaus Weihrauch, *Computability* (Berlin: Springer-Verlag, 1987), 107.

41. See Leon Brillouin, *Science and Information Theory*, 2nd ed. (New York: Academic Press, 1962), 267–269, where he makes this point beautifully. Brillouin quotes a delightful passage from Edgar Allen Poe, who, commenting as far back as 1836 on Babbage's inference engine, understood clearly that deterministic systems are incapable of attaining to "the intellect of man." Brillouin concludes thus: "[A] machine does not create any new information, but it performs a very valuable transformation of known information. It would be very interesting to find some measure of this transformation and to compute its value, but up to now no method has been discovered to evaluate this work." Brillouin wrote this back in the 1950s. Since then such measures for the transformation of information have been developed. They are called *complexity measures*. Indeed, an entire new discipline has developed since Brillouin's prescient observation, to wit, *complexity theory*. For an introduction to complexity theory see Dembski, *The Design Inference*, ch. 4.

42. Emile Borel, *Probabilities and Life*, trans. M. Baudin (New York: Dover, 1962), 28.

43. See Dembski, *The Design Inference*, sec. 6.5.

44. Bernd-Olaf Küppers, *Information and the Origin of Life* (Cambridge, Mass.: MIT Press, 1990), 59.

45. Richard Dawkins, *The Blind Watchmaker*, 139.

46. See Stewart Ethier and Thomas Kurtz, *Markov Processes: Characterization and Convergence* (New York: John Wiley, 1986), 49–50.

47. Ibid., 50.

48. In showing that a stochastic process cannot generate CSI, I broke the stochastic

process $f$ into a random and then a deterministic component. This is mathematically legitimate and involves no loss of generality. Nonetheless, for readers unfamiliar with stochastic processes and who think this breaking of stochastic processes into random and deterministic components might constitute some sleight of hand on my part, it is possible to do the analysis in one shot as in the deterministic case. Thus one can define the universal composition function $U$ on $\Omega_1 \times \Omega_2 \times \Omega_3$ into $\Omega_4$ where $\Omega_4$ is the reference class that contains the CSI $j$ and $\Omega_1 \times \Omega_2 \times \Omega_3$ is the Cartesian product of $\Omega_1$ (which contains $i$—the state of affairs antecedent to $j$), $\Omega_2$ (which contains $\omega$—the random element antecedent to $j$), and $\Omega_3$ (which is a function space that contains $f$—the stochastic process that acts on $i$ and $\omega$ to yield $j$). $\Omega_1 \times \Omega_2 \times \Omega_3$ comes with a probability measure that under the action of $U$ induces a probability measure on $\Omega_4$. Moreover, the inverse image of $j$ under the action of $U$ constitutes CSI in the product space $\Omega_1 \times \Omega_2 \times \Omega_3$. To establish this rigorously involves some tedious bookkeeping. The analysis in the body of the text is conceptually simpler.

49. See Samuel Karlin and Howard Taylor, A *First Course in Stochastic Processes*, 2nd ed. (New York: Academic Press, 1975) and by the same authors A *Second Course in Stochastic Processes* (New York: Academic Press, 1981).

50. Medawar, *The Limits of Science*, 78–82.

51. For a thorough exploration of how design might enter the natural order, see Del Ratzsch, *Nature, Design, and Science: The Status of Design in Natural Science*, in SUNY Series in Philosophy and Biology (Albany, N.Y.: SUNY Press, 2001).

52. See Timothy Lenoir, *The Strategy of Life: Teleology and Mechanics in Nineteenth Century German Biology* (Dordrecht: Reidel, 1982).

53. Dawkins, *The Blind Watchmaker*, 13, 316; Daniel Dennett, *Darwin's Dangerous Idea* (New York: Simon & Schuster, 1995), 153.

54. Douglas Robertson goes so far as to claim that the defining feature of intelligence is the "creation of new information," by which is properly understood the creation of complex specified information. See Douglas S. Robertson, "Algorithmic Information Theory, Free Will, and the Turing Test," *Complexity* 4(3) (1999): 25–34.

55. The textual transmission of the New Testament is a wonderful place to begin for understanding the problems facing textual critics. See Bruce Metzger, *The Text of the New Testament: Its Transmission, Corruption, and Restoration* (Oxford: Oxford University Press, 1992).

56. Keith Devlin, *Logic and Information* (Cambridge: Cambridge University Press, 1991), 2.

57. See Nicholas Georgescu-Roegen, *The Entropy Law and the Economic Process* (Cambridge, Mass.: Harvard University Press, 1971) and C. Bianciardi, A. Donati, and S. Ulgiati, "Complete Recycling of Matter in the Frameworks of Physics, Biology, and Ecological Economics," *Ecological Economics* 8 (1993): 1–5.

58. Victor F. Weisskopf, "The Frontiers and Limits of Science," *American Scientist* 65(4) (July-August 1977): 405–411.

59. Letters to the Editor, *American Scientist* 65(6) (November-December 1977): 674–675.

60. Polanyi, "Life Transcending Physics and Chemistry" and "Life's Irreducible Structure."

61. Letters to the Editor, *American Scientist*, 676.

62. See Harold J. Morowitz, *Beginnings of Cellular Life: Metabolism Recapitulates Biogenesis* (New Haven, Conn.: Yale University Press, 1992), 77. See also Ilya Prigogine's work on nonequilibrium thermodynamics, e.g., Ilya Prigogine and Isabelle Stengers, *Order Out of Chaos* (New York: Bantam, 1984), 140–145.

63. See Per Bak, *How Nature Works: The Science of Self-Organized Criticality* (New York: Copernicus, 1996) and Stuart Kauffman, *Investigations* (New York: Oxford University Press, 2000), especially ch. 8.

64. Kauffman, *Investigations*, 160.

65. Ibid., 8.

66. Stuart Kauffman, *The Origins of Order: Self-Organization and Selection in Evolution* (Oxford: Oxford University Press, 1993), xvi.

67. Francis Crick and Leslie E. Orgel, "Directed Panspermia," *Icarus* 19 (1973): 341–346; Fred Hoyle and Chandra Wickramasinghe, *Evolution from Space* (New York: Simon & Schuster, 1981), 1–33, 130–141; Fred Hoyle, *Cosmology and Astrophysics* (Ithaca, N.Y.: Cornell University Press, 1982), 1–65.

68. John R. Pierce, *An Introduction to Information Theory: Symbols, Signals and Noise*, 2nd rev. ed. (New York: Dover, 1980), 199.

69. Ibid.

70. Leo Szilard, "Über die Entropieverminderung in einem thermodynamischen System bei Eingriff intelligenter Wesen," *Zeitschrift für Physik* 53 (1929): 840–856. For the English translation see John A. Wheeler and Wojciech H. Zurek, eds., *Quantum Theory and Measurement* (Princeton: Princeton University Press, 1983), 539–548.

71. Charles H. Bennett, "Demons, Engines and the Second Law," *Scientific American* 257(5) (November 1987): 116.

72. Wojciech H. Zurek, "Algorithmic Information Content, Church-Turing Thesis, Physical Entropy, and Maxwell's Demon," in *Complexity, Entropy and the Physics of Information*, ed. W. H. Zurek (Reading, Mass.: Addison-Wesley, 1990), 74. IGUSes are now part of the popular science literature. For instance, science reporter Tom Siegfried (with the *Dallas Morning News*) titles one of the chapters in his popular exposition on quantum computing and information "IGUSes." See Tom Siegfried, *The Bit and the Pendulum: From Quantum Computing to M Theory—The New Physics of Information* (New York: Wiley, 2000), ch. 8.

73. Y. M. Guttmann, *The Concept of Probability in Statistical Physics* (Cambridge: Cambridge University Press, 1999), 2–3.

74. Dretske, *Knowledge and the Flow of Information*, 26.

CHAPTER FOUR

~

# Evolutionary Algorithms

## 4.1  METHINKS IT IS LIKE A WEASEL

How did life originate? No convincing answer has been given to date. Notwithstanding optimistic claims to the contrary, the origin of life remains a thoroughly intractable problem for science. Many scientists, like Ian Stewart, are optimistic:

> The origin of life no longer appears to be a particularly difficult problem. We know that—at least on this planet—the key ingredient is DNA. Life's basis is molecular. What we need is an understanding of complex molecules: how they might have arisen in the first place, and how they contribute to the rich tapestry of living forms and behavior. It turns out that the main scientific issue is *not* the absence of any plausible explanation for the origin of life—which used to be the case—but an embarrassment of riches. There are many plausible explanations; the difficulty is to choose among them. That surfeit causes problems for the question "How *did* life begin on Earth?" but not for the more basic issue, which is "*Can* life emerge from nonliving processes?"[1]

Stewart's assessment of the origin-of-life problem is too easy. Indeed, the embarrassment of riches that Stewart cites should itself give us pause. It is usually hard enough to come up with just one good theory to account for a phenomenon (cf. the state of mechanics prior to Newton). An embarrassment of riches points not to the solution of a problem but to vague gestures at a solution.[2]

Paul Davies is less sanguine about resolving the problem of life's origin.

In *The Fifth Miracle* Davies goes so far as to suggest that any laws capable of explaining the origin of life must be radically different from any scientific laws known to date.[3] The problem, as he sees it, with currently known scientific laws, like the laws of chemistry and physics, is that they cannot explain the key feature of life that needs to be explained.[4] That feature is *specified complexity*. As Davies puts it: "Living organisms are mysterious not for their complexity *per se*, but for their tightly specified complexity."[5] Nonetheless, once life (or more generally some self-replicator) arrives on the scene, Davies thinks there is no problem accounting for specified complexity. Indeed, he thinks the Darwinian mechanism of natural selection and random variation is fully adequate to account for specified complexity once replicators are here: "Random mutations plus natural selection are one surefire way to generate biological information, extending a short random genome over time into a long random genome. Chance in the guise of mutations and law in the guise of selection form just the right combination of randomness and order needed to create 'the impossible object.' The necessary information comes, as we have seen, from the environment."[6]

The problem with invoking the Darwinian mechanism to explain specified complexity at the origin of life is the absence of any identifiable replicator to which the mechanism might apply. Theodosius Dobzhansky was therefore right when he remarked that "prebiological natural selection is a contradiction in terms."[7] Indeed, the Darwinian mechanism of natural selection and random variation is simply not available until after the origin of life and specifically of self-replicating systems. Once life has started and self-replication has begun, however, the Darwinian mechanism is usually invoked to explain the specified complexity of living things. According to Stuart Kauffman this is the majority position within the biological community: "Biologists now tend to believe profoundly that natural selection is the invisible hand that crafts well-wrought forms. It may be an overstatement to claim that biologists view selection as the sole source of order in biology, but not by much. If current biology has a central canon, you have now heard it."[8]

In this chapter I will argue that the problem of explaining specified complexity is even worse than Davies makes out in *The Fifth Miracle*. Not only have we yet to explain specified complexity at the origin of life, but the Darwinian mechanism fails to explain it for the subsequent history of life as well. To see that the Darwinian mechanism is incapable of generating specified complexity, it is necessary to consider the mathematical underpinnings of that mechanism, to wit, evolutionary algorithms. By an evolutionary algorithm I mean any well-defined mathematical procedure that generates contingency via some chance process and then sifts it via some law-like process.

The Darwinian mechanism, simulated annealing, training neural nets, and genetic algorithms all fall within this broad construal of evolutionary algorithms.[9]

Given the popular enthusiasm for evolutionary algorithms, to claim that they are incapable of generating specified complexity may seem misconceived. But consider a well-known example by Richard Dawkins in which he purports to show how an evolutionary algorithm can generate specified complexity.[10] He starts with the following target sequence, a putative instance of specified complexity:

METHINKS•IT•IS•LIKE•A•WEASEL

(he considers only capital Roman letters and spaces, spaces represented here by bullets—thus 27 possibilities at each location in a symbol string 28 characters in length).

If we tried to attain this target sequence by pure chance (for example, by randomly shaking out Scrabble pieces), the probability of getting it on the first try would be around 1 in $10^{40}$, and correspondingly it would take on average about $10^{40}$ tries to stand a better than even chance of getting it.[11] Thus, if we depended on pure chance to attain this target sequence, we would in all likelihood be unsuccessful. As a problem for pure chance, attaining Dawkins's target sequence is an exercise in generating specified complexity, and it becomes clear that pure chance simply is not up to the task. (Technically, Dawkins's target sequence is not long enough for its probability to fall below the 1 in $10^{150}$ universal probability bound or correspondingly for its complexity to surpass the 500-bit universal complexity bound. Dawkins's target sequence therefore does not qualify as complex specified information in the strict sense—see sections 2.8 and 3.9. Nonetheless, for practical purposes the complexity is sufficient to illustrate specified complexity.)

But consider next Dawkins's reframing of the problem. In place of pure chance, he considers the following evolutionary algorithm: (1) Start out with a randomly selected sequence of 28 capital Roman letters and spaces, such as

WDL•MNLT•DTJBKWIRZREZLMQCO•P;

(2) randomly alter all the letters and spaces in the current sequence that do not agree with the target sequence; and (3) whenever an alteration happens to match a corresponding letter in the target sequence, leave it and randomly alter only those remaining letters that still differ from the target sequence.

In very short order this algorithm converges to Dawkins's target sequence.

In *The Blind Watchmaker*, Dawkins provides the following computer simulation of this algorithm:[12]

```
 (1)  WDL•MNLT•DTJBKWIRZREZLMQCO•P
 (2)  WDLTMNLT•DTJBSWIRZREZLMQCO•P
              . . .
(10)  MDLDMNLS•ITJISWHRZREZ•MECS•P
              . . .
(20)  MELDINLS•IT•ISWPRKE•Z•WECSEL
              . . .
(30)  METHINGS•IT•ISWLIKE•B•WECSEL
              . . .
(40)  METHINKS•IT•IS•LIKE•I•WEASEL
              . . .
(43)  METHINKS•IT•IS•LIKE•A•WEASEL
```

Thus, Dawkins's simulation converges to the target sequence in 43 steps. In place of $10^{40}$ tries on average for pure chance to generate the target sequence, it now takes on average only 40 tries to generate it via an evolutionary algorithm.

Although Dawkins and fellow Darwinists use this example to illustrate the power of evolutionary algorithms,[13] in fact it raises more problems than it solves. For one thing, choosing a prespecified target sequence as Dawkins does here is deeply teleological (the target here is set prior to running the evolutionary algorithm and the evolutionary algorithm here is explicitly programmed to end up at the target). This is a problem because evolutionary algorithms are supposed to be capable of solving complex problems without invoking teleology (indeed, most evolutionary algorithms in the literature are programmed to search a space of possible solutions to a problem until they find an answer—not, as Dawkins does here, by explicitly programming the answer into them in advance). For the sake of argument let us therefore assume that when it comes to biology, nature somehow selects targets without introducing teleology into the Darwinian mechanism, thus allowing us to set aside the teleological problem raised by Dawkins's example.[14]

A more serious problem then remains. We can see it by posing the following question: Given Dawkins's evolutionary algorithm, what besides the target sequence can this algorithm attain? Think of it this way. Dawkins's evolutionary algorithm is proceeding along; what are the possible terminal points of this algorithm? Clearly, the algorithm is always going to converge on the target sequence (with probability 1 for that matter). An evolutionary algorithm acts as a *probability amplifier*. Whereas it would take pure chance on average $10^{40}$ tries to attain Dawkins's target sequence, his evolutionary algorithm on average attains it in the logarithm of that number, that is, on average in only 40 tries and with virtual certainty in a few hundred tries. Since these "waiting times" for attaining the target sequence are inversely proportional to the probability of attaining the target in a given number of tries, the probability of Dawkins's evolutionary algorithm attaining the target sequence is much bigger than the probability of a purely chance-driven process. Dawkins's evolutionary algorithm therefore acts as a probability amplifier.

But a probability amplifier is also a *complexity diminisher*. For something to be complex, there must be many live possibilities that could take its place. Increasingly numerous live possibilities correspond to increasing improbability of any one of those possibilities. Complexity and probability therefore vary inversely—the greater the complexity, the smaller the probability.[15] It follows that Dawkins's evolutionary algorithm, by vastly increasing the probability of getting the target sequence, vastly decreases the complexity inherent in that sequence. As the sole possibility that Dawkins's evolutionary algorithm can attain, the target sequence in fact has minimal complexity (i.e., the probability is 1 and the complexity, as measured by the usual information measure, is 0).[16] Evolutionary algorithms are therefore incapable of generating true complexity. And since they cannot generate true complexity, they cannot generate true specified complexity either.

In general, then, evolutionary algorithms generate not true specified complexity but at best the *appearance of specified complexity*. This claim is reminiscent of one made by Richard Dawkins. On the opening page of *The Blind Watchmaker* he states, "Biology is the study of complicated things that give the appearance of having been designed for a purpose."[17] Just as the Darwinian mechanism does not generate actual design but only its appearance, so too the Darwinian mechanism does not generate actual specified complexity but only its appearance. But this raises an obvious question, namely, whether there might not be a fundamental connection between intelligence or design on the one hand and specified complexity on the other. As we have seen in the previous chapters, there is indeed.

Dawkins, fellow Darwinists, and even some non-Darwinists are quite taken with the METHINKS•IT•IS•LIKE•A•WEASEL example and see it as illustrating the power of evolutionary algorithms to generate specified complexity.[18] Closer investigation, however, reveals that the evolutionary algorithm Dawkins uses to produce this target sequence has not so much generated specified complexity as merely shuffled it around. As we shall see, invariably when evolutionary algorithms appear to generate specified complexity, what they actually do is smuggle in preexisting specified complexity. Indeed, evolutionary algorithms are inherently incapable of generating specified complexity. The rest of this chapter is devoted to justifying that claim.

## 4.2  Optimization

Generating specified complexity via an evolutionary algorithm is typically understood as an optimization problem. Optimization is one of the main things mathematicians do for a living. What is the shortest route connecting ten cities? What is the fastest algorithm to sort through a database? What is the most efficient way to pack objects of one geometrical shape inside another (e.g., bottles in boxes)?[19] Optimization quantifies superlatives like "shortest," "fastest," and "most efficient" and then attempts to identify states of affairs to which those superlatives apply.

As in ordinary life, so too in mathematics optimization always involves constraints. Indeed, it is fair to say that all optimization is *constrained optimization*.[20] Constrained optimization is the art of compromise among competing objectives. Suppose you own a sausage factory. You want to produce premium sausages. But that requires premium ingredients, and these cost extra. As a good capitalist you want to minimize costs and maximize profits. Nonetheless, skimping on ingredients too much will reduce the quality of your sausages sufficiently so that people will not buy them at any cost. Thus, you have to find the right balance between quality of sausages and cost of sausages. Despite an epidemic of books titled *Having It All* (fashion publisher Helen Gurley Brown as far as I know started this dismal trend), it is impossible to have it all. What you can have is a preponderance of some things to the exclusion of others. The trick of optimization is to figure out how to stack the preponderance so that it best achieves one's objectives.

The first step in mathematically formulating an optimization problem is therefore to identify the relevant collection of possible solutions to a problem, or what we will call the *reference class of possible solutions* (compare this reference class with the reference class of possibilities that in chapter 3 formed the backdrop for generating information; the similarity of usage is not coincidental—indeed, the usage here is a special case of the usage there). The reference class of possible solutions is never uniquely determined, and its choice reflects the constraints associated with an optimization problem. For instance, with the sausage factory, an industrial systems engineer will ignore possible solutions where a gullible public buys vast quantities of poor quality sausages at grossly inflated prices. The factory owner might wish the world such that it include these solutions as live possibilities, but "having it all" is not an option, and the systems engineer hired by the factory owner to optimize production and profits will make sure that such solutions are omitted from the reference class of possible solutions.[21]

Constraints inherent in an optimization problem are thus in part represented mathematically in the choice of reference class of possible solutions. Once that reference class has been identified, the next step is to identify a *univalent* measure of optimality. Our initial measures of optimality are typically *multivalent*, that is, there are several things we are trying to optimize simultaneously. For instance, with the sausage factory, the owner wants to produce as many sausages as possible, to sell as many as are produced, to keep the costs as low as possible, and to generate maximal sales with minimal expenditures in advertising and marketing. But ultimately the factory owner is going to want to maximize profits. Profit, then, is the univalent measure of optimization that ties together all these disparate measures of optimization that constitute the owner's initial multivalent measure of optimality.

Substituting a univalent measure of optimality for a multivalent one is another way of factoring constraints into an optimization problem. A multivalent measure of optimality pits competing measures of optimality against each other, so that optimizing one measure is apt to impair another. For instance, with the sausage factory, quality of ingredients and cost of ingredients are in competition. Selecting a univalent measure of optimality adjudicates among the competing claims inherent in a multivalent measure of optimality. Only after a univalent measure of optimality has been identified can an optimization problem be said to be well-defined. With a univalent measure of optimality, we can say that one solution is better than another because the values assigned are directly comparable. Profit, for instance, can be measured in dollars, and dollar amounts are comparable. Multivalent measures, on the other hand, by their very nature defeat comparability and thus leave an optimization problem ill-defined.

Univalent measures of optimality need to assume values that are comparable so that for any pair of values one can decide which is preferable. Dollar amounts representing profits, for instance, are not just comparable but also come with a preferred ordering according to which "more is better." It follows that real numbers (i.e., numbers like 0, 1, −53, the square root of two, and pi) are ideally suited for representing the values assumed by a univalent measure of optimality. Any pair of distinct real numbers are directly comparable with one strictly greater than the other. Consequently, the "greater than" and "less than" relations for real numbers are capable of representing the preferability of one solution over another. A solution is then optimal

provided that it assumes the extreme-most value in the preferred direction—
that is, either a maximum or a minimum.

But which is it, a maximum or a minimum? The answer depends on
whether we regard "less as better" or "more as better." When our univalent
measure of optimality measures profits in dollar amounts, then "more is better." But when it measures costs, be it anything from cash outlays to computation times that we want to minimize, then "less is better." Now it turns
out that the techniques for minimizing or maximizing real-valued functions
are entirely equivalent (real-valued functions being the ones that represent
univalent measures of optimality). This is because minimizing a real-valued
function is logically equivalent to maximizing its negative (think of credits
and debits: maximizing credits is equivalent to minimizing debits, debits and
credits being negatives of each other). It follows that we can transform any
optimization problem into a maximization problem.

Transforming a minimization into a maximization problem is simply a
matter of inverting the direction of the univalent measure we are trying to
optimize (i.e., highs becomes lows and lows becomes highs). There are other
ways we can transform a univalent measure of optimality without changing
the underlying optimization problem. Since we represent such measures as
real-valued functions, we can change the scaling as well as shift all values in
the positive direction so that none remains negative. Such transformations
leave unchanged the preference structure of the original univalent measure
of optimality. Applied successively, such transformations enable us to represent univalent measures of optimality as nonnegative real-valued functions
that are optimized by being maximized. This simplification is useful for studying evolutionary algorithms.[22]

Before closing this section, I need to offer two qualifications. First, the
sausage example was too easy. In that example profit too easily trumped
other measures of optimality. Multicriteria optimization is currently a hot
topic of research, and the problem of balancing objectives that it tries to
resolve is much more difficult than the sausage example would indicate.[23]
There is often no way to combine multiple competing measures of optimality
into a straightforward, singularly appropriate univalent measure of optimality. Nonetheless, my point remains that until some form of univalence is
achieved, optimization cannot begin. The second qualification is this. By
optimization I do not just mean finding the one true global optimum for a
single fitness function. That is the prototype case. Yet often our search techniques do not take us to a global optimum but to a solution that is "good
enough" for our purposes. As Melanie Mitchell puts it, "The goal is to *satis-
fice* (find a solution that is good enough for one's purposes) rather than to

optimize (find the best possible solution)."[24] Nonetheless, I shall continue to employ the language of optimization inasmuch as satisficing presupposes searching for the best possible solution. As for fitness functions that vary with time, they are readily dealt with once we understand the prototype case (see section 4.10).

## 4.3   Statement of the Problem

We are now in a position to state what it would mean for an evolutionary algorithm to generate specified complexity. Generating specified complexity via an evolutionary algorithm can be understood as the following optimization problem. We are given a reference class of possible solutions known as the *phase space*. Possible solutions within the phase space are referred to as *points*. The univalent measure defined with reference to the phase space and needing to be optimized is known as the *fitness function* (also *fitness measure* or *fitness landscape*). The fitness function is a nonnegative real-valued function that is optimized by being maximized (i.e., fitness can never fall below zero and is regarded as a benefit for which "more is better"). The task of an evolutionary algorithm is to locate a possible solution where the fitness function attains at least a certain level of fitness. The set of possible solutions where the fitness function attains at least that level of fitness will be called the *target*.[25]

Think of it this way. Imagine that the phase space is a vast plane (i.e., two-dimensional surface) and the fitness function is a vast hollowed-out mountain range over the plane (complete with low-lying foothills and incredibly high peaks). The task of an evolutionary algorithm is by moving around in the plane to get to some point under the mountain range where it attains at least a certain height (like 10,000 feet). The collection of all such places on the plane where the mountain range attains at least that height (here 10,000 feet) is the target. Thus the job of the evolutionary algorithm is by navigating the phase space to find its way into the target.

This can be formulated mathematically. Let us refer to the phase space as $\Omega$ and to the fitness function as f. The function f takes values only in the nonnegative reals, and by optimization here we mean maximizing f. Thus we want to find a point in that subset of $\Omega$ where f attains at least a certain level $a^*$ ($a^*$ being a positive real number). Let us refer to this subset as the target T. This subset is represented mathematically as $T = \{x \in \Omega \mid f(x) \geq a^*\}$, that is, the set of all x in $\Omega$ such that $f(x)$ is at least $a^*$. Our task, then, is to find some point q in T, that is, a point q in the phase space $\Omega$ satisfying $f(q) \geq a^*$. Note that the target T is the rejection region induced by the fitness function f, now treated as a rejection function; moreover, T is identical to the extremal set $T^{a*}$ (see sections 2.2 and 2.5). Indeed, the entire theoretical apparatus of chapter 2 applies to this discussion.

The phase space Ω (which we are picturing as a giant plane) invariably comes with some additional topological structure, typically given by a metric or distance function. A metric d on Ω assigns to any pair of points the distance separating those points. Metrics satisfy the following three conditions: (1) for all x in Ω, $d(x,x) = 0$ (identity); (2) for all x and y in Ω, $d(x,y) = d(y,x)$ (symmetry); and (3) for all x, y, and z in Ω, $d(x,z) \leq d(x,y) + d(y,z)$ (triangle inequality).[26] In plain English this means that (1) any point in the phase space has zero distance from itself; (2) the distance between two points does not depend on the order in which one considers them (e.g., flying from Atlanta to Dallas is the same distance as flying from Dallas to Atlanta); and (3) the direct distance between two points is never bigger than the distance of going through some intermediate point.

The topological structure induced by a metric tells us how points in the phase space are related geometrically to nearby points. Though typically huge, phase spaces tend to be finite (strictly finite for problems represented on computer and topologically finite, or what topologists call "compact," in general). Moreover, such spaces typically come with a uniform probability that is adapted to the topology of the phase space. What this means is that Ω possesses a uniform probability measure U adapted to the metric on Ω so that geometrically congruent pieces of Ω get assigned identical probabilities (see section 2.2).[27]

If you think of the phase space as a giant (but not infinite) plane, this means that if you get out your tape measure and measure off, say, a three-by-five-foot area in one part of the phase space, the uniform probability will assign it the same probability as a three-by-five-foot area in another portion of the phase space. All the spaces to which evolutionary algorithms have till now been applied do indeed satisfy these two conditions of having a finite topological structure (i.e., they are compact) and possessing a uniform probability (i.e., U). Moreover, this uniform probability is what typically gets used to estimate the complexity or improbability of the target (e.g., the improbability of landing in that region of a plane where a supervening mountain range attains at least a certain height, like 10,000 feet).

For instance, in Dawkins's METHINKS•IT•IS•LIKE•A•WEASEL example (see section 4.1), the phase space consists of all sequences 28 characters in length comprising upper case Roman letters and spaces (spaces being represented by bullets). A uniform probability on this space assigns equal probability to each of these sequences—the probability value is approximately 1 in $10^{40}$ and signals a highly improbable state of affairs. It is this improbability that corresponds to the complexity of the target sequence and which by its explicit identification specifies the sequence and thus renders it an instance of specified complexity (though as pointed out in section 4.1, we are being somewhat loose in this example about the level of complexity required for specified complexity—technically, the level of complexity should correspond to the universal probability bound of 1 in $10^{150}$).

In general, given a phase space whose target is those places where the fitness function attains a certain height (e.g., at least 10,000 feet), the (uniform) probability of randomly choosing a point from the phase space and landing in the target will be extremely small. In Dawkins's example, the target equals the character string METHINKS•IT•IS•LIKE•A•WEASEL and the improbability is 1 in $10^{40}$. For nontoy examples, the improbability is typically much less than the universal probability bound of 1 in $10^{150}$ described in section 1.5.[28] Note that if the probability of the target were not small, a random search through the phase space would suffice to find a point in the target, and there would be no need to construct an evolutionary algorithm to find it.

We therefore suppose that the target is just a tiny portion of the whole phase space. More precisely, the (uniform) probability of the target in relation to the entire phase space is exceedingly small. Also, we suppose that the target, in virtue of its explicit identification, is specified (certainly this is the case in Dawkins's example, where the target coincides with the character string METHINKS•IT•IS•LIKE•A•WEASEL and Dawkins explicitly identified this string prior to running his evolutionary algorithm). Thus it would seem that for an evolutionary algorithm to find such a point in the target would be to generate specified complexity.

But let us look deeper. Consider a typical evolutionary algorithm in search of a target. An evolutionary algorithm is a stochastic process, that is, an indexed set of random values with precisely given probabilistic dependencies among those various values.[29] The values an evolutionary algorithm assumes occur in phase space. An evolutionary algorithm moves around phase space some finite number of times and stops as soon as it reaches the target. More specifically, an evolutionary algorithm $\mathbf{E}$ is a stochastic process indexed by the natural numbers 1, 2, 3, . . . and taking values in the phase space $\Omega$.[30] Since we need $\mathbf{E}$ to locate the target in a manageable number of steps, we fix a natural number m and consider only those values of $\mathbf{E}$ indexed by 1 through m (we refer to m as the "sample size"). Truncating $\mathbf{E}$ after a certain point prevents the evolutionary algorithm from cycling endlessly through phase space. Consequently, even though $\mathbf{E}$ is defined on all the natural numbers, we limit ourselves to considering only $E_1$, $E_2$, . . . , $E_m$ (i.e., the values of $\mathbf{E}$ indexed by the numbers 1, 2, . . . , m).

The evolutionary algorithm starts at some point $E_1$ in the phase space (usually chosen at random). Then it moves to $E_2$. Then to $E_3$. And so on all the way up to $E_m$. For $\mathbf{E}$ successfully to find the target (in other words, to find a point where the fitness function attains at least a certain value—e.g., a place under the mountain range where the mountain attains at least 10,000 feet) then means that within a manageable number of steps (i.e., m), $\mathbf{E}$ is very likely to land in the target. Simply put, the evolutionary algorithm has to get into the target with high probability in a relatively short number of steps. Alternatively, the probability is high (i.e., reasonably close to 1) that some one of $E_1$, $E_2$, . . . , $E_m$ intersects the target T. In the Dawkins example, $E_m$ rapidly converged to METHINKS•IT•IS•LIKE•A•WEASEL for m around 40.

An evolutionary algorithm needs to improve on pure random sampling and blind search.[31] Pure random sampling treats the phase space like a giant urn from which items are drawn at random according to the uniform probability. Each random draw in effect constitutes a brand-new attempt by pure chance to locate a point in the target. It follows that a random sample from $\Omega$ of size k will have a better-than-even chance of containing a point in the target only if k is close to the reciprocal of the (uniform) probability of the target.[32] Since in most cases of interest that probability is going to be less than the universal probability bound of 1 in $10^{150}$, it follows that k will need to be on the order of $10^{150}$. This number is enormous and far exceeds the number of computations conceivable in the known universe (note that quantum computation is not going to render this number any more tractable because the points in phase space need to be made explicit in any random sampling scheme, implying decoherence and thus preventing the exploitation of quantum superposition[33]).

Blind search is no better than pure random sampling. Whereas pure random sampling at each stage selects at random a point from the entire phase space, blind search at each stage selects at random a point within a certain fixed proximity of the previously selected point (the initial starting point being chosen randomly over the entire phase space as with pure random sampling). Whereas pure random sampling looks like sampling from a giant urn, blind search looks like a drunken walk. Indeed, probabilists refer to it as a "random walk."[34] Since random walks can easily double back on themselves and thus lead to inefficiencies in search, the probability of locating the target for a blind search is no better than for pure random sampling. If, therefore, pure random sampling has little hope of locating the target for a sample size of m, then blind search has even less. (Note that pure random sampling is actually a special case of blind search in which the neighborhood of proximity from which points are selected is always the entire phase space. Consequently, I focus in the sequel on blind search, the more general category, rather than on pure random sampling, the more restrictive category.)

Let us now return to the evolutionary algorithm **E**. We are going to allow ourselves a certain number of steps m for **E** to land in the target. Clearly m will have to be much less than $10^{150}$ if we are going to program **E** on a computer and have any hope of **E** landing in the target. With the sample size m fixed, we can determine the probability that **E** will land in any subset of phase space within m steps. In that case, given the target $T = \{x \in \Omega \mid f(x) \geq a^*\}$, the probability that at least one of $E_1, E_2, \ldots, E_m$ intersects T needs to be large enough so that we can reasonably expect the evolutionary algorithm to land in T within m steps. In other words, the probability that at least one of $E_1, E_2, \ldots, E_m$ intersects T needs to be reasonably close to 1. For instance, in the Dawkins example, for m = 100 and the target sequence METHINKS•IT•IS•LIKE•A•WEASEL and **E** the cumulative selection algorithm Dawkins constructed, the probability of **E** attaining the target in m = 100 steps is close to 1.

Thus, even though the target has exceedingly small probability for random samples of size m (in this case the m points in phase space are selected with respect to the uniform probability measure **U**), the probability of the evolutionary algorithm **E** getting into the target in m steps is no longer small (this time the m points in phase space are selected with respect to the evolutionary algorithm **E**). Indeed, if the evolutionary algorithm is doing its job, its probability of locating the target in m steps will be quite large. And since for the present discussion complexity and improbability are equivalent notions, the target, though complex and specified with respect to the uniform probability **U**, remains specified but no longer complex with respect to **E**.

But does this not mean that the evolutionary algorithm has generated specified complexity after all? No. At issue here is the *generation* of specified complexity and not its *reshuffling*. To appreciate the difference, let us be clear about a condition **E** must satisfy if it is to count as a genuine evolutionary algorithm (i.e., a legitimate correlative of the Darwinian mutation-selection mechanism). It is not, for instance, legitimate for **E** to survey the fitness landscape induced by the fitness function, see where in the phase space it attains a global maximum, and then head in that direction. That would be teleology. No, **E** must be able to navigate its way to the target either by randomly choosing points from the phase space or by using those as starting points and then selecting other points in the phase space based solely on the topology of the phase space and without recourse to the fitness function except to evaluate the fitness function at individual points of the phase space already traversed by **E**.

We can think of it this way: $E_1$, the first point selected by the evolutionary algorithm, is selected randomly from the phase space (i.e., with respect to the uniform probability on the phase space). The fitness function can then be evaluated at $E_1$ (in our running analogy, we can determine how high the mountain range is at that point $E_1$). Given only $E_1$, the fitness function's height at $E_1$, and the topology of the phase space, the evolutionary algorithm **E** next selects $E_2$. Then the height of the fitness function can be evaluated at $E_2$. Then $E_3$ is selected based only on $E_1$, $E_2$, the height of the fitness function at these two points, and the topology of the phase space. And so on. Consequently, the evolutionary algorithm **E** must be independent of the fitness function except for those points that **E** has hitherto traversed and then only insofar as the fitness function is evaluated at those points.[35]

Certainly this means that the evolutionary algorithm **E** has to be highly constrained in its use of the fitness function. But there is more: the success of **E** in hitting the target depends crucially on the structure of the fitness function. If, for instance, the fitness function is totally flat and close to zero whenever it is outside the target, then it fails to discriminate between points outside the target and so cannot be any help guiding an evolutionary algorithm into the target. For such a fitness function, the probability of the evolutionary algorithm landing in the target is no better than the probability of pure random sampling or blind search landing in the target, which as we know are inadequate to get us there (an eventuality we have dismissed out of hand—the target simply has too small a probability for pure random sampling or blind search to have any hope of success). Let us now turn in detail to the fitness function and the crucial role its structure plays in guiding an evolutionary algorithm into the target.

## 4.4 Choosing the Right Fitness Function

To understand how a fitness function guides an evolutionary algorithm into a target, let us consider a slight modification of Dawkins's METHINKS•IT•IS•LIKE•A•WEASEL example. Recall that the phase space in his example comprised all sequences of 28 capital Roman letters and spaces (spaces being represented by bullets), and that the target was the sentence ME-THINKS•IT•IS•LIKE•A•WEASEL. In this example the essential feature of the fitness function was that it assigned higher fitness to sequences having more characters in common with the target sequence. There are many ways to represent such a fitness function mathematically, but perhaps the simplest is simply to count the number of characters identical with the target sequence. Such a fitness function assumes integer values between 0 and 28, assigning 0 to sequences with no coinciding characters and 28 solely to the target sequence.

In modifying Dawkins's example, let us keep the phase space, target, and fitness function the same, but alter slightly the algorithm. In Dawkins's original example, the evolutionary algorithm started with a random sequence of 28 characters, and then at each stage randomly modified all the characters that did not coincide with the corresponding character in the target sequence. As we saw in section 4.1, this algorithm converged on the target sequence with probability 1 and on average yielded the target sequence after about 40 iterations.

Dawkins's original algorithm requires at each iteration an explicit character-by-character comparison of a given sequence with the target sequence. This is highly artificial, giving the algorithm complete access to the target at each step in its execution. Not only does this algorithm introduce a teleology foreign to the natural world (and certainly to the Darwinian mechanism), but it also introduces a teleology foreign to the evolutionary algorithms actually used by working computer scientists. For a computer scientist programming an evolutionary algorithm, the sequence METHINKS•IT•IS• LIKE•A•WEASEL would presumably be the solution to some problem. Yet there is no point to solving a problem whose solution is already staring us in the face. What is needed, then, is to come up with an evolutionary algorithm that solves the problem of finding METHINKS•IT•IS•LIKE•A•WEASEL by searching the phase space *without* explicit recourse to the target.

In his choice of evolutionary algorithm Dawkins made his job too easy. I want therefore to modify his example by introducing a slightly different but more realistic evolutionary algorithm. In place of an evolutionary algorithm that randomly varies only those characters that do not already coincide with the corresponding characters in the target sequence, let us consider the following algorithm, which we will call **E**:

**E**  Start with an entirely random sequence of 28 letters and spaces. Next pick a position at random in the sequence (this can be conceived as picking a random number between 1 and 28). Then randomly alter the character in that position. If the new sequence has higher fitness than the old, keep it and discard the old. Otherwise stay with the old. Repeat this process.

As before, fitness is determined by how close a sequence is to the target sequence.

Unlike the evolutionary algorithm in Dawkins's original example, which made constant and explicit reference to the target, **E** searches for the target solely on the basis of the phase space and the fitness function. Moreover, **E** employs the fitness function only at points previously identified in the phase space. **E** therefore avoids smuggling in any obvious teleology. Yet, as in Dawkins's original example, **E** converges with probability 1 on the target sequence, though somewhat more slowly. One can show mathematically that **E** takes on average about two orders of magnitude longer to converge on the target than Dawkins's original algorithm (an order of magnitude is a power of 10). Dawkins's algorithm converged on average in 40 iterations. Thus, whereas Dawkins's algorithm converges in decades, **E** converges in millennia—a negligible difference on evolutionary time scales.

It would seem, then, that **E** has generated specified complexity after all. To be sure, not in the sense of generating a target sequence that is inherently improbable for the algorithm (as with Dawkins's original example, the evolutionary algorithm here converges to the target sequence with probability 1). Nonetheless, with respect to the original uniform probability on the phase space, which assigned to each sequence a probability of around 1 in $10^{40}$, **E** appears to have done just that, to wit, generate a highly improbable specified event, or what we are calling specified complexity. What's more, unlike Dawkins's original algorithm, **E** is doing everything on the up and up. Indeed, to guide it to an otherwise complex specified target, the evolutionary algorithm is using nothing more than the phase space and the fitness function. Moreover, it is using these in ways that give it no unfair access to the target. If specified complexity is being smuggled in where instead it is supposed to be generated, it is not due to any fault of **E**.

Even so, the problem of generating specified complexity has not been solved. Indeed, it is utterly misleading to say that **E** has *generated* specified complexity. What the algorithm **E** has done is take advantage of the specified complexity inherent in the fitness function and utilized it in searching for and then locating the target sequence. Any fitness function that assigns higher fitness to sequences the more characters they have in common with the target sequence is hardly going to be arbitrary. Indeed, it is going to be *highly specific* and *carefully adapted* to the target. Indeed, its definition is going to require more complex specified information than the original target.

This is easily seen in the present example. Given the sequence ME-THINKS•IT•IS•LIKE•A•WEASEL, we adapted the fitness function to this sequence so that the function assigns the number of places where an arbitrary sequence agrees with it. But note that in the construction of this fitness function, there is nothing special about the sequence METHINKS•IT•IS•LIKE•A•WEASEL. Any other character sequence of 28 letters and spaces would have served equally well. Given any target sequence whatsoever, we can define a fitness function that assigns the number of places where an arbitrary sequence agrees with it. Moreover, given this fitness function, our evolutionary algorithm **E** will just as surely converge to the new target as previously it converged to METHINKS•IT•IS•LIKE•A•WEASEL.

It follows that every sequence corresponds to a fitness function specifically adapted for conducting **E** to it. Every sequence is therefore potentially a target for **E**, and the only thing distinguishing targets is the choice of fitness function. But this means that the problem of finding a given target sequence has been *displaced* to the new problem of finding a corresponding fitness function capable of locating the target. Our original problem was finding a certain target sequence within phase space. Our new problem is finding a certain fitness function within the entire collection of fitness functions—and one that is specifically adapted for locating the original target.

The collection of all fitness functions has therefore become a new phase space in which we must locate a new target (the new target being a fitness function capable of locating the original target in the original phase space). But this new phase space is far less tractable than the original phase space. The original phase space comprised all sequences of capital Roman letters and spaces 28 characters in length. This space contained about $10^{40}$ elements. The new phase space, even if we limit it as we have here to integer-valued functions with values between 0 and 28, will have at least $10^{10^{40}}$ elements (i.e., one followed by $10^{40}$ zeros).[36] If the original phase space was big, the new one is a monster.

To say that **E** has generated specified complexity within the original phase space is therefore really to say that **E** has borrowed specified complexity from a higher-order phase space, namely, the phase space of fitness functions. And since this phase space is always much bigger and much less tractable than the original phase space (the increase is exponential in the cardinality of the original space), it follows that **E** has in fact not generated specified complexity at all but merely shifted it around.

We have here a particularly vicious regress. For the evolutionary algorithm **E** to generate specified complexity within the original phase space presupposes that specified complexity was first generated within the higher-order phase space of fitness functions. But how was this prior specified complexity generated? Clearly, it would be self-defeating to claim that some higher-order evolutionary algorithm on the higher-order phase space of fitness functions generated specified complexity; for then we face the even more difficult problem of generating specified complexity from a still higher-order phase space (i.e., fitness functions over fitness functions over the original phase space).

This regress, in which evolutionary algorithms shift the problem of generating specified complexity from an original phase space to a higher-order phase space holds not just for Dawkins's METHINKS•IT•IS•LIKE•A• WEASEL example but in general. Complexity theorists are aware of this regress and characterize it by what are called "No Free Lunch theorems."[37] The upshot of these theorems is that averaged over all possible fitness functions, no search procedure outperforms any other. It follows that any success an evolutionary algorithm has in outputting specified complexity must ultimately be referred to the fitness function that the evolutionary algorithm employs in conducting its search. The No Free Lunch theorems dash any hope of generating specified complexity via evolutionary algorithms. Before turning to these theorems, however, we need to consider blind search more closely.

## 4.5   Blind Search

Joseph Culberson begins his discussion of the No Free Lunch (NFL) theorems with the following vignette:

> In the movie *UHF*, there is a marvelous scene that every computing scientist should consider. As the camera slowly pans across a small park setting, we hear a voice repeatedly asking "Is this it?" followed each time by the response "No!" As the camera continues to pan, it picks up two men on a park bench, one of them blind and holding a Rubik's cube. He gives it a twist, holds it up to his friend and the query–response sequence is repeated. This is blind search.[38]

This scene is humorous for the same reason it is instructive. There are an enormous number of possible configurations of the Rubik's cube. Of these only one constitutes a solution (viz., the configuration where each face displays exactly one color). Within the reference class of all possible configurations, the solution therefore constitutes an instance of specified complexity.[39] The scene is instructive because it illustrates specified complexity. This is also why it is humorous, for the two men on the park bench will long expire before reaching a solution.

In terms of the formal apparatus developed in the previous sections, blind search can be characterized as follows. Imagine two interlocutors, Alice and Bob. Alice has access to a reference class of possible solutions to a problem, call it $\Omega$. Bob not only has access to $\Omega$, but also to the set of actual solutions, call it T (T is a subset of $\Omega$; both $\Omega$ and T are nonempty). For any possible solution x in $\Omega$, Bob is able to tell Alice whether x is in T (i.e., whether x is in fact a solution). Alice now successively selects possible solutions $x_1$, $x_2$, . . . , $x_m$ from $\Omega$, at each step querying Bob (note that Alice limits herself to a finite search with at most m steps—she does not have infinite resources to continue the search indefinitely). Bob then truthfully answers whether each proposed solution is in fact in T. The search is successful if one of the proposed solutions lands in T. Computer scientists call this blind search.[40]

Two questions now arise: (1) When is blind search likely to succeed? and (2) What supplemental information might assist Alice in her search? We consider the first of these questions in this section and the second in the next section. First, when is blind search likely to succeed? Clearly, blind search is *guaranteed* to succeed if $\Omega$ has no more than m possible solutions, for then Alice can exhaust all the elements of $\Omega$. Actually, a slightly stronger result obtains: If the possible solutions in $\Omega$ that lie outside T (i.e., the possible solutions that are not also actual solutions) do not exceed m−1, then Alice need simply make sure that $x_1$, $x_2$, . . . , $x_m$ are all distinct to guarantee that at least one of them lands within T. (Alice would here be exploiting what is known as a "pigeonhole principle.")

Once the set of possible solutions that are not actual solutions is at least m, however, we are in the realm of probabilities. In that case Alice cannot simply exhaust the elements of $\Omega$ to guarantee a solution in T. Rather, Alice must draw samples of size m from $\Omega$ according to some probability distribution. But which probability distribution? Given any nonempty subset of $\Omega$ whatsoever, it is always possible to construct a probability distribution on $\Omega$ that concentrates all the probability on that set.[41] In other words, there is always a probability distribution that assigns a probability of 1 to any nonempty subset. Think of a die, for instance. It can always be loaded to land on some prescribed face. Hence even though there are five other faces, none of them will ever land. All the probability here is concentrated on a single arbitrarily chosen face. In general, then, one can guarantee that a random sample $x_1$, $x_2$, . . . , $x_m$ will contain a solution in T provided the probability of T is 1 (indeed, every one of the xs will in this instance be in T).

But this makes things too easy. Instead of concentrating all the probability on the solution set T, the usual move is to spread the probability over the reference class $\Omega$ as diffusely as possible. Such a probability distribution is called a uniform probability and typically presupposes a topological structure induced by a metric d (the metric characterizes what it means for a probability to be spread out "diffusely").[42] We have seen uniform probabilities already and denoted them by $U$ (see section 4.3). For now the important thing about $U$ is that it be a probability distribution defined over $\Omega$ and that it assign probability strictly less than 1 to $T$ (i.e., $U(T) < 1$). Once $U(T)$ is strictly less than 1, random sampling from $\Omega$ with respect to $U$ cannot guarantee landing in $T$ (i.e., the probability of landing in $T$ will be strictly less than 1). Even so, depending on the size of $U(T)$ and the sample size m, the probability of landing in $T$ can assume any value strictly less than 1 (though m may have to get extremely large for the probability to get close to 1).[43]

But what happens if $U(T)$ is so small that for any sample size we might reasonably take, the probability that a random sample lands in the target remains small? This was the problem with doing a blind search with the Rubik's cube—the blind man on the park bench cannot give the cube enough random twists to stand a reasonable chance of solving the puzzle. The probability that a random sample of size m lands in a target $T$ with probability value $p = U(T)$ is

$$1 - (1-p)^m.$$

The calculation demonstrating this claim is straightforward and can be found in any elementary probability text.[44] And as noted in section 4.3, this probability gives an upper bound for the probability that a blind search with sample size m can locate $T$.

Suppose, now, that m is less than $1/p$ by, say, two orders of magnitude (i.e., $100m < 1/p$ or equivalently $mp < 1/100$; an order of magnitude is a power of 10 so that two orders of magnitude is $10^2 = 100$). Then $1 - (1-p)^m$ is approximately mp:

$$1 - (1-p)^m \approx mp.$$

This follows from expanding $(1-p)^m$ via the binomial theorem (the symbol $\approx$ here means "approximately equal"). Thus, as a rule of thumb, so long as m is considerably less than $1/p$ (say by at least two orders of magnitude), mp is the probability of a random sample of size m landing in a target of probability p. But if m is considerably less than $1/p$, then mp is itself close to zero, and therefore the probability of a random sample of size m landing in the target is close to zero. With the Rubik's cube puzzle, for instance, the probability p of solving the puzzle with a single random twist is so small that for any reasonable number of twists that one might give it—let m denote an upper bound—the product mp will be tiny and the puzzle will very likely remain unsolved.

To sum up, given a sample of size m and a target of probability p, if m is considerably less than $1/p$ (say two or more orders of magnitude), then mp provides an upper bound on the probability that a blind search of sample size m will locate the target. What's more, mp will in that case be small, and the blind search of sample size m will therefore very likely fail to land in T. And since for most real-world problems the reference class $\Omega$ of possible solutions is huge compared to the target T, the probability $p = U(T)$ of the target will typically be minuscule. Consequently, for any sample size m that might reasonably be taken, m will be much less than $1/p$, mp will be close to zero, and a blind search for T will be almost sure to fail.

## 4.6  The No Free Lunch Theorems

If blind search is destined to fail, how, then, to make a search succeed? More to the point, what additional information needs to supplement a blind search to make it successful? To answer this question, let us return to the exchange between Alice and Bob. Bob, recall, has a full grasp of the target T and for any proposed solution x in $\Omega$ is able to answer whether x is in T. Alice, on the other hand, knows only $\Omega$ and whatever information Bob is willing to divulge about T. We may assume that in knowing $\Omega$, Alice knows $\Omega$'s topological structure (as induced by the metric d) as well as the uniform probability U. We may also assume that Alice knows enough about the problem in question to ascertain the probability $p = U(T)$. Finally, we assume that m is an upper bound on the number of possible solutions in $\Omega$ that Alice can verify with Bob, and that mp, the approximate probability for locating the target by random sampling given a sample size m, is so small that Alice has no hope of ever attaining T via a blind search.

Alice and Bob are playing a game of "m questions" in which Bob divulges too little information for Alice to have any hope of winning the game. Alice therefore needs some *additional information* from Bob. But what additional information? Bob could just inform Alice of the exact location of T and be done with it. But that would be too easy. If Alice is playing the role of scientist and Bob the role of nature, then Bob needs to make Alice drudge and sweat to locate T—nature, after all, does not divulge her secrets easily. Alice and Bob are operating here with competing constraints. Bob wants to give Alice the minimum information she needs to locate T. Alice, on the other hand, wants to make maximal use of whatever information Bob gives her to ensure that her m questions are as effective as possible in locating the target.

Let us therefore suppose that Bob identifies some additional information j and makes it available to Alice. This information is supposed to help Alice locate T. Bob, however, is not simply going to hand j over to Alice. Rather, he is going to wait for Alice to propose a possible solution x in $\Omega$, and then inform Alice of what j has to say about x. Moreover, Alice is going to be able to propose at most m such candidates from $\Omega$ since m sets an upper bound on Alice's sample size.

We have therefore a new protocol for the interchange between Alice and Bob. Before, Bob would only tell Alice whether a candidate solution belonged to the target (i.e., for any x identified by Alice, Bob would only tell her whether x belonged to T). Now, for any candidate solution that Alice proposes, Bob will tell her what this additional information has to say about it (i.e., for any x that Alice proposes, Bob will tell her what j has to say about x). Thus, instead of handing the full information j over to Alice, Bob is going to hand over only that aspect of j pertaining to x, which we will refer to as *the item of information* j(x). The item of information j(x) might be the distance of x from T, the color of x, the weight of x, the fitness of x, the number of elements of $\Omega$ directly adjacent to x, some combination of these, etc., etc. (the possibilities are limitless).

We therefore have a new game of "m questions" in which the answer to each question is not *whether* some proposed solution x belongs to T but rather *what* the information j has to say specifically about x—an item of information we denote by j(x). It follows that all the action in this new game of "m questions" centers on the information j. Is j enough to render Alice's m-step search for T successful? And if so, what characteristics must j possess?

The fundamental point of the No Free Lunch (NFL) theorems is that it does not matter what characteristics j possesses. Instead, what matters is the reference class of possibilities to which j belongs and from which j is drawn. Precisely because j is information, it must by definition belong to some reference class of possibilities. Information always presupposes multiple live possibilities, one of which has been selected to the exclusion of others. Recall the quote from Robert Stalnaker in section 3.1: "Content requires contingency. To learn something, to acquire information, is to rule out possibilities. To understand the information conveyed in a communication is to know what possibilities would be excluded by its truth."[45] Additionally, recall the quote from Fred Dretske in that same section: "Information theory identifies the amount of information associated with, or generated by, the occurrence of an event (or the realization of a state of affairs) with the reduction in uncertainty, the elimination of possibilities, represented by that event or state of affairs."[46]

Let us therefore refer to the reference class of possibilities from which j is drawn as the *information-resource space* and denote it by J. More informally, we refer to J as the *informational context*. J constitutes the totality of informational resources that might assist Alice in locating T. We now suppose that Bob has access to **J**, selects some possibility j from it, and makes j available to Alice to help her locate T. The reason we speak of No Free Lunch theorems (plural) is to distinguish the different types of information-resource spaces **J** from which Bob might select j.

What can **J** look like? We have already seen in section 4.4 where **J** is the class of fitness functions on $\Omega$ (i.e., the nonnegative real-valued functions on $\Omega$) and j is an individual fitness function (which we previously denoted by f). Since $\Omega$ is a topological space, **J** could as well be the continuous fitness functions on $\Omega$. If $\Omega$ is a differentiable manifold, **J** could be the differentiable fitness functions on $\Omega$. **J** could even be a class of temporally indexed fitness functions on $\Omega$ so that when Alice proposes a possible solution x, j(x) is not merely the fitness of x simpliciter but its fitness at time t when Alice proposes x. Such temporally indexed fitness functions have yet to find widespread use but seem more appropriate than ordinary fitness functions for modeling fitness, which in any realistic environment is not likely to be static (see section 4.10).

Alternatively, **J** need not involve any fitness functions whatsoever. **J** could be a group action on $\Omega$, that is, an algebraic group the elements of which form bijective mappings of $\Omega$ to itself (i.e., each j in **J** is a one-to-one and onto function from $\Omega$ to itself).[47] **J** could be a class of dynamical systems, that is, a class of temporally indexed functions from $\Omega$ to itself that describe the flow of particles through the phase space $\Omega$.[48] The possibilities for such information-resource spaces are limitless, and each such **J** has its own No Free Lunch theorem.

Suppose now that we are given a phase space $\Omega$, a metric d on $\Omega$, a uniform probability **U** on $\Omega$ induced by d, a target T that is a nonempty subset of $\Omega$, an information-resource space **J**, information j in **J**, and a maximal sample size m. An evolutionary algorithm **E** on $\Omega$ is a stochastic process indexed by the natural numbers 1, 2, 3, . . . and taking values in the phase space $\Omega$ satisfying the following condition: For any k, the value $E_k = x_k$ depends only on the phase space $\Omega$, the metric d, the uniform probability **U**, and the values of $E_i = x_i$ and $j(x_i)$ for $i < k$. In other words, **E** depends on the information j only at those points in phase space that **E** has hitherto traversed and then only insofar as j is evaluated at those points (see section 4.3).

A generic NFL theorem now takes the following form: It sets up a performance measure **M** that characterizes how effectively an evolutionary algorithm **E** locates a target T within m steps using the information j. This measure can be denoted by **M(E,T,m,j)** (for instance, **M(E,T,m,j)** can be the probability that **E** locates T within m steps using **j**). Next **M(E,T,m,j)** needs to be averaged over all **j** in **J**. This new averaged performance measure can be denoted by **M(E,T,m,J)**. A generic NFL theorem now states that **M(E,T,m,J)** is independent of the evolutionary algorithm **E**—in other words, it is the same for all evolutionary algorithms. And since blind search always constitutes a perfectly valid evolutionary algorithm, this means that the average performance of any evolutionary algorithm **E** is no better than blind search.[49]

The significance of the No Free Lunch theorems is that an information-resource space **J** does not, and indeed cannot, privilege a target T. Instead, **J** contains information that is equally adept at guiding an evolutionary algorithm to other targets in the phase space. The information-resource spaces associated with NFL theorems typically possess the following symmetry relation: for any two targets T and T′ within Ω of roughly the same probability, there exists information **j** and **j**′ in **J** such that **j** guides the evolutionary algorithm successfully into T if and only if **j**′ guides it successfully into T′ (success being determined by whether the evolutionary algorithm locates the target without exceeding the sample size m).

We saw this in section 4.4. Recall there the evolutionary algorithm **E** that located the target phrase METHINKS•IT•IS•LIKE•A•WEASEL. We first defined a fitness function that to each sequence of 28 letters and spaces (spaces being represented by bullets) assigned the number of characters coinciding with the target phrase. We then defined **E** in relation to this fitness function:

**E**   Start with an entirely random sequence of 28 letters and spaces. Next pick a position at random in the sequence (this can be conceived as picking a random number between 1 and 28). Then randomly alter the character in that position. If the new sequence has higher fitness than the old, keep it and discard the old. Otherwise stay with the old. Repeat this process.

As it turned out, **E** converged to the target phrase with high probability once the sample size m reached three or four orders of magnitude.

The fitness function is of course the additional information that turns this search from a blind to a constrained search. Let us therefore denote by **j** the fitness function that assigns to each sequence the number of characters coinciding with METHINKS•IT•IS•LIKE•A•WEASEL. Additionally, let us denote by **J** the class of all fitness functions defined with respect to such sequences. It is then clear that in **j**'s formulation there is nothing special about the sequence METHINKS•IT•IS•LIKE•A•WEASEL. Any other character sequence of 28 letters and spaces would have served equally well. Given any target sequence whatsoever, we can define a fitness function **j**′ that assigns the number of places where an arbitrary sequence agrees with it. Moreover, given this fitness function, the evolutionary algorithm **E** will just as surely converge to the new target as previously it converged to ME-THINKS•IT•IS•LIKE•A•WEASEL.

In general, the information-resource spaces **J** that arise in NFL theorems are sufficiently rich so that if **J** contains information for guiding an evolutionary algorithm into one target, then it also contains information for guiding it into any other target. Consequently, there are no privileged targets with respect to **J**, and the only thing distinguishing targets is the choice of **j** in **J**. But this means that the problem of finding a given target has been displaced to the new problem of finding the information **j** capable of locating that target. Our original problem was finding a certain target within phase space. Our new problem is finding a certain **j** within the information-resource space **J**.

The information-resource space **J** has therefore become a higher-order phase space in which we must locate a new target (the new target being information in **J** needed to locate the original target in the original phase space). But this new phase space is generally far less tractable than the original phase space (in general the increase in complexity going from a phase space to an information-resource space is exponential; see section 4.4 where we considered a concrete example). To say that an evolutionary algorithm has generated specified complexity within the original phase space is therefore really to say that it has borrowed specified complexity from a higher-order phase space, namely, the information-resource space. And since in practice this new phase space (i.e., **J**) is much bigger and much less tractable than the original phase space, it follows that the evolutionary algorithm has in fact not generated specified complexity at all but merely shifted it around.

## 4.7 The Displacement Problem

The essential difficulty in generating specified complexity with an evolutionary algorithm can now be stated quite simply. An evolutionary algorithm is supposed to find a target within phase space. To do this successfully, however, it needs more information than is available to a blind search. But this additional information is situated within a wider informational context (what in the last section we called the information-resource space). And locating that additional information within the wider context is no easier than locating the original target within the original phase space. Evolutionary algorithms therefore displace the problem of generating specified complexity but do not solve it. I call this *the displacement problem.*

Think of it this way. There is an island with buried treasure. You can scour the island trying to find the buried treasure. Alternatively, you can try to find a map that tells you where the treasure is buried. Once such a map is in hand, finding the treasure is no problem. But how to find such a map? Suppose such a map exists but is mixed among a huge assortment of other maps. Finding the right map within that huge assortment will then be no easier than simply searching the island directly. The huge assortment of maps is the informational context. In general, an informational context is no easier to search than the original phase space.

There is no way around the displacement problem. This is not so say that there have not been attempts to get around it. But invariably we find that when specified complexity seems to be generated for free, it has in fact been front-loaded, smuggled in, or hidden from view. I want, then, in this section to review some attempts to get around the displacement problem and uncover just where the displaced information resides once it goes underground.

First off, let us be clear that the No Free Lunch theorems that underwrite the displacement problem apply with perfect generality—NFL applies to *any* information that might supplement a blind search, and not just to fitness functions. Usually the NFL theorems are stated in terms of fitness functions over phase spaces (indeed, that is how I motivated the discussion in the earlier sections of this chapter). Thus, in the case of biological evolution, one can try to mitigate the force of NFL by arguing that evolution is nonoptimizing. Joseph Culberson, for instance, asks, "If fitness is supposed to be increasing, then in what nontrivial way is a widespread species of today more fit than a widespread species of the middle Jurassic?"[50] But NFL theorems can just as well be formulated for informational contexts that do not comprise fitness functions. The challenge facing biological evolution, then, is to avoid the force of NFL when evolutionary algorithms also have access to information other than fitness functions. Merely arguing that evolution is nonoptimizing is therefore not enough. Rather, one must show that finding the information that guides an evolutionary algorithm to a target is substantially easier than finding the target directly through a blind search.

Think of it this way. In trying to locate a target, you can sample no more than m points in phase space. What's more, your problem is sufficiently complex that you will need additional information to find the target. That information resides in a broader informational context (what we have called the information-resource space). If searching through that broader informational context is no easier than searching through the original phase space, then you are no better off going with an evolutionary algorithm than going with a straight blind search. Moreover, you cannot arbitrarily truncate your informational context simply to facilitate your search, for any such truncation will itself be an act of ruling out possibilities, and that by definition means an intrusion of novel information, and in particular of specified complexity. In effect, you will be smuggling in what you are claiming to discover.[51]

To resolve the displacement problem therefore requires an answer to the following question: How can the informational context be simplified sufficiently so that finding the information needed to locate a target is easier than finding the target using blind search? There is only one way to do this without arbitrarily truncating the informational context, and that is for the phase space itself to constrain the informational context. Structures and regularities of the phase space must by themselves be enough to constrain the selection of points in the phase space and thus facilitate locating the target. The move here, then, is from contingency to necessity; from evolutionary algorithms to dynamical systems; from Darwinian evolution to complex self-organization. Stuart Kauffman's approach to biological complexity epitomizes this move, focusing on autocatalytic reactions that reliably settle into complex behaviors and patterns (see his chapter "Order for Free" in *At Home in the Universe*).[52]

Nonetheless, even this proposed resolution of the displacement problem fails. Yes, structures and regularities of the phase space can simplify the informational context so that finding the information needed to locate a target is easier than finding the target using blind search. But whence those structures and regularities in the first place? Structures and regularities are constraints. And constraints, by their very specificity, could always have been otherwise. A constraint that is not specific is no constraint at all. Constraints are constraints solely in virtue of their specificity—they permit some things and rule out others. But in that case different constraints could fundamentally alter what is permitted and what is ruled out. Thus, the very structures and regularities that were supposed to eliminate contingency, information, and specified complexity merely invite them back in.

To see this, consider a phase space with an additional structure or regularity that simplifies the informational context, thereby facilitating the task of finding the information needed to locate a target. What sorts of structures or regularities might these be? Are they topological? A topology defines a class of open sets on the phase space.[53] If those open sets in some way privilege the target, then by permuting the underlying points of the phase space, a new topology can be generated that privileges any other target we might choose (that is why mathematicians refer to topology as "point-set topology"). What's more, the totality of topologies associated with a given phase space is vastly more complicated than the original phase space. Searching a phase space by exploiting its topology therefore presupposes identifying a suitable topology within a vast ensemble of topologies. As usual, the displacement problem refuses to go away.

Or suppose instead that the constraint to be exploited for locating a target constitutes a dynamical system, that is, a temporally indexed flow describing how points move about in phase space. Dynamical systems are the stuff of "chaos theory" and underlie all the wonderful fractal images, strange attractors, and self-similar objects that have so captured the public imagination.[54] Now, if the dynamical system in question helps locate a target, it is fair to ask what other dynamical systems the phase space is capable of sustaining and what other targets they are capable of locating. In general, the totality of different possible dynamical systems associated with a phase space will be far more complicated than the original phase space (if, for instance, the phase space is a differentiable manifold, then any vector field induces a differentiable flow whose tangents are the vectors of the vector field; the totality of different possible flows will in this case be immense, with flows going in all conceivable directions).[55] Searching a phase space by exploiting a dynamical system therefore presupposes identifying a suitable dynamical system within a vast ensemble of dynamical systems. As usual, the displacement problem refuses to go away.

Exploiting constraints on a phase space to locate a target is therefore merely another way of displacing information. Not only does it not solve the displacement problem; its applicability is quite limited. Many phase spaces are homogeneous and provide no help in locating targets. Consider for instance a phase space comprising all possible character sequences from a fixed alphabet (such phase spaces model not only written texts but also polymers—e.g., DNA, RNA, and proteins). Such phase spaces are perfectly homogeneous, with one character string geometrically interchangeable with the next. Whatever else the constraints on such spaces may be, they provide no help in locating targets. Rather, external semantic information (in the case of written texts) or functional information (in the case of polymers) is needed to locate a target.[56]

To sum up, there is no getting around the displacement problem. Any output of specified complexity requires a prior input of specified complexity. In the case of evolutionary algorithms, they can yield specified complexity only if they themselves are carefully front-loaded with the right information (typically via a fitness function) and thus carefully adapted to the problem at hand. In other words, all the specified complexity we get out of an evolutionary algorithm has first to be put into its construction and into the information that guides the algorithm. Evolutionary algorithms therefore do not generate or create specified complexity, but merely harness already existing specified complexity.

How, then, does one generate specified complexity? There is only one known generator of specified complexity, and that is intelligence.[57] In every case where we know the causal history underlying an instance of specified complexity, an intelligent agent was involved (see sections 1.6 and 1.7). Most human artifacts, from Shakespearean sonnets to Dürer woodcuts to Cray supercomputers, are specified and complex. For a signal from outer space to convince astronomers that extraterrestrial life is real, it too will have to be complex and specified, thus indicating that the extraterrestrial is not only alive but also intelligent (hence the search for extraterrestrial *intelligence*—SETI).[58] Thus, to claim that natural laws, even radically new ones as Paul Davies suggests, can produce specified complexity is to commit a category mistake. It is to attribute to laws something they are intrinsically incapable of delivering.[59] Indeed, all our evidence points to intelligence as the sole source for specified complexity.

## 4.8   Darwinian Evolution in Nature

We need now to step back and consider carefully what the displacement problem means for Darwinian evolution as it occurs in nature. Darwinists are unlikely to see the displacement problem as a serious threat to their theory. I have argued that evolutionary algorithms like the one in Dawkins's METHINKS•IT•IS•LIKE•A•WEASEL example fail to generate specified complexity because they smuggle it in during construction of the fitness function. Now, if evolutionary algorithms modeled, say, the stitching together of monomers to generate some initial self-replicating polymer, strict Darwinists would admit the relevance of the displacement problem (to paraphrase Theodosius Dobzhansky, to speak of generating an initial replicator via a Darwinian selection mechanism is a contradiction in terms because that very mechanism presupposes replication). Darwinists, however, are principally interested in modeling evolutionary progress once a replicator has come into existence, and here they argue the displacement problem is irrelevant.

Why? According to Richard Dawkins, nature's criterion for optimization is not an arbitrarily chosen distant target but survival and reproduction, and these are anything but arbitrary. As he puts it in *The Blind Watchmaker*, commenting on the METHINKS•IT•IS•LIKE•A•WEASEL example:

> Although the monkey/Shakespeare model is useful for explaining the distinction between single-step selection and cumulative selection, it is misleading in important ways. One of these is that, in each generation of selective "breeding," the mutant "progeny" phrases were judged according to the criterion of resemblance to a distant ideal target, the phrase METHINKS IT IS LIKE A WEASEL. Life isn't like that. Evolution has no long-term goal. There is no long-distance target, no final perfection to serve as a criterion for selection. . . . In real life, the criterion for selection is always short-term, either simple survival or, more generally, reproductive success. . . . The "watchmaker" that is cumulative natural selection is blind to the future and has no long-term goal.[60]

The Darwinist therefore objects that "real life" Darwinian evolution can in fact generate specified complexity without smuggling it in after all. The fitness function in biological evolution follows directly from differential survival and reproduction, and this, according to the Darwinist, *can* legitimately be viewed as a "free lunch." In biological systems the replicator (i.e., the living organism) will sample different variants via mutation, and then the fitness function freely bestowed by differential survival and reproduction will select those variants that constitute an improvement, which within Darwinism is defined by being better at surviving and reproducing. No specified complexity is required as input in advance.

If this objection is conceded, then the only way to show that the Darwinian mechanism cannot generate specified complexity is by demonstrating that the gradients of the fitness function induced by differential survival and reproduction are not sufficiently smooth for the Darwinian mechanism to drive large-scale biological evolution. To use another Dawkins metaphor, one must show that there is no gradual way to ascend "Mount Improbable."[61] This is a separate line of argument and one that I shall take up in the next chapter. Here, however, I want to show that this concession need not be granted and that the displacement problem does indeed undercut Darwinism.

Things are not nearly as simple as taking differential survival and reproduction as brute givens and from there concluding that the fitness function induced by these is likewise a brute given. Differential survival and reproduction by themselves do not guarantee that anything interesting will happen. Consider, for instance, Sol Spiegelman's work on the evolution of polynucleotides in a replicase environment. Leaving aside that the replicase protein is supplied by the investigator (from a viral genome), as are the activated mononucleotides needed to feed polynucleotide synthesis, the problem here and in experiments like it is the steady attenuation of information over the course of the experiment. As Brian Goodwin notes:

> In a classic experiment, Spiegelman in 1967 showed what happens to a molecular replicating system in a test tube, without any cellular organization around it. The replicating molecules (the nucleic acid templates) require an energy source, building blocks (i.e., nucleotide bases), and an enzyme to help the polymerization process that is involved in self-copying of the templates. Then away it goes, making more copies of the specific nucleotide sequences that define the initial templates. But the interesting result was that these initial templates did not stay the same; they were not accurately copied. They got shorter and shorter until they reached the minimal size compatible with the sequence retaining self-copying properties. And as they got shorter, the copying process went faster. So what happened with natural selection in a test tube: the shorter templates that copied themselves faster became more numerous, while the larger ones were gradually eliminated. This looks like Darwinian evolution in a test tube. But the interesting result was that this evolution went one way: toward greater simplicity. Actual evolution tends to go toward greater complexity, species becoming more elaborate in their structure and behavior, though the process can also go in reverse, toward simplicity. But DNA on its own can go nowhere but toward greater simplicity. In order for the evolution of complexity to occur, DNA has to be within a cellular context; the whole system evolves as a reproducing unit.[62]

My point here is not that Darwinian evolution in a test tube should be regarded as disconfirming evidence for Darwinian evolution in nature. Rather, it is that if the Darwinian mechanism of differential survival and reproduction is what in fact drives full-scale biological evolution in nature, then the fitness function induced by that mechanism has to be very special. Indeed, many prior conditions need to be satisfied for the fitness function to take a form consistent with the Darwinian mechanism being the principal driving force behind biological evolution. Granted, the fitness function induced by differential survival and reproduction in nature is nonarbitrary. But that does not make it a free lunch either.

Think of it this way. Suppose we are given a phase space $\Omega$ of replicators that replicate according to a Darwinian mechanism of differential survival and reproduction. Suppose this mechanism induces a fitness function f. Given just this information, we do not know if evolving within this phase space will over time lead to anything interesting. In the case of Spiegelman's experiment, it did not—Darwinian evolution led to increasingly simpler replicators. In real life, however, Darwinian evolution is said to lead to vast increases in the complexity of replicators, with all cellular organisms tracing their lineage back to a common unicellular ancestor. Let us grant this. The phase space Ω then comprises a vast array of DNA-based self-replicating cellular organisms and the Darwinian mechanism of differential survival and reproduction over this phase space induces a fitness function f that underwrites full-scale Darwinian evolution.[63] In other words, the fitness function f is consistent not only with the descent of all organisms from a common ancestor (i.e., common descent), but also with the Darwinian mechanism accounting for the genealogical interrelatedness of all organisms. Now suppose this is true. What prior conditions have to be satisfied for f to be the type of fitness function that allows a specifically Darwinian form of evolution to flourish?

For starters, Ω had better be nonempty, and that presupposes raw materials like carbon, hydrogen, and oxygen. Such raw materials, however, presuppose star formation, and star formation in turn presupposes the fine-tuning of cosmological constants. Thus for f to be the type of fitness function that allows Darwin's theory to flourish presupposes all the anthropic principles and cosmological fine-tuning that lead many physicists to see design in the universe.[64] Yet even with cosmological fine-tuning in place, many additional conditions need to be satisfied. The phase space Ω of DNA-based self-replicating cellular organisms needs to be housed on a planet that is not too hot and not too cold. It needs a reliable light source. It needs to have a sufficient diversity of minerals and especially metals. It needs to be free from excessive bombardment by meteors. It needs not only water but enough water. Michael Denton's *Nature's Destiny* is largely devoted to such specifically terrestrial conditions that need to be satisfied if biological evolution on earth is to stand any chance of success.[65]

But there is more. Cosmology, astrophysics, and geology fail to exhaust the conditions that a fitness function must satisfy if it is to render not just biological evolution but a specifically Darwinian form of it the grand success we see on planet earth. Clearly, DNA-based replicators need to be robust in the sense of being able to withstand frequent and harsh environmental insults (this may seem self-evident, but computer simulations with artificial life forms tend to be quite sensitive to unexpected perturbations and thus lack the robustness we see in terrestrial biology). What's more, the DNA copying mechanism of such replicators must be sufficiently reliable to avoid error catastrophes. Barring a high degree of reliability the replicators will go extinct or wallow interminably at a low state of complexity (basically just enough complexity to avoid the error catastrophe).[66]

Perhaps most importantly, the replicators must be able to increase fitness and complexity in tandem. In particular, fitness must not be positively correlated with simplicity. This last requirement may seem easily purchased, but it is not. Stephen Jay Gould, for instance, argues in *Full House* that replication demands a certain minimal level of complexity below which things are dead (i.e., no longer replicate). Darwinian evolution is thus said to constitute a random walk off a reflecting barrier, the barrier constituting a minimal complexity threshold for which increases in complexity always permit survival but decreases below that level entail death. Enormous increases in complexity are thus said to become not only logically possible but also highly probable.[67]

The problem with this argument is that in the context of Darwinian evolution such a reflecting barrier tends also to be an absorbing barrier (i.e., there is a propensity for replicators to stay close to, if not right at, the minimal complexity threshold). As a consequence, such replicators will over the course of evolution remain simple and never venture into high degrees of complexity. Simplicity by definition always entails a lower cost in raw materials (be they material or computational) than increases in complexity, and so there is a inherent tendency in evolving systems for selection pressures to force such systems toward simplicity (or as it is sometimes called, *elegance*).

Fitness functions induced by differential survival and reproduction are more naturally inclined to place a premium on simplicity and regard replicators above a certain complexity threshold as too cumbersome to survive and reproduce. The Spiegelman example is a case in point. Thomas Ray's Tierra simulation gave a similar result, showing how selection acting on replicators in a computational environment also tended toward simplicity rather than complexity—unless parameters were set so that selection could favor larger sized organisms (complexity here corresponding to size).[68] This is not to say that the Darwinian mechanism automatically takes replicating systems toward a minimal level of complexity, but that if it does not, then some further conditions need to be satisfied, conditions reflected in the fitness function.

Vast is the catalogue of conditions that the fitness function induced by differential survival and reproduction needs to satisfy if the spectacular diversity of living forms that we see on earth is properly to be attributed to a Darwinian form of evolution. Such a catalogue is going to require a vast amount of specified complexity, and this specified complexity will be reflected in the fitness function that, as Darwinists rightly note, is nonarbitrary but, as Darwinists are reluctant to accept, is also not a free lunch. Throw together some replicators, subject them to differential survival and reproduction, perhaps add a little game theory to the mix (à la Robert Wright),[69] and there is no reason to think you will get anything interesting, and certainly not a form of Darwinian evolution that is worth spilling any ink over. Thus I submit that even if Darwinian evolution is the means by which the panoply of life on earth came to be, the underlying fitness function that constrains biological evolution would not be a free lunch and not a brute given, but a finely crafted assemblage of smooth gradients that presupposes much prior specified complexity.

In the next chapter we will see why there is no good reason to think that the gradients are smooth. But even if the gradients are smooth, there is no reason to think that the Darwinian mechanism is the driving force behind evolution. Smooth gradients are a necessary condition for Darwinian evolution to take place. But they are hardly a sufficient condition. Even if the gradients of a fitness function are smooth, the portions of phase space that the fitness function renders optimal could be thoroughly dull and of no biological significance. Smooth gradients tell us that an evolutionary algorithm is able to optimize a fitness function possessing those gradients. Smooth gradients do not tell us whether optimizing that fitness function leads to anything interesting.[70]

## 4.9  Following the Information Trail

The No Free Lunch theorems are essentially bookkeeping results. They keep track of how well evolutionary algorithms do at optimizing fitness functions over a phase space. The fundamental claim of these theorems is that when averaged across fitness functions, evolutionary algorithms cannot outperform blind search. The significance of these theorems is that if an evolutionary algorithm actually proves successful at locating a complex specified target, the algorithm has to exploit a carefully chosen fitness function. This means that any complex specified information in the target had first to reside in the fitness function.

The No Free Lunch theorems underscore the fundamental limits of the Darwinian mechanism. Up until their proof, it was thought that because the Darwinian mechanism could account for all of biological complexity, evolutionary algorithms (i.e., their mathematical underpinnings) must be universal problem solvers. The No Free Lunch theorems show that evolutionary algorithms, apart from careful fine-tuning by a programmer, are no better than blind search and thus no better than pure chance. Consequently, these theorems cast doubt on the power of the Darwinian mechanism to account for all of biological complexity. Granted, the No Free Lunch theorems are bookkeeping results. But bookkeeping can be very useful. It keeps us honest about the promissory notes our various enterprises—science being one of them—can and cannot make good. In the case of Darwinism we are no longer entitled to think that the Darwinian mechanism can offer biological complexity as a free lunch.

I summarize the results of this chapter thus far because even though the theory developed here is clear and ought to be uncontroversial, often it fails to be applied in practice and gets so misrepresented that what NFL denies actually seems to be affirmed. It is a very human impulse to look for magical solutions to circumvent mathematical impossibilities. The theory of accounting tells us that Ponzi schemes cannot work. The theory of probability tells us that games of chance whose expected gain favors not us but the casino can only lead to our loss in the long run. Nonetheless, Ponzi schemes and casino gambling continue to be big business. Likewise, in biology, even though computational theory is clear that evolutionary algorithms cannot generate complex specified information, by suitably shuffling information around one often gets the impression that evolutionary algorithms can in fact generate CSI and that CSI is a free lunch after all. Invariably what is involved here is a shell game in which the shells are adroitly moved so that one loses track of just which shell contains the elusive pea. The pea here is complex specified information. The task of the bookkeeper is to follow the information trail so that it is properly accounted for and not magically smuggled in.

As an example of smuggling in complex specified information that is purported to be generated for free, consider the work of Thomas Schneider. Schneider heads a laboratory focusing on molecular information theory. Schneider's laboratory belongs to the LECB (Laboratory of Experimental and Computational Biology) at the National Cancer Institute. Schneider is well-versed in Shannon's theory of information, regularly applies it in his research, and devotes considerable space to it on his website.[71] In the summer of 2000 he published an article in *Nucleic Acids Research* titled "Evolution of Biological Information."[72] In that paper he identified a computational phase space consisting of all sequences 256 letters in length constructed from a four-letter alphabet (cf. the four nucleotide bases). The phase space therefore consisted of $4^{256}$ sequences, or approximately $10^{154}$ sequences. Starting with an evolutionary algorithm acting on a randomly chosen sequence from the phase space, Schneider then purported to generate an information-rich sequence corresponding to a finely tuned genetic control system in which one part of the genome codes for proteins that precisely bind to another part of the genome. To model genetic control, Schneider divided his 256-letter computational genomes essentially in half, treating the first half as what he called a "weight matrix" and the second half as binding sites. The optimization task of his evolutionary algorithm was to get the weight matrix to match up suitably with the binding sites. Here the weight matrix corresponded to translation and protein folding of natural biological systems, and the binding sites corresponded to locations on DNA where these proteins would then bind.

The details here are not that important. What is important is the discrepancy between what Schneider thinks his computer simulation establishes and what it in fact establishes. Schneider thinks that he has generated biologically relevant information for free, or as he puts it, "from scratch." Early in his article he writes, "The necessary information should be able to evolve from scratch."[73] Later in the article he claims to have established precisely that: "The program simulates the process of evolution of new binding sites from scratch."[74] According to Schneider the advantage of his simulation over other simulations that attempt to generate biologically relevant information (like Richard Dawkins's biomorphs program and Thomas Ray's Tierra environment) is that Schneider's program "starts with a completely random genome, and no further intervention is required."[75] Schneider gives his readers to believe that he has decisively confirmed the full sufficiency of the Darwinian mechanism to account for biological information. Accordingly, he claims his model "addresses the question of how life gains information, . . . [and] shows explicitly how this information gain comes about from mutation and selection, without any other external influence."[76]

Schneider himself would quibble with the previous paragraph. It is not that he would deny that information has been generated "from scratch"—he has affirmed that clearly enough. It is that he would refuse to equate information being generated from scratch with information being a free lunch. As he wrote in response to an earlier criticism of mine:

> The phrase "for free" does not appear in the paper [i.e., "Evolution of Biological Information"]. The claim in [that paper] is that the information appears under replication, mutation and selection, commonly known as "evolution." It is not for free! Half of the population DIES every generation! . . . "From scratch" does *not* mean the same thing as "for free." "From scratch" refers (obviously) to the *initial condition of the genome* which is random in this case. . . . That is, there is no *measurable* information in the binding sites at the beginning of the simulation. "For free" would mean "without effort," and . . . there is quite a bit of effort and (virtual) pain for the gains observed.[77]

Schneider is here engaged in some semantic hair-splitting. Darwinists frequently cite the huge cost of Darwinian evolution, pointing to the deaths of countless organisms as natural selection's price for our climb up the evolutionary peak (if humans are not your preferred organism, take your pick). The issue is not whether Darwinian evolution incurs a cost according to some method or other of denominating currency. The issue is whether in the currency of information, and CSI in particular, Darwinian evolution incurs a cost. It does not, and Schneider agrees that it does not. For Schneider, information is a product of evolution and not a cost paid to make Darwinian evolution work.

Semantics aside, the question remains whether Schneider has in fact successfully answered the charge that the Darwinian mechanism is inadequate to generate biological information and in particular CSI? In reading Schneider's article, and more generally when confronting Darwinian scenarios that purport to generate CSI for free, I always go back to my days as a graduate student in mathematics teaching undergraduates trigonometry. When it came time to grade their tests, I always had to watch that they did not trick me by purporting to establish a trigonometric equality when in fact they did not have a clue why one trigonometric expression was equal to another. What students would do is write one expression at the top of the page, the other at the bottom of the page. Then they would manipulate the top expression, transforming it line by line down the middle of the page. Next they would manipulate the bottom expression, transforming it line by line up the middle of the page. In the middle of the page the transformed top and bottom expressions would happily meet, offering no clue how they were related. My challenge was to find where the unwarranted leap occurred (i.e., where the transformation from one expression to the other could no longer be justified).

I find myself in a similar position analyzing Schneider's article and Darwinian scenarios like his. Schneider purports to have generated biologically relevant information, and thus CSI, for free (or "from scratch" as he prefers). The No Free Lunch theorems, however, tell us this is not possible. Where, then, has he smuggled in CSI? The precise place where he smuggles it in is not hard to find if one knows what to look for. Here is the crucial paragraph in his article:

> The organisms [i.e., the computational sequences in phase space] are subjected to rounds of selection and mutation. First, the number of mistakes made by each organism in the population is determined. Then the half of the population making the least mistakes is allowed to replicate by having their genomes replace ("kill")

the ones making more mistakes. (To preserve diversity, no replacement takes place if they are equal.) At every generation, each organism is subjected to one random point mutation in which the original base is obtained one-quarter of the time.[78]

Within this crucial paragraph, the crucial sentence is: "The number of mistakes made by each organism in the population is determined." Who or what determines the number of mistakes? Clearly, Schneider had to program any such determination of number of mistakes into his simulation. Moreover, the determination of number of mistakes is the key defining feature of his fitness function. For this function optimal fitness corresponds to minimal number of mistakes.

We have seen all this before, to wit, in Richard Dawkins's METHINKS IT IS LIKE A WEASEL simulation (see section 4.1). To be sure, Schneider's simulation is more subtle. But the parallels are unmistakable. Like Dawkins's simulation, Schneider's simulation starts with a randomly given "genome" and requires no further intervention. Unlike Dawkins's simulation, Schneider's does not identify an explicitly given target sequence. Even so, it identifies target sequences implicitly through the choice of fitness function. Moreover, by tying fitness to number of mistakes, Schneider guarantees that the gradients of his fitness function rise gradually and thus that his evolutionary algorithm converges in short order to an optimal computational sequence (optimality being defined in relation to his fitness function). Although once the algorithm starts running there is no intervention on the part of the investigator, it is not the case that Schneider did not intervene crucially in structuring the fitness function. He did, and this is where he smuggled in the CSI that he claimed to obtain from scratch.

Schneider has responded to this criticism. His response is in two parts. Ironically the parts cancel each other. First Schneider contends that there is no fitness function: "I generally do not find 'fitness' to be a useful concept. In the ev program [his computer simulation] is no fitness function and the word 'fitness' does not appear in the paper. Unlike most biologists I dispense with the concept of a fixed 'fitness function.' . . . At best there is only 'relative fitness' in a changing environment. That is, whomever [sic] makes the fewest mistakes in the current environment is likely to survive."[79] This last statement is a tautology. It says that the survivors are the fittest (according to some apparently inexpressible notion of "relative fitness") and that the fittest are the survivors. Furthermore, it sidesteps the issue of who or what counts the mistakes by suggesting that there really is no fitness function but that there are nevertheless mistakes to be counted and that nature has no problem counting those mistakes. But if nature has no problem counting mistakes and somehow scoring the count monotonically with respect to survivability, how can Schneider argue that there is no fitness function?

The second part of Schneider's response is therefore to admit that the counting of mistakes does occur after all (though he refuses to refer to the counting function as a fitness function). Nevertheless, he is quick to deny that this counting of mistakes is in any way artificial: "Counting of the number of mistakes matches what happens in nature, as described above. I only claim that the ev simulation matches what happens in nature in essential points. If Dembski finds that this produces information, then he will understand that the simulation shows that *information can be generated in nature solely by replication, mutation and selection.*"[80] Yet if the counting of mistakes matches what happens in nature in essential points, then the obvious conclusion is that nature is chock-full of design and that replication, mutation, and selection are merely instruments for expressing that design. Indeed, the error-counting function in Schneider's evolutionary algorithm is anything but natural. Rather, it is fully contrived to make his simulation achieve the desired end. It is an instance of complex specified information. Consequently, if Schneider's simulation matches nature in essential points, his insertion of complex specified information into his simulation must mirror the insertion of complex specified information into nature. Schneider resists this conclusion. Indeed, he refuses to distinguish information simpliciter from complex specified information (he even asserts that there is no distinction).[81] Schneider's refusal to recognize this distinction is hardly an argument against it. There is indeed a fundamental distinction between these two types of information, and Schneider's error-counting function (I will forgo calling it a fitness function) is where he inserts complex specified information into his evolutionary algorithm.

Schneider's choice of error-counting function is the most obvious place where he smuggles in CSI. But there are others. In the *Nucleic Acids Research* article we have been discussing, he does not list the source code for the program underlying his simulation. For that code he refers readers to the relevant web address. The source code is revealing and shows that Schneider had to do a lot of fine-tuning to his evolutionary algorithm to make his simulation come out right. For instance, in the crucial paragraph from his article that I quoted above, Schneider remarks parenthetically: "To preserve diversity [of organisms], no replacement takes place if [the number of mistakes is] equal." Schneider's Pascal source code (which appears not in his *Nucleic Acids Research* article but at a separate web location) reveals why: "SPECIAL RULE: if the bugs have the same number of mistakes, reproduction (by replacement) does not take place. This ensures that the quicksort algorithm does not affect who takes over the population. (1988 October 26) Without this, the population quickly is taken over and evolution is extremely slow!"[82] Schneider is here fine-tuning his evolutionary algorithm to obtain the results he wants. All such fine-tuning amounts to investigator interference smuggling in complex specified information.[83]

In the case of computer simulations, following the information trail and finding the place where complex specified information was smuggled in is usually not difficult. I predict it will become more difficult in the future as this shell game becomes more sophisticated, involving more shells and quicker movements of the shells. But just as accounts where profits and losses cannot be squared with receivables contain an error in addition or subtraction somewhere, so simulations that claim to generate complex specified information from scratch contain an unwarranted insertion of preexisting complex specified information. With simulations all that is needed is to follow the information trail and find the point of insertion. That may be complicated, but the entire trail is surveyable and will eventually yield to sustained analysis—it is not as though we are missing any crucial piece of the puzzle.

The same cannot be said for actual biological examples. Consider, for instance, a proposed counterexample to my claim that evolutionary algorithms cannot generate specified complexity. This counterexample was much discussed on the Internet in February and March 2000.[84] The counterexample concerns the gene T-urf13 and its protein product URF13. The two are found in the mitochondria of Texas cytoplasmic-male-sterile maize (cms-T). URF13 is a protein 113 amino acids long. It forms three membrane-spanning alpha helices and a channel in the mitochondrial membrane. The problem is that T-urf13, the gene that encodes URF13, was produced by recombining non-protein-coding gene segments only. What's more, most of the sequence is homologous to a nearby ribosomal RNA gene (rrn26). It therefore appears that a biologically functional 113-amino acid protein formed de novo, and thus that biological specified complexity can arise purely by natural causes after all.[85]

But let us consider this claim more closely. First off, T-urf13 appears not to be doing Texas cytoplasmic-male-sterile maize any good—its protein product URF13 renders the plant sterile and increases its susceptibility to fungal toxins. What's more, any time one strings together a sequence of amino acids, one is likely to obtain some three-dimensional structure that includes alpha helices since these are easy to form. URF13's function is therefore deleterious and not all that well defined. Also its form does not appear carefully adapted to its function. URF13 has 113 amino acids. It is therefore one of $20^{113}$ possible proteins sequences of length 113. Since $20^{113}$ is approximately $10^{147}$, URF13's improbability of 1 in $10^{147}$ does not fall below the universal probability bound of $10^{-150}$. What's more, the minimal functional size of URF13 is 83 amino acids since the last 30 are not needed for function. Since $20^{83}$ is approximately $10^{107}$, the improbability of URF13 is now at 1 in $10^{107}$. This is still uncomfortably small, but well above the universal probability bound. We can increase this probability still further by considering the mutational stability of these 83 amino acids.[86] Some swapping of amino acids retains function, thereby increasing the probability of proteins performing the same function as URF13. At issue is not the individual improbability of URF13 but the improbability of getting it or some homologous sequence that performs the same function (see section 5.10).

It seems then that we have averted the challenge of URF13 to the naturalistic generation of complex specified information, though just barely. But what if we came upon a longer protein that was more specific and did its host organism some evident good? What if that protein resulted from a gene that in turn resulted from recombining portions of DNA all of which were non-protein-coding gene segments? What if any way we sliced it, the improbabilities computed turned out to be less than the universal probability bound? Would that demonstrate that CSI had been naturalistically generated? No. First off, there is no reason to think that non-protein-coding gene segments themselves are truly random—as noted above, T-urf13, which is composed of such segments, is homologous to ribosomal RNA. So it is not as though these segments were produced by sampling an urn filled with loosely mixed nucleic acids. What's more, it is not clear that recombination is itself truly random. All recombinations that are supposed to confirm the naturalistic generation of CSI do, after all, occur within a cellular context. The CSI—if indeed it is CSI—that we see in genes produced from recombining non-protein-coding gene segments could just be CSI that had gone underground and now has been reconstituted. Unlike computer simulations, following the information trail for actual biological systems in the wild is rarely possible and depends on contingencies that may forever lie beyond the veil of history. Nonetheless, the mathematics underlying CSI is clear—you cannot get it via chance and necessity. This does not mean that we reflexively trust mathematics over biology. But it does mean giving both their due.

I want next to consider still another challenge that purports to show how CSI can be obtained naturalistically. The challenge in this instance focuses neither on computer simulations nor on actual biological systems in the wild, but on carefully controlled experiments with biopolymers. Here is the challenge as it has been put to me in several unsolicited emails over the Internet:

> For selection to produce some innovation that is both complex and specific would demolish your hypothesis. In fact selection can do just that. Consider in vitro RNA evolution [N.B.: the actual type of biopolymer used is unimportant; RNA is the fashion these days]. Using only a random pool of RNAs (none of them designed), we can select for RNAs that perform a certain highly specified function. They can be selected to bind to any molecule of choice with high specificity or to catalyze a highly specific reaction. This is molecular specified information, by anyone's definition. We have thus empirically seen that highly specific information can be generated in a molecule without designing the molecule. Information theory just has to catch up with what we know from experiment.
>
> At the beginning of a SELEX experiment, for instance, you have a random pool of RNAs that cannot do much at all. At the end you have a pool of RNAs that can perform a complex specified function, such as catalyze a specific reaction or bind a specific molecule. In other words, there is an increase in net CSI through the course of the experiment. The pool of molecules you get at the end of the experiment were never designed. To the contrary, the scientist has no clue as to the identity of their sequence or structure. An extensive effort usually follows a SELEX experiment to characterize the evolved RNA. The RNA must be sequenced, and in some cases it is crystallized and the structure is solved. Only then does the scientist know what was created, and how it performs its complex specific function.[87]

In no way do SELEX, ribozyme engineering, or similar experimental techniques falsify the Law of Conservation of Information or circumvent the No Free Lunch theorems. In SELEX experiments large pools of randomized RNA molecules are formed by intelligent synthesis and not by chance—there is no natural route to RNA. These molecules are then sifted chemically by various means for catalytic function. What's more, the catalytic function is specified by the investigator. Those molecules showing some activity are isolated and become templates for the next round of selection. And so on, round after round. At every step in both SELEX and ribozyme (catalytic RNA) engineering experiments generally, the investigator is carefully arranging the outcome, even if he or she does not know the specific sequence that will emerge. It is simply irrelevant that the investigator is ignorant of the identity and structure of the evolved ribozyme and must determine it after the experiment is over. The investigator first had to specify a precise catalytic function, next had to specify a fitness measure gauging degree of catalytic function for a given biopolymer, and finally had to run an experiment optimizing the fitness measure. Only then does the investigator obtain a biopolymer exhibiting the catalytic function of interest. In all such experiments the investigator is inserting CSI right and left, most notably in specifying the fitness measure that gauges degree of catalytic function. Once it is clear what to look for, following the information trail in such experiments is straightforward.[88]

I want now to step back and consider why researchers who employ evolutionary algorithms might be led to think that these algorithms generate specified complexity as a free lunch. The mathematics, as we have seen, is against specified complexity arising de novo from any nontelic process. What's more, the three counterexamples considered in this section that purport to show how specified complexity can arise as a free lunch are readily refuted once one follows the information trail and, as it were, audits the books. Even so, there is something oddly compelling and almost magical about the way evolutionary algorithms find solutions to problems where the solutions are not like anything we have imagined.[89] A particularly striking example is the "crooked wire genetic antennas" of Edward Altshuler and Derek Linden.[90] The problem these researchers solved with evolutionary (or genetic) algorithms was to find an antenna that radiates equally well in all directions over a hemisphere situated above a ground plane of infinite extent. Contrary to expectations, no wire with a neat symmetric geometric shape solves this problem. Instead, the best solutions to this problem look like zigzagging tangles.[91] What's more, evolutionary algorithms find their way through all the various zigzagging tangles—most of which do not work—to one that actually does. This is remarkable. Even so, the fitness function that prescribes optimal antenna performance is well-defined and readily supplies the complex specified information that an optimal crooked wire genetic antenna seems to acquire for free.

Perhaps the most subtle example I know of an evolutionary algorithm that appears to generate specified complexity for free is the evolutionary checker program of Kumar Chellapilla and David Fogel. As James Glanz reported in the *New York Times*, "Knowing only the rules of checkers and a few basics, and otherwise starting from scratch, the program must teach itself how to play a good game without help from the outside world—including from the programmers."[92] The program is an evolutionary algorithm that searches a space of checker-playing neural nets. In the initial work of Chellapilla and Fogel in 1999, their program found neural nets that play checkers one or two notches below the level of expert.[93] Since then, "with longer evolutionary trials and the inclusion of a preprocessing layer to let the neural network learn that the game is played on a two-dimensional board, rather than a one-dimensional 32-element vector," the program found neural nets that attain the level of expert.[94] The program is therefore able to find neural nets that play checkers at a level far superior to most humans. What is remarkable about this program is that it attained such a high level of play without having to be explicitly programmed with expert knowledge like the world champion chess program Deep Blue or the world champion checker program Chinook.[95]

But did the evolutionary checker program of Chellapilla and Fogel find its expert checker-playing neural nets without commensurate input from prior intelligence? To be sure, a good deal of knowledge was inserted into the representation of the neural nets. For instance, a preprocessing layer of ninety-one neurons took inputs from each square subregion of the checker board ($3 \times 3$, $4 \times 4$, etc.). The preprocessing was therefore adapted specifically to the two-dimensional geometry of the board—a natural enough move, but a constraining choice nonetheless. Even so, apart from how it represented neural nets, the program seemed not to be incorporating any special knowledge or prior input of intelligence. The program was run for 840 generations. Each generation consisted of a tournament involving 30 neural nets. Within a given generation, each neural net played 5 games as red (the color that moves first) against randomly selected (with replacement) opponents playing white (thus on average each net played 10 games). Scoring assigned $+1$ to a win, 0 for a draw, and $-2$ for a loss. The program kept the top 15 neural nets at each generation and made 15 offspring (1 per parent) by randomly varying all weights as well as each neural net's king value.[96]

Chellapilla and Fogel's evolutionary checker program therefore appears to have generated expert checker-playing neural nets without a commensurate input of prior intelligence. Indeed, everything seems to be happening locally and without any top-down control. There is not even a fitness function defined over the entire space of checker-playing neural nets. Instead, each collection of 30 neural nets gets its own local fitness function that assigns fitness depending on how a neural net fares in a tournament with other neural nets (tournament pairings being random as described in the last paragraph). At the end of each tournament, 15 winners out of 30 contestants are selected—not enough of a reduction of possibilities to generate specified complexity in a single tournament. Yet over successive tournaments qua generations, specified complexity does appear to be generated. Since specified complexity is not a free lunch, there must be an insertion of prior specified complexity. But where was it inserted?

The answer is simple though not obvious: Prior specified complexity was inserted via the *coordination* of local fitness functions. It is important to understand that there is nothing requiring one local fitness function defined for 30 neural nets to match up with another local fitness function defined for another 30 neural nets. A local fitness function, as it were, hands off winning neural nets satisfying its criterion of success to another local fitness function imposing the same criterion of success. Chellapilla and Fogel kept the criterion of winning constant from one set of neural nets to the next. But this was a choice on their part. To be sure, it was the appropriate choice given that they were trying to optimze checker playing. But it was a choice nonetheless. Indeed, it was a choice that inserted an enormous amount of specified complexity (the space of all possible combinations of local fitness functions from which they chose their coordinated set of local fitness functions is enormous). Also, their choice is without a natural analogue. Chellapilla and Fogel kept constant their criterion for "tournament victory." For biological systems, the criterion for "tournament victory" will vary considerably depending on who is playing in the tournament.

In closing this section I want to draw a pair of lessons. Indeed, both intelligent design and evolutionary algorithms have a lesson to learn from each other. The No Free Lunch theorems show that for evolutionary algorithms to output CSI they had first to receive a prior input of CSI. And since CSI is reliably linked to intelligence, evolutionary algorithms, insofar as they output CSI, do so on account of a guiding intelligence. The lesson, then, for evolutionary algorithms is that any intelligence these algorithms display is never autonomous but always derived. On the other hand, evolutionary algorithms do produce remarkable solutions to problems—solutions that in many cases we would never have imagined on our own. Having been given some initial input of CSI, evolutionary algorithms as it were mine that CSI and extract every iota of value from it. The lesson, then, for intelligent design is that natural causes can synergize with intelligent causes to produce results far exceeding what intelligent causes left to their own abstractions might ever accomplish. Too often design is understood in a deterministic sense in which every aspect of a designed object has to be preordained by a designing intelligence. Evolutionary algorithms underwrite a nondeterministic conception of design in which design and nature operate in tandem to produce results that neither could produce by itself.[97] I close with a quote by Michael Polanyi very much in this spirit (see also section 6.5):

It is true that the teleology rejected in our day is understood as an overriding cosmic purpose necessitating all the structures and occurrences in the universe in order to accomplish itself. This form of teleology is indeed a form of determinism— perhaps even a tighter form of determinism than is provided for by a materialistic, mechanistic atomism. However, since at least the time of Charles Saunders Peirce and William James a looser view of teleology has been offered to us—one that would make it possible for us to suppose that some sort of intelligible directional tendencies may be operative in the world without our having to suppose that they *determine* all things. Actually it is possible that even Plato did not suppose that his "Good" *forced* itself upon all things. As Whitehead has pointed out, Plato tells us that the Demiurge, looking toward the Good, "persuades" an essentially free matter to structure itself, to some extent, in imitation of the Forms. Plato appeared to Whitehead to have modeled the cosmos on a struggle to achieve the Good in the somewhat recalcitrant media of space and time and matter, a struggle well known to all souls with purposes and ends and aims. Whether or not it is true that Plato did this, certainly Whitehead modeled his *own* cosmos very much this way.[98]

## 4.10   Coevolving Fitness Landscapes

There is yet one remaining exit strategy for trying to circumvent the displacement problem, and that is Stuart Kauffman's proposal of coevolving fitness landscapes. Kauffman fully appreciates the challenge that the displacement problem (in the guise of NFL) raises for evolution:

> The no-free-lunch theorem says that, averaged over all possible fitness landscapes, no search procedure outperforms any other. . . . In the absence of any knowledge, or constraint, on the fitness landscape, on average, any search procedure is as good as any other. But life uses mutation, recombination, and selection. These search procedures seem to be working quite well. Your typical bat or butterfly has managed to get itself evolved and seems a rather impressive entity. The no-free-lunch theorem brings into high relief the puzzle. If mutation, recombination, and selection only work well on certain kinds of fitness landscapes, yet most organisms are sexual, and hence use recombination, and all organisms use mutation as a search mechanism, where did these well-wrought fitness landscapes come from, such that evolution manages to produce the fancy stuff around us?[99]

According to Kauffman, "No one knows."[100] Nonetheless, Kauffman offers a proposal for how such well-wrought fitness landscapes (or fitness functions as we have been calling them) might have come about.

Before describing Kauffman's proposal, I want to reiterate what I stressed in section 4.8, namely, that it is a separate and prior question whether the fitness functions upon which the Darwinian mechanism operates exercise sufficient control over the evolutionary process to account for all of biological complexity. In the next chapter I will argue that they do not, and in particular that the irreducible complexity of certain biochemical systems argues decisively against the gradients of these fitness functions (or fitness landscapes) being smooth enough to make the Darwinian mechanism the driving force behind evolution. Note that in offering such an argument I do not challenge evolution as such but the sufficiency of the Darwinian mechanism to account for it. Even so, in this section we assume for the sake of argument that nature provides us with the right fitness functions to make a specifically Darwinian form of evolution completely successful—in other words, we assume the fitness functions are just what they need to be for the Darwinian mechanism to account for all of evolutionary change.

Kauffman is therefore right to observe that the crucial problem is how nature happened to provide just the right fitness functions to make evolution work. First off, let us be clear that Kauffman provides no solution to this problem. Kauffman's proposal of coevolving fitness landscapes is a proposal for where to look for a solution but is not itself a solution. Kauffman admits this: "The strange thing about the theory of evolution is that everyone thinks he understands it. But we do not. A biosphere, or an econosphere, self-consistently coconstructs itself according to principles we do not yet fathom."[101] What then is Kauffman's proposal? I quote his proposal at length both because it is instructive and because it is sufficiently detailed to allow itself to be critiqued and refuted:

> The no-free-lunch theorem led me to wonder about the following: We organisms use mutation, recombination, and selection in evolution, and we pay twofold fitness for sex and recombination to boot. But recombination is only a useful search procedure on smooth enough fitness landscapes where the high peaks snuggle rather near one another. In turn, this led me to wonder where such nice fitness landscapes arise in evolution, for not all fitness landscapes are so blessedly smooth. Some are random. Some are anticorrelated. In turn, this led me to think about and discuss natural games, or ways of making a living. Since ways of making a living evolve with the organisms making those livings, we got to the winning games are the games the winners play. Which led me to suggest that those ways of making a living that are well searched out and exploited by the search mechanisms organisms happen to use—mutation, recombination, and selection—will be ways of making a living that are well-populated by organisms and similar species. Ways of making a living that cannot be well searched out by organisms and their mutation recombination search procedures will not be well populated. So we came to the reasonable conclusion that a biosphere of autonomous agents is a self-consistently self-constructing whole, in which agents, ways of making a living, and ways of searching for how to make a living all work together to coconstruct the biosphere. Happily, we are picking the problems we can manage to solve. Of course, if we could not solve our chosen ways of making livings, we would be dead.[102]

Thus, according to Kauffman, organisms undergo biological evolution by "tuning landscape structure (ways of making a living) and coupling between landscapes."[103] We can think of Kauffman's proposal as follows. For an ordinary evolutionary algorithm $E$, the phase space $\Omega$ and fitness function f are fixed. In this case the evolutionary algorithm is successful provided it reaches a target T where f attains a certain level of fitness. According to Kauffman this static view of evolutionary algorithms needs to be replaced by a dynamic view in which the fitness function changes as the evolutionary algorithm runs through the phase space. In this case the evolutionary algorithm $E$ starts at $E_1$ in $\Omega$ conditional upon an initial fitness function $f_1$. Because $E_1$ happened, $E_1$'s environment is now changed. Consequently, what it means for some organism to be fit in that new environment is no longer determined by $f_1$ but by a new fitness function $f_2$. Conditional upon $f_2$, $E$ now produces $E_2$. This in turn changes the environment and induces a new fitness function $f_3$. And so on. $E$ is regarded as successful if after m steps $E_m$ has reached a place where $f_m$ attains a certain level of fitness. This is what Kauffman means by coevolving fitness landscapes.

Although such coevolving fitness landscapes seem to complicate our understanding of evolutionary algorithms and open the door to new mechanisms for generating specified complexity, in fact they introduce nothing new and fail to resolve the displacement problem. Suppose we are given a coevolving fitness landscape that is defined with respect to the evolutionary algorithm $E$, the phase space $\Omega$, and the fitness functions $f_i$ for $1 \leq i \leq m$. We can then define a new coevolutionary phase space $\Omega'$ as the Cartesian product $\Omega \times J$ where $\Omega$ is the old phase space and $J$ is the set of all fitness functions on $\Omega$. Moreover, we can then define a new evolutionary algorithm $E'$ on $\Omega \times J$ such that $E'$ takes values $(E_i, f_i)$ characterizing the coevolution of the previous evolutionary algorithm and fitness functions. The problem of coevolving fitness landscapes now becomes the problem of optimizing a higher-order fitness function, call it F, on $\Omega \times J$ that assigns fitness for ordered pairs (x,f) of $\Omega \times J$ according to how successful the first element x in $\Omega$ is at "making a living" with respect to the second element, the fitness function f. But this F sits in some information-resource space $J'$ for the phase space $\Omega \times J$ and itself exhibits specified complexity with respect to $J'$. We are therefore back to an ordinary evolutionary algorithm with all the problems of displacement that such algorithms raise.

In section 4.8 we saw that the catalogue of conditions is vast that a fitness function induced by differential survival and reproduction needs to satisfy if the spectacular diversity of living forms that we see on earth is properly to be attributed to a Darwinian form of evolution. How much more vast, then, is the catalogue of conditions that a higher-order fitness function induced by the coevolution of fitness landscapes needs to satisfy if in evolving, organisms "make their living" by exploiting coevolving fitness landscapes? Such a catalogue is going to require a vast amount of specified complexity, and this specified complexity will be reflected in the higher-order fitness function (i.e., F) defined on the coevolutionary phase space (i.e., $\Omega \times J$). This fitness function, as with the one induced by differential survival and reproduction, is nonarbitrary (as Kauffman puts it, "if we could not solve our chosen ways of making livings, we would be dead"[104]) but, as Kauffman seems less ready to admit, is also not a free lunch. Throw together communities of autonomous agents in Kauffman's sense and let them evolve to optimize "ways of making a living," and there is no reason to expect you will get anything interesting or anything that grows in complexity over time. Most "ways of making a living" stress dull routine and simplicity, stripping away frills and avoiding costly increases in complexity (in contrast to the emergence of sexual reproduction). Thus I submit that even if coevolution of fitness landscapes is the means by which the panoply of life on earth came to be, any higher-order fitness function on a coevolutionary phase space that facilitates biological evolution would not be a free lunch and not a brute given, but a finely crafted assemblage of peaks, valleys, and inclines that together presuppose much prior specified complexity.

Kauffman's dynamic view of evolutionary algorithms in terms of coevolving fitness landscapes is therefore mathematically equivalent to the old static view of evolutionary algorithms in which the fitness landscape (or fitness function) is fixed. Kauffman's proposed exit strategy therefore lands us right back at the problem that motivated his exit strategy in the first place, namely, How does one account for the well-wrought fitness functions that are needed to make evolutionary algorithms work? He has not gotten rid of this problem. By letting fitness functions coevolve with evolving trajectories through the original phase space $\Omega$, he has in fact changed the phase space and imposed on it a higher-order fitness function that characterizes how successful organisms are at making a living with respect to the coevolving lower-order fitness functions. All the problems with evolutionary algorithms that adopt a fixed fitness function therefore remain. Well-wrought fitness functions that make for interesting evolutionary pathways invariably exhibit specified complexity. Coevolving fitness landscapes merely displace the specified complexity to a higher order fitness function.

In closing I want to preclude one last loophole that Kauffman seems to have left himself. Throughout his work on coevolving fitness landscapes Kauffman stresses that the phase spaces relevant to biology are not finitely prestatable.[105] By this he means that we cannot explicitly state or identify all the biological possibilities that might emerge in the natural world over time. This limitation seems to leave open the possibility of evolutionary algorithms generating specified complexity after all by allowing not just fitness landscapes but also the very phase spaces themselves to evolve. The idea of evolving phase spaces, however, seems misconceived (in fairness to Kauffman, he does not talk about evolving phase spaces; I am merely trying to preclude a possible direction where his ideas might lead). The phase spaces scientists identify in their research are mathematical constructs that depend on the state of scientific knowledge. On the other hand, nature is a given that does not allow infinite free play but places definite constraints on the arrangements that mass-energy can take and what such arrangements can do. The phase spaces relevant to biology reflect nature or various aspects of nature. Thus, while it may be legitimate to say that the phase spaces we use to represent nature evolve (how felicitous this manner of speaking is, is another matter), nature herself allows only limited possibilities, and these do not evolve. Such inherent limitations constrain all our phase spaces that model nature and give us no reason to think that our assessment of the inability of evolutionary algorithms to generate complex specified information may need to be revised in light of further knowledge about such phase spaces. It follows that coevolving fitness landscapes, with or without finitely prestatable phase spaces, offer no resolution of the displacement problem and thus no account of how complex specified information is generated for free.

# Notes

1. Ian Stewart, *Life's Other Secret: The New Mathematics of the Living World* (New York: John Wiley, 1998), 48.

2. More realistic assessments of the origin of life problem can be found in the work of Robert Shapiro and Hubert Yockey. See Robert Shapiro, *Origins, A Skeptics Guide to the Creation of Life on Earth* (New York: Summit Books, 1986); and Hubert Yockey, *Information Theory and Molecular Biology* (Cambridge: Cambridge University Press, 1992), chs. 8, 9, and 10. See also Charles Thaxton, Walter Bradley, and Roger Olsen, *The Mystery of Life's Origin: Reassessing Current Theories* (New York: Philosophical Library, 1984); and Gordon Mills and Dean Kenyon, "The RNA World: A Critique," *Origins & Design* 17(1) (1996): 9–14.

3. Davies claims that we are "a very long way from comprehending" how life originated. "This gulf in understanding is not merely ignorance about certain technical details, it is a major conceptual lacuna. . . . My personal belief, for what it is worth, is that a fully satisfactory theory of the origin of life demands some radically new ideas." Paul

Davies, *The Fifth Miracle: The Search for the Origin and Meaning of Life* (New York: Simon & Schuster, 1999), 17.

4. Ibid., 115–122. See also Michael Polanyi, "Life Transcending Physics and Chemistry," *Chemical and Engineering News* (21 August 1967): 55–66; and Michael Polanyi, "Life's Irreducible Structure," *Science* 113 (1968): 1308–1312.

5. Davies, *Fifth Miracle*, 112. Consider also the following claim by Leslie Orgel: "Living organisms are distinguished by their specified complexity. Crystals such as granite fail to qualify as living because they lack complexity; mixtures of random polymers fail to qualify because they lack specificity." In Leslie Orgel, *The Origins of Life* (New York: John Wiley, 1973), 189.

6. Davies, *Fifth Miracle*, 120. Consider also from that same book: "Natural selection . . . acts like a ratchet, locking in the advantageous errors and discarding the bad. Starting with the DNA of some primitive ancestor microbe, bit by bit, error by error, the increasingly lengthy instructions for building more complex organisms came to be constructed" (42). Or, "The environment feeds the information into the genetic message via natural selection" (57).

7. Theodosius Dobzhansky, Discussion of G. Schramm's paper, in *The Origins of Prebiological Systems and of Their Molecular Matrices*, ed. S. W. Fox (New York: Academic Press, 1965), 310.

8. Stuart Kauffman, *At Home in the Universe: The Search for the Laws of Self-Organization and Complexity* (New York: Oxford University Press, 1995), 150. Note that Kauffman himself dissents from this majority view.

9. The definition of evolutionary algorithms given here is more general than is customary. For a popular exposition of the types of search strategies included here under evolutionary algorithms, consult Peter Coveney and Roger Highfield, *Frontiers of Complexity: The Search for Order in a Chaotic World* (New York: Fawcett Columbine, 1995). For the connection between organic evolution and evolutionary algorithms, see Thomas Bäck, *Evolutionary Algorithms in Theory and Practice: Evolution Strategies, Evolutionary Programming, Genetic Algorithms* (New York: Oxford University Press, 1996), ch. 1.

The training of neural nets can be assimilated to evolutionary algorithms as follows: Let $\Omega$, the phase space, be the set of all possible graphs (simple, directed, or multiply connected) on a fixed number of N nodes with different weights attached to the edges connecting the nodes (in the simplest case the weights are all either 0 or 1). Now put a fitness function over $\Omega$ that assesses how well each graph (i.e., neural net) performs some task (e.g., performs a visual recognition task). Training the neural net then means running an evolutionary algorithm to optimize the fitness function over the phase space.

10. Richard Dawkins, *The Blind Watchmaker* (New York: Norton, 1986), 47–48.

11. For an event of probability p to occur at least once in k trials has probability $1 - (1-p)^k$—see Geoffrey Grimmett and David Stirzaker, *Probability and Random Processes* (Oxford: Clarendon, 1982), 38. If p is small and $k = 1/p$, then this probability is greater than $1/2$. But if k is much smaller than $1/p$, this probability will be quite small (i.e., close to 0). These probabilities will be considered at much greater length later in this chapter. For now it is enough to be aware as a rule of thumb that an event of probability p requires around $1/p$ trials to become reasonably probable.

12. Dawkins, *Blind Watchmaker*, 48.

13. See Bernd-Olaf Küppers, "On the Prior Probability of the Existence of Life," in *The Probabilistic Revolution*, vol. 2, eds. L. Krüger, G. Gigerenzer, and M. S. Morgan (Cambridge, Mass.: MIT Press, 1987), 365–369. According to Küppers (367), simulation experiments like Dawkins's show that "meaningful information can indeed arise from a meaningless initial sequence by way of random variation and selection. Since the appearance of mutants is, on the genetic level, completely indeterminate, the process of natural selection lays down a general gradient of evolution, but not the detailed path by which the local maximum will be reached."

14. According to Bernd-Olaf Küppers ("On the Prior Probability of the Existence of Life," 367–368), "Darwinian theory predicts a priori only the emergence of information in general, but not the detailed structure of this information. In consequence, the neo-Darwinistic view of the origin of life attempts not to reconstruct the historical course of this process, but simply to uncover its fundamental laws and principles that can be expressed in the language of physics."

15. Complexity as I am using it here is in the information-theoretic or Shannon sense. There are lots and lots of different complexity measures. Seth Lloyd records over thirty (see John Horgan, *The End of Science* [New York: Broadway Books, 1996], 303, n. 11). John Horgan regards this abundance of complexity measures as a bad thing (194–198), but it is not. Having many "flavors of complexity" does not subjectivize the notion. Just as we need many types of measures in daily life (volumes, densities, weights, lengths, times, etc.), so we need many different complexity measures to measure the diverse types of complication associated with diverse structures.

16. The (Shannon) information $I(A)$ associated with an event A is by definition $-\log_2 P(A)$, where $P(A)$ is the probability of that event and the logarithm is taken to the base 2. It follows that $I(A)$ equals zero if and only if $P(A)$ equals one.

17. Dawkins, *Blind Watchmaker*, 1.

18. In this regard I have already cited Küppers, "On the Prior Probability of the Existence of Life," 367. Recently Jeffrey Satinover, who does not seem wedded to Darwinism, has offered a variant of Dawkins's METHINKS•IT•IS•LIKE•A•WEASEL example. In *The Quantum Brain: The Search for Freedom and the Next Generation of Man* (New York: Wiley, 2001), 89–92, Satinover purports to demonstrate the power of evolutionary algorithms by showing how such an algorithm could generate the target phrase MONKEYS-WROTESHAKESPEARE. Satinover's algorithm is quite similar to Dawkins's except that Satinover utilizes a few more techniques from the evolutionary algorithms toolchest (specifically crossover and mating). In thus jazzing up Dawkins's algorithm, Satinover requires on average ninety iterations instead of Dawkins's forty to produce the target phrase. But Satinover's target phrase was there from the start: "We define the fitness of a sequence as the sum of the distances of each character (on a keyboard) from the correct one. . . ." (90) Thus the "correct sequence" was there all the time, and the fitness function was defined specifically with reference to that sequence. It is remarkable how Dawkins's example gets recycled without any indication of the fundamental difficulties that attend it.

19. See, for example, Dervis Karaboga and Duc Truon Pham, *Intelligent Optimization Techniques: Genetic Algorithms, Tabu Search, Simulated Annealing, and Neural Networks* (New York: Springer-Verlag, 2000); Toshihide Ibaraki, *Resource Allocation Problems: Algorithmic Approaches* (Cambridge, Mass.: MIT Press, 1988); John Horton Conway and N. J. A. Sloane, *Sphere Packings, Lattices, and Groups*, 3rd ed. (New York: Springer-Verlag, 1998).

20. For instance, Henry Petroski, writing about the optimization of design, notes, "All design involves conflicting objectives and hence compromise, and the best designs will always be those that come up with the best compromise." Quoted from *Invention by Design: How Engineers Get from Thought to Thing* (Cambridge, Mass.: Harvard University Press, 1996), 30.

21. Actually, one is usually safer erring on the side of including too many possibilities in the reference class, and then weeding them out later. In the sausage example, however, the optimization technique of choice will be linear programming, and for this optimization procedure the possible solutions always constitute a tightly constrained convex set in hyperspace, thereby omitting the type of "far-out" solutions indicated in the text. Linear programming is one of the most widely used of optimization techniques. For a thorough mathematical treatment of linear programming consult David Gale, *The Theory of Linear Economic Models* (New York: McGraw-Hill, 1960). For a less technical account see Frederick S. Hillier and Gerald J. Lieberman, *Introduction to Operations Research*, 5th ed. (New York: McGraw-Hill, 1990).

22. This discussion of real-valued functions falls under what mathematicians call "real analysis." For an introduction to this field, see Walter Rudin, *Principles of Mathematical Analysis*, 3rd ed. (New York: McGraw-Hill, 1976).

23. For a bibliography of the vast literature on multicriteria optimization, see http://ubmail.ubalt.edu/~harsham/refop/Refop.htm (last accessed 25 June 2001).

24. Melanie Mitchell, *An Introduction to Genetic Algorithms* (Cambridge, Mass.: MIT Press, 1996), 124.

25. The precise terminology here is unimportant and varies according to discipline. "Phase space," for instance, occurs in the study of dynamical systems—see Morris Hirsch and Stephen Smale, *Differential Equations, Dynamical Systems, and Linear Algebra* (New York: Academic Press, 1974) or I. P. Cornfeld, S. V. Fomin, and Ya. G. Sinai, *Ergodic Theory* (New York: Springer-Verlag, 1982). Within the genetic algorithms literature the phase space is usually referred to as the "population," takes the form of a data structure, and is referred to generically as the "search space"—see David E. Goldberg, *Genetic Algorithms in Search, Optimization, and Machine Learning* (Reading, Mass.: Addison-Wesley, 1989), 60; and Mitchell, *An Introduction to Genetic Algorithms*, 6, 8. Moreover, the elements of phase space are often referred to as "chromosomes" (Goldberg, 60; Mitchell, 8). Fitness functions are also referred to as "fitness measures" or "fitness landscapes." Within the field of operations research these functions are called "objective functions"—see Hillier and Lieberman, *Introduction to Operations Research*. The target or solution set tends not to be explicitly named as such in the genetic algorithms literature, being instead merely identified as where in the phase space a fitness function attains a certain level of fitness.

26. See Albert Wilansky, *Topology for Analysis* (Malabar, Fla.: Krieger, 1983), 12.

27. For a general account of uniform probability see my article "Uniform Probability," *Journal of Theoretical Probability* 3(4) (1990): 611–626.

28. A universal probability bound is a level of improbability that precludes specified events below that level from occurring by chance in the observable universe. Emile Borel proposed $10^{-50}$ as a bound below which probabilities could be neglected universally (i.e., neglected across the entire observable universe). See Emile Borel, *Probabilities and Life*, trans. M. Baudin (New York: Dover, 1962), 28 and Eberhard Knobloch, "Emile Borel as a Probabilist," in *The Probabilistic Revolution*, vol. 1, eds. L. Krüger, L. J. Daston, and M. Heidelberger, 215–233 (Cambridge, Mass.: MIT Press, 1987), 228. In *The Design Inference* (Cambridge: Cambridge University Press, 1998) I justify a more stringent universal probability bound of $10^{-150}$ based on the number of elementary particles in the observable universe, the duration of the observable universe, and the Planck time. See Dembski, *The Design Inference*, sec. 6.5. Universal probability bounds also come up in the cryptographic literature, setting a level of probability at which a cryptosystem is judged secure despite all conceivable computational resources that might be arrayed against it. See Kenneth W. Dam and Herbert S. Lin (eds.), *Cryptography's Role in Securing the Information Society* (Washington, D.C.: National Academy Press, 1996), 380, n. 17, where a universal probability bound of $10^{-95}$ is computed. Universal probability bounds presuppose limited probabilistic resources. With unlimited probabilistic resources (e.g., multiple universes) anything, however improbable, becomes certain. Unlimited probabilistic resources have their own problems, however. In particular, they permit attributing any event whatsoever to chance—see section 2.8.

29. Compare section 3.8. For an overview of stochastic processes see Samuel Karlin and Howard Taylor, *A First Course in Stochastic Processes*, 2nd ed. (New York: Academic Press, 1975) and by the same authors *A Second Course in Stochastic Processes* (New York: Academic Press, 1981).

30. The indexing set for a stochastic process need not be confined to the natural numbers. Often the indexing set denotes time and is represented by nonnegative real numbers. For a general treatment of stochastic processes consult the references in the previous note.

31. Though note that pure random sampling and blind search are both evolutionary algorithms. They are just not particularly effective ones for most purposes.

32. In general, an event of probability p has probability $1 - (1-p)^k$ of occurring at least once in k trials. This number approaches 1/2 only as k approaches 1/p, but remains close to 0 as k falls far short of 1/p. See Grimmett and Stirzaker, *Probability and Random Processes*, 38.

33. See Colin P. Williams and Scott H. Clearwater, *Explorations in Quantum Computing* (New York: Springer-Verlag, 1998), 213–218.

34. For a thorough treatment of random walks, see Frank Spitzer, *Principles of Random Walk*, 2nd ed. (New York: Springer-Verlag, 1976).

35. Joseph C. Culberson puts it this way: "Evolutionary algorithms (EAs) are often touted as 'no prior knowledge' algorithms. This means that we expect EAs to perform

without special information from the environment. Similar claims are often made for other adaptive algorithms." From "On the Futility of Blind Search: An Algorithmic View of 'No Free Lunch'," *Evolutionary Computation* 6(2) (1998): 109–127.

36. The set of functions on a given set tends to go up exponentially in size. Thus if a given set has cardinality N (i.e., has N elements), the set of all functions from this set to another set will have cardinality $M^N$ for M the cardinality of the other set. See D. van Dalen, H. C. Doets, and H. de Swart, *Sets: Naïve, Axiomatic and Applied* (Oxford: Pergamon, 1978), 60.

37. The first of these theorems were proven in 1996 by Wolpert and Macready. See David H. Wolpert and William G. Macready, "No Free Lunch Theorems for Optimization," *IEEE Transactions on Evolutionary Computation* 1(1) (1997): 67–82.

38. Culberson, "Futility of Blind Search," 109.

39. With this example as well as with others in this chapter I am being lax about the level of complexity needed to qualify as CSI. Technically, the level of complexity needs to attain at least the universal probability bound of $10^{-150}$ or the corresponding universal complexity bound of 500 bits. But for the purposes of illustration I am allowing less stringent bounds.

40. This account of blind search is consistent with the one given in section 4.3.

41. The simplest way to concentrate a probability measure on a nonempty set is by defining what is called a unit or point mass for some point in the set. A point mass assigns probability 1 to any set containing that point and 0 to any set that does not. See Heinz Bauer, *Probability Theory and the Elements of Measure Theory*, trans. R. B. Burckel, 2nd English ed. (London: Academic Press, 1981), 20.

42. See Dembski, "Uniform Probability."

43. No matter how small $U(T)$ is, provided it does not equal zero, if the sample size m is sufficiently large, the probability of landing in T can be made arbitrarily close to 1, though never exactly 1.

44. See Grimmett and Stirzaker, *Probability and Random Processes*, 38 for their discussion of the geometric distribution.

45. Robert Stalnaker, *Inquiry* (Cambridge, Mass.: MIT Press, 1984), 85.

46. Fred Dretske, *Knowledge and the Flow of Information* (Cambridge, Mass.: MIT Press, 1981), 4.

47. See, for instance, M. A. Naimark and A. I. Stern, *Theory of Group Representations*, trans. E. Hewitt (New York: Springer-Verlag, 1982) or Persi Diaconis, *Group Representations in Probability and Statistics* (Hayward, Calif.: Institute of Mathematical Statistics, 1988).

48. See Hirsch and Smale, *Differential Equations, Dynamical Systems, and Linear Algebra*.

49. For this generic way of formulating NFL theorems, see Culberson, "On the Futility of Blind Search," 111–112.

50. Ibid., 125.

51. The essential idea behind information is the reduction of possibilities from a reference class of possibilities. That is why information theorists define information as the

*reduction* or *resolution of uncertainty.* See John R. Pierce, *An Introduction to Information Theory: Symbols, Signals and Noise,* 2nd ed. (New York: Dover, 1980), 24.

52. Kauffman, *At Home in the Universe,* ch. 4.

53. James R. Munkres, *Topology: A First Course* (Englewood Cliffs, N.J.: Prentice-Hall, 1975), 76.

54. See Heinz-Otto Peitgen, Harmut Jürgens, and Dietmar Saupe, *Chaos and Fractals: New Frontiers of Science* (New York: Springer-Verlag, 1992).

55. William M. Boothby, *An Introduction to Differentiable Manifolds and Riemannian Geometry* (New York: Academic Press, 1975), chap. 4. For a quick overview of vector fields in the simpler context of Euclidean n-space, see the first few (very short) chapters of V. I. Arnold, *Ordinary Differential Equations,* trans. R. A. Silverman (Cambridge, Mass.: MIT Press, 1978).

56. Stephen Meyer has argued this point convincingly. See his article "DNA by Design: An Inference to the Best Explanation for the Origin of Biological Information," *Rhetoric and Public Affairs* 1(4) (1998): 519–556.

57. See Douglas Robertson, "Algorithmic Information Theory, Free Will, and the Turing Test," *Complexity* 4(3) (1999): 25–34. Robertson argues that the defining feature of agents with free will is their ability to create (complex specified) information.

58. Recall the crucial signal in the movie *Contact* that convinced the radio astronomers that they had indeed established "contact" with an extraterrestrial intelligence, namely, a long sequence of prime numbers—see section 1.3.

59. Davies, *The Fifth Miracle,* 17. The subtitle of Stuart Kauffman's *At Home in the Universe* demonstrates quite plainly this impulse to explain specified complexity in terms of laws: *The Search for the Laws of Self-Organization and Complexity.* Note that Kauffman refers explicitly to "the search" for such laws. At present they remain unknown. See also Roger Penrose's *The Emperor's New Mind* (Oxford: Oxford University Press, 1989) and *Shadows of the Mind* (Oxford: Oxford University Press, 1994). Penrose hopes to unravel the problem of human consciousness through unknown quantum-theoretical laws. There are no proposals for what laws that generate specified complexity might look like, much less how they might actually be formulated. The point of this chapter is to argue that no such laws can exist.

60. Dawkins, *Blind Watchmaker,* 50.

61. Richard Dawkins, *Climbing Mount Improbable* (New York: Norton, 1996).

62. Brian Goodwin, *How the Leopard Changed Its Spots: The Evolution of Complexity* (New York: Scribner's, 1994), 35–36.

63. More precisely, f needs to be an evolving fitness function indexed by time. My argument, however, remains intact. See section 4.10.

64. See Paul Davies, *The Mind of God: The Scientific Basis for a Rational World* (New York: Touchstone, 1992), ch. 8, titled "Designer Universe."

65. Michael Denton, *Nature's Destiny: How the Laws of Biology Reveal Purpose in the Universe* (New York: Free Press, 1998).

66. See Stuart Kauffman, *The Origins of Order: Self-Organization and Selection in Evolution* (Oxford: Oxford University Press, 1993), 95–101; see also Manfred Eigen and Peter

Schuster, *The Hypercycle: A Principle of Natural Self-Organization* (New York: Springer, 1979).

67. Stephen Jay Gould, *Full House: The Spread of Excellence from Plato to Darwin* (New York: Harmony Books, 1996), 169–173.

68. See http://www.isd.atr.co.jp/~ray/tierra (last accessed 10 June 2001).

69. See Robert Wright, *Nonzero: The Logic of Human Destiny* (New York: Pantheon, 2000), 4–5.

70. With all this talk of smooth gradients, it is worth offering a definition. Smooth gradients are most easily defined in terms of what mathematicians call a Lipschitz condition, in which distances between points in phase space proportionately constrain distances between their corresponding fitness values. Thus, for a metric d on the phase space $\Omega$, the fitness function f can be defined as smooth provided there is some positive real number k such that for all x and y in $\Omega$, $|f(x) - f(y)| \leq k \cdot d(x,y)$ (known as a Lipschitz condition). The smaller k, the smoother the fitness function f. Although mathematicians often use "smooth" to refer to functions that are infinitely differentiable, in the study of evolutionary algorithms it is more appropriate to adopt this Lipschitz characterization of smoothness, especially since many of the phase spaces we deal with in biology and computation are discrete and thus do not admit differentiability. Typically, however, these spaces do have have a metric and therefore admit a Lipschitz condition. See Tom M. Apostol, *Mathematical Analysis*, 2nd ed. (Reading, Mass.: Addison-Wesley, 1974), 121.

71. http://www.lecb.ncifcrf.gov/~toms (last accessed 10 June 2001).

72. Thomas D. Schneider, "Evolution of Biological Information," *Nucleic Acids Research* 28(14) (2000): 2794–2799.

73. Ibid., 2794.

74. Ibid., 2796.

75. Ibid.

76. Ibid., 2797.

77. Thomas D. Schneider, "Rebuttal to William A. Dembski's Posting," (6 June 2001): http://www.lecb.ncifcrf.gov/~toms/paper/ev/dembski/rebuttal.html (last accessed 8 June 2001). Schneider was responding to my piece "America's Obsession with Design: A Response to Wolfhart Pannenberg," *Metaviews* 3294 (5 June 2001): http://www.metanexus.net (last accessed 26 June 2001).

78. Schneider, "Evolution of Biological Information," 2795.

79. Schneider, "Rebuttal to William A. Dembski's Posting."

80. Ibid.

81. Ibid. Schneider writes, "Shannon used the term 'information' in a precise mathematical sense and that is what I use. I will assume that the extra words 'complex specified' are jargon that can be dispensed with."

82. http://www.lecb.ncifcrf.gov/~toms/delila/ev.html (last accessed 10 June 2001).

83. Schneider objects to me referring to such insertions into his simulation program as "fine-tuning." See Schneider, "Rebuttal to William A. Dembski's Posting" as well as his article "Effect of Ties on the Evolution of Information by the Ev Program," (7 June 2001): http://www.lecb.ncifcrf.gov/~toms/paper/ev/dembski/claimtest.html (last accessed

8 June 2001). Call such insertions what you will. They amount to an insertion of knowledge that within Darwinism has no natural analogue.

84. In the remainder of this section I shall be drawing from comments by "Mike Gene" and "DNAunion" (both pseudonyms) on the Access Research Network Intelligent Design Discussion group at http://www.arn.org/ubb/Forum1/HTML/000066.html (last accessed 10 June 2001).

85. See R. E. Dewey, C. S. Levings III, and D. H. Timothy, "Novel Recombinations in the Maize Mitochondrial Genome Produce a Unique Transcriptional Unit in the Texas Male-Sterile Cytoplasm," *Cell* 44(3) (14 February 1986): 439–449.

86. See Erich Bornberg-Bauer and Hue Sun Chan, "Modeling Evolutionary Landscapes: Mutational Stability, Topology, and Superfunnels in Sequence Space," *Proceedings of the National Academy of Sciences* 96(19) (14 September 1999): 10689–10694.

87. Adapted from one of many emails like it that I have received. SELEX refers to "systematic evolution of ligands by exponential enrichment." In 1990 the laboratories of J. W. Szostak (Boston), L. Gold (Boulder), and G. F. Joyce (La Jolla) independently developed this technique, which permits the simultaneous screening of more than $10^{15}$ polynucleotides for different functionalities. See S. Klug and M. Famulok, "All You Wanted to Know about SELEX," *Molecular Biology Reports* 20 (1994): 97–107. See also Gordon Mills and Dean Kenyon, "The RNA World: A Critique," *Origins & Design* 17(1) (1996): 9–14.

88. I am indebted to Paul Nelson for helping me see how the formal mathematical theory developed in this and the previous chapter connects to current experimental work with biopolymers.

89. For a survey of the diverse problems to which evolutionary algorithms have been applied and for many of which these algorithms have generated unexpected solutions see Melanie Mitchell, *An Introduction to Genetic Algorithms*, 15–16.

90. Edward E. Altshuler and Derek S. Linden, "Design of Wire Antennas Using Genetic Algorithms," 211–248 in *Electromagnetic Optimization by Genetic Algorithms*, eds. Y. Rahmat-Samii and E. Michielssen (New York: Wiley, 1999). I am indebted to Karl Stephan for pointing me to this example. See Karl Stephan, "Evolutionary Computing and Intelligent Design," *Zygon* (2001): in review.

91. Altshuler and Linden, "Design of Wire Antennas Using Genetic Algorithms," fig. 22.

92. James Glanz, "It's Only Checkers, but the Computer Taught Itself," *New York Times* (25 July 2000): on the web at http://www.cognitivetherapy/neural_checkers.html (last accessed 10 June 2001).

93. Kumar Chellapilla and David B. Fogel, "Co-Evolving Checkers Playing Programs Using Only Win, Lose, or Draw," *SPIE's AeroSense'99: Applications and Science of Computational Intelligence II* (Orlando, Fla.: 5–9 April 1999). SPIE is the International Society for Optical Engineering.

94. Personal communication from David B. Fogel, 27 February 2001.

95. Deep Blue's defeat of Gary Kasparov in 1997 is widely known. For an account of Chinook, see J. Schaeffer, R. Lake, P. Lu, and M. Bryant, "Chinook: The World Man-

Machine Checkers Champion," *AI Magazine* 17 (1996): 21–29. Since the world champion programs did require expert knowledge, again the question arises how far the scope and power of evolutionary algorithms extends. While evolutionary algorithms seem well-suited for honing functions of extant systems, they seem less adept at constructing integrated systems that require multiple parts to achieve novel functions (see chapter 5). Also, as an optimization technique, evolutionary algorithms seem not to have caught on with the professionals who do optimization for a living. INFORMS, the Institute for Operations Research and the Management Sciences (http://www.informs.org), the professional organization of the OR (operations research) community has an annual conference that on average features around 450 presentations. Of these only a handful (three or four but not much more) are devoted to evolutionary algorithms. Other optimization methods, like hill-climbing and barrier methods, are much more widely used and discussed. The OR community is well-aware of evolutionary algorithms. Thus the failure of this problem-solving technique to catch on within the OR community is reason to be skeptical of the technique's general scope and power.

96. Personal communication from David B. Fogel, 3 June 2001. I am extremely grateful to David Fogel for walking me through his most fascinating work. He spent more time educating me about it than I deserve. My criticisms of his work are at the level of interpretation—how to make sense of it within a broader theoretical context. The work itself inspires my full admiration.

97. This I take to be the take-home lesson of Roger Lewin and Birute Regine's *The Soul at Work: Embracing Complexity Science for Business Success* (New York: Simon & Schuster, 2000). For a business to thrive, a framework within which the business operates must be designed. Yet once that framework is designed and in place, the business must not be micromanaged but allowed to follow its "natural course."

98. Michael Polanyi and Harry Prosch, *Meaning* (Chicago: University of Chicago Press, 1975), 162–163. Although the synergizing of intelligence and nature can be understood from the perspective of Whiteheadian process theology, it is also possible to take a more traditional theological view. Consider, for instance, the Eastern Orthodox view on synergy as described in Timothy Ware, *The Orthodox Church* (London: Penguin, 1963), 226–227. Here Ware cites John Chrysostom and Cyril of Jerusalem in support of a synergy between a transcendent intelligence (in this case the Christian God) and nature (in particular, human nature). For a more sustained treatment of synergy from the Eastern Orthodox perspective, see Philip Sherrard, *Human Image: World Image* (Ipswich, U.K.: Golgonooza Press, 1992), especially chapter 7.

99. Stuart Kauffman, *Investigations* (New York: Oxford University Press, 2000), 19.
100. Ibid., 18.
101. Ibid., 20.
102. Ibid., 239.
103. Ibid., 160.
104. Ibid., 239.
105. Ibid., 130–139.

~

# CHAPTER FIVE

# The Emergence of Irreducibly
# Complex Systems

## 5.1   The Causal Specificity Problem

In its heyday alchemy was a comprehensive theory of transmutation addressing not only transformations of base into precious metals but also transformations of the soul up and down the great chain of being. Alchemy was not just
a physics but also a metaphysics. Alchemy as metaphysics attracts interest to
this day—as in writings about the soul and personal identity by mystics like
Franz Bardon and mystically inclined psychologists like Carl Jung.[1] But to
include alchemy within the natural sciences is nowadays regarded as irretrievably misguided. The scientific community rejects alchemy as supersti-
tion and commends itself for having successfully debunked it.

For scientists the problem with alchemy is that it fails to specify the processes by which transmutations are supposed to take place. A well-known
Sidney Harris cartoon illustrates this point. The cartoon shows two scientists
viewing a chalkboard. The chalkboard displays some fancy equations, a gap,
and then some more fancy equations. In the gap are written the words:
"Then a miracle occurs." Pointing to the gap, one scientist remarks to the
other, "I think you need to be more explicit here." This is the problem
with alchemy. To characterize a transformation scientifically, it needs to be
specified explicitly. Alchemy never did this. Instead it continually offered
promissory notes promising that some day it would make the transformation
explicit. None of the promissory notes was ever kept. Indeed, the much
sought after philosopher's stone remains to be found.[2]

Officially, the scientific community rejects alchemy and has rejected it since the rise of modern science.[3] Unofficially, however, the scientific community has had a much harder time eradicating it. Indeed, I will argue that alchemical thinking pervades those fields concerned with the emergence of complex systems. This is not to deny that complex systems emerge out of simple systems (houses, for instance, are built out of simple components like bricks). But unless the process by which a complex system emerges from simpler systems is specified, emergence remains an empty word. And given that such specificity is often lacking, much (though by no means all) of what is described as the emergence of complex systems is alchemy by another name.

Alchemy followed a certain logic, and it is important to see the fallacy inherent in that logic. The problem with alchemy was not its failure to *understand* the causal process responsible for a transformation. It is not alchemy, for instance, to assert that a certain one-dimensional polypeptide will fold into the three-dimensional conformation of a functional protein. How polypeptides fold to form proteins is an open problem in biology. Three-dimensional proteins *emerge*, one might say, from suitably sequenced one-dimensional polypeptides (in suitable cellular contexts). This happens repeatedly and reliably. We can describe the transformation, but as yet we cannot explain how the transformation takes place. Ignorance about the underlying mechanism responsible for a transformation does not make the transformation alchemical.

Things transform into other things. Sometimes we can explain the process by which the transformation occurs. At other times we cannot. Sometimes the process requires an intelligent agent, sometimes no intelligent agent is required. Thus, a process that arranges randomly strewn Scrabble pieces into meaningful English sentences requires a guiding intelligence. On the other hand, the process by which water crystallizes into ice requires no guiding intelligence—lowering the temperature sufficiently is all that is needed. It is not alchemy that transforms water into ice. Nor is it alchemy that transforms randomly strewn Scrabble pieces into meaningful sentences. Nor, for that matter, is it alchemy that transforms a one-dimensional polypeptide into a functional protein, and that despite our ignorance about the mechanisms governing protein folding.

What, then, is the problem with alchemy? Alchemy's problem is its lack of *causal specificity*. Causal specificity means specifying a cause sufficient to account for an effect in question. Often we can specify the cause of an effect even if we cannot explain how the cause produces the effect. For instance, I may know from experience that shaking a closed container filled with a gas will cause the temperature of the gas to rise. Thus, by specifying the causal antecedents (i.e., a closed container filled with gas and my shaking it), I account for the container's rise in temperature. Nonetheless, I may have no idea why the temperature rises. Boltzmann's kinetic theory tells me that the temperature of the gas rises because temperature corresponds to average kinetic energy of the particles constituting the gas, and by shaking the container I impart additional kinetic energy to the particles. Boltzmann's theory enables me to explain why the temperature goes up. Even so, I do not need Boltzmann's theory to specify a cause that accounts for the temperature going up. For that, it is enough that I specify the causal antecedents (i.e., a closed container filled with gas and my shaking it).

Alchemy eschews causal specificity. Consider the stereotypical example of alchemical transformation, the transmutation of lead into gold. There is no logical impossibility that prevents potions and furnaces from acting on lead and turning it into gold. It may just be that we have overlooked some property of lead that in combination with the right ingredients allows it to be transformed into gold. But the alchemists of old never specified the precise causal antecedents that would bring about this transformation. Consequently, they lacked any compelling evidence that the transformation was even possible. Note, modern-day particle physicists can transform lead into gold with their particle accelerators, smashing the lead into more elementary constituents and then reconstituting them as gold. But here the causal antecedents are specified and differ plainly from those considered by the alchemists (particle accelerators were not part of the alchemists' tool chest).

Causal specificity was evident in the examples considered earlier: Water cooled below zero degrees Celsius is sufficient to account for it turning to ice. A random collection of Scrabble pieces left in the hands of a literate, nonhandicapped English speaker is sufficient to account for the Scrabble pieces spelling a coherent English sentence. A given sequence of l-amino acids joined by peptide bonds within a cellular context is sufficient to account for it folding into a functional protein, say cytochrome c. In each of these cases the causal antecedent is specified and accounts for the effect in question. We may not be able to explain how the cause that was specified produces its effect, but we know that it does so nonetheless.

But how do we get from causal antecedents like lead, potions, and furnaces and end up with gold? The alchemists' conviction was that if one could find just the right ingredients to combine with lead, lead would transform into gold. Thereafter the transformation could be performed at will and the alchemist who discovered the secret of transmutation would be rich (until, that is, the secret got out and gold became so common that it too became a base metal). Discovering the secret of transmutation was the alchemist's deepest hope. The interesting question for this discussion, however, is the alchemist's reason for that hope. Why were alchemists so confident that the transmutation from base into precious metals could even be effected? From our vantage we judge their enterprise a failure and one that had no possibility of success (contemporary solid state physics giving the coup de grace). But why were they unshaken in their conviction that with the few paltry means at their disposal (particle accelerators not being among them), they could transform base into precious metals? Put another way, why, lacking causal specificity, did they think the transformation could be effected at all?

Without causal specificity, one has no empirical justification for affirming that a transformation can be effected. At the same time, without causal specificity, one has no empirical justification for denying that a transformation can be effected. There is no way to demonstrate with complete certainty that Dr. Jekyll cannot transform into Mr. Hyde by some *unspecified* process. Lack of causal specificity leaves one without the means to judge whether a desired transformation can or cannot be effected. Any conviction about the desired transformation being possible, much less inevitable, must therefore derive from something other than a causal analysis. But from where?

Enter metaphysics. It is no secret that the motivation behind alchemy was never scientific (as we use the term nowadays) but metaphysical. Alchemy is a corollary of Neoplatonic metaphysics. Neoplatonism held to a great chain of being in which all reality emanates from God and ultimately returns to God. The great chain of being is strictly hierarchical so that for any two distinct items in the chain one is higher than the other. Now consider lead and gold. Gold is higher on the chain than lead (lead is a base metal, gold is a precious metal). Moreover, since everything is returning to God, lead is returning to God and on its way to God will pass through gold. Consequently, there is a natural pull for lead to get to gold on its way to God. The alchemist's task is therefore not to violate nature, but simply to help nature along. All lead needs is a fillip to achieve gold. The modest means by which alchemists hoped to achieve the transformation of lead into gold thus seemed entirely reasonable (in particular, no particle accelerators would be required).

Here, then, is the fallacy in alchemy's logic. Alchemy relinquishes causal specificity, yet confidently asserts that an unspecified process will yield a desired transformation. Lacking causal specificity, the alchemist has no empirical grounds for holding that the desired transformation can be effected. Even so, the alchemist remains convinced that the transformation can be effected because prior metaphysical beliefs ensure that some process, though for now unspecified, *must* effect the desired transformation. In short, metaphysics guarantees the transformation even if the empirical evidence is against it.

Alchemy continues to flourish, though nowadays it goes by the name *emergence*. Whereas classical alchemy was concerned with transforming base into precious metals, emergence is concerned with transforming simple into complex systems. Now I do not want to give the impression that emergence is a disreputable concept. The concept has applications that are entirely innocent. Consider, for instance, Bénard cell convection. In Bénard cell convection, boiling liquid organizes into columns of hexagonal cells. Bossomaier and Green elaborate: "In some columns liquid travels up from the bottom of the vessel, while in adjacent columns liquid travels down. Just think how extraordinary this is: there are no inherent boundaries in the liquid and these columns have formed spontaneously. Furthermore, they have a clear geometrical shape, again something in no sense obvious from the initial setup."[4]

In Bénard cell convection an unexpected global behavior emerges from the joint action of simple localized effects. No central planning governs the joint action of these simple localized effects. Rather, the unexpected global behavior comes about simply by having the right pieces in place (for Bénard cell convection heating a thin sheet of liquid in a frying pan is enough to produce the phenomenon). Emergence in this nonproblematic sense occurs in everything from the self-organization of dynamical systems to the self-regulation of ecosystems to the self-optimization of market economies.

In each of these cases emergence is nonproblematic. Why? Because of causal specificity. Bénard cell convection, for instance, happens repeatedly and reliably so long as the appropriate fluid is sufficiently heated in the appropriate vessel. We may not understand what it is about the properties of the fluid that makes it organize itself into hexagonal cells, but the causal antecedents that produce the hexagonal cells are clearly specified. So long as we have causal specificity, emergence is a perfectly legitimate concept.

But what about emergence without causal specificity? Consider, for instance, the origin of life. For most of the scientific community, the presumption is that life organized itself through chemical means apart from any designing intelligence. Yet, unlike the causal specificity in Bénard cell convection, origin of life researchers have yet to specify the purely chemical pathways that supposedly lead to life. Despite a vast literature on the origin of life, causally specific proposals for just what those purely chemical pathways might be are sorely absent. Which is not to say there have not been any proposals. In fact, there are too many of them. RNA worlds, clay templates, hydrothermal vents, and numerous other naturalistic scenarios have all been proposed to account for the emergence of life. Yet none of these scenarios is detailed enough to be seriously criticized or tested. In short, they all lack causal specificity.

The logic of emergence parallels the logic of alchemy. Emergence, like alchemical transformation, is a relational notion. To say that something emerges is to say what it emerges from (e.g., gold emerges from lead plus some other things). "X emerges" is an incomplete sentence. It needs to be completed by reading "X emerges from Y." Moreover, the claim that X emerges from Y remains vacuous until one specifies Y and can demonstrate that Y is sufficient to account for X. Lowering the temperature of water below zero degrees Celsius is causally specific and adequately accounts for the freezing of water. On the other hand, a complete set of the building materials for a house do not suffice to account for a house—additionally what is needed is an architectural plan (drawn up by an architect) as well as assembly instructions (executed by a contractor) to implement the plan. Likewise, in the origin of life, it does no good simply to have the building blocks for life (e.g., nucleotide bases or amino acids). The means for organizing those building blocks into a coherent system (i.e., a living organism) need to be specified as well.

Given the pervasive lack of causal specificity in origin-of-life studies, what confidence have we that purely physical causes are even up to the task of originating life? If we take seriously the parallel between emergence and alchemy, then we should be looking for a prior metaphysical commitment that ensures that purely physical causes, though for now unspecified, *must* effect the desired transformation. In the case of alchemy, the prior metaphysical commitment was Neoplatonism. In the case of emerging complex systems, the prior metaphysical commitment is naturalism. Naturalism is the view that purely physical causes undirected by any guiding intelligence at base govern the world. Given naturalism as a prior metaphysical commitment, it follows that life and complex systems in general must emerge from purely physical causes. But that commitment, like the alchemists' commitment to Neoplatonism, is highly problematic.

It is important to be clear why this parallel with alchemy poses a stumbling block for complexity theory. In reference to the origin of life, for instance, proponents of naturalism are apt merely to note that life is here, life was not always here, and so some transformation from nonlife to life had to occur. Life has emerged even if we cannot quite spell out the precise causal antecedents for life. The origin of life is a great unsolved problem, and complexity theory is valiantly trying to resolve it. For me to compare the emergence of complex systems with alchemy will therefore strike the naturalist as misconceived if not downright churlish.

To see why this dismissal is too easy, consider what it means to say that life has, as the naturalist claims, emerged from purely physical causes. Because the origin of life is an open problem, the reference to "purely physical causes" lacks, to be sure, causal specificity. But there is a deeper problem, and that is the imposition of an arbitrary restriction. The problem with claiming that life has emerged from purely physical causes is not that it admits ignorance about an unsolved problem, but that it restricts the possible solutions to that problem; namely, it requires that solutions limit themselves to purely physical causes. This is an arbitrary and metaphysically driven restriction. Life has emerged from purely physical causes. How do we know that? In general, to hypothesize that X emerges from Y remains pure speculation until the process that takes Y to X is causally specified. Until then, to impose restrictions on the types of causal factors that may or may not be employed in Y is arbitrary and certain to frustrate scientific inquiry.

In this respect complexity theory is even more culpable than alchemy. Alchemy sought to transform lead into gold, but left open the means by which the transformation could be effected (though in practice alchemists hoped the transformation could be effected through the modest technical means at their disposal). Complexity theory, on the other hand, seeks to transform nonlife into life, but—when biased by naturalism—excludes any place for intelligence or teleology in the transformation. Such a restriction is gratuitous given complexity theory's lack of causal specificity in accounting for the origin of life. Perhaps naturalism will eventually be vindicated and the great open problems of complexity theory will submit to purely naturalistic solutions. But in the absence of causal specificity, there is no reason to let naturalism place such restrictions on our scientific theorizing. It is restrictions like these—typically unspoken, metaphysically motivated, and at odds with free scientific inquiry—that need to be resisted and exposed. Science must not degenerate into applied naturalistic philosophy. At its best, science is a free inquiry into *all* the possibilities that nature might have to offer. Design remains very much a live possibility.

The origin of life is just one example of emergence without causal specificity. The emergence of consciousness from neurophysiology is another. Nonetheless, the one I want to focus on especially in this chapter is the emergence of increasingly complex life forms from simpler life forms. Although the Darwinian mutation-selection mechanism is supposed to handle such cases of emergence, I shall argue that the Darwinian mechanism encounters the same failure of causal specificity endemic to alchemy. The lesson of alchemy is clear: Causal specificity cannot be redeemed in the coin of metaphysics, be it Neoplatonic or naturalistic.

## 5.2   The Challenge of Irreducible Complexity

There is no question that the Darwinian selection mechanism constitutes a fruitful idea for biology, and one that continues to spur interesting research. Even so, Darwinism is more than just this mechanism; it is the totalizing claim that this mechanism accounts for all the diversity of life. The evidence simply does not support this claim. What evidence there is indicates that the mechanism can account for small-scale changes in organisms, like insects developing insecticide resistance or bacteria developing antibiotic resistance. For such small-scale changes, the Darwinian mechanism possesses the causal specificity requisite for a successful scientific theory. If, for instance, one subjects a certain population of bacteria to a certain regimen of antibiotics, then in due course one will witness a specific pattern of antibiotic resistance. Darwinian theory gives a scientifically fruitful, empirically adequate, and causally specific account of such small-scale changes.

On the other hand, it is completely unfounded to assert that this mechanism can account for the unlimited plasticity of organisms to diversify across all apparent boundaries. In making this claim, I am not challenging the genealogical interrelatedness of all organisms, or what is called common descent. Rather, I am simply noting that large-scale evolutionary changes in which novel information-rich structures are added to an organism cannot legitimately be derived from the Darwinian selection mechanism. To do so is to extrapolate the Darwinian theory beyond its evidential base, which consists entirely of small-scale organismal changes.[5] This is always a danger in science—to think that one's theory encompasses a far bigger domain than it actually does. This happened with Newtonian mechanics—physicists had thought that Newton's laws provided a total account of the constitution and dynamics of the universe. Maxwell, Einstein, and Heisenberg each showed that the proper domain of Newtonian mechanics was far more constricted. So too, the proper domain of the Darwinian selection mechanism is far more constricted than most Darwinists would like to admit.

Indeed, the following problems have proven utterly intractable not only for the Darwinian selection mechanism, but also for any other undirected natural process proposed to date: the origin of life, the origin of the genetic code, the origin of multicellular life, the origin of sexual reproduction, the scarcity of transitional forms in the fossil record, the biological Big Bang that occurred in the Cambrian era, the development of complex organ systems, and the formation of irreducibly complex molecular machines.[6] These are just a few of the more serious difficulties that confront every theory of biological evolution that posits only undirected natural processes. I want in this chapter to focus on the last of these, namely, the formation of irreducibly complex molecular machines. As we shall see, causal specificity collapses as soon as the Darwinian selection mechanism attempts to account for such systems.

Highly intricate molecular machines play an integral part in the life of the cell and are increasingly attracting the attention of the biological community. For instance, in February 1998 the premier biology journal *Cell* devoted a special issue to "macromolecular machines." All cells use complex molecular machines to process information, build proteins, and move materials across their membranes. Bruce Alberts, president of the National Academy of Sciences, introduced this issue with an article titled "The Cell as a Collection of Protein Machines." In it he remarked,

> We have always underestimated cells. . . . The entire cell can be viewed as a factory that contains an elaborate network of interlocking assembly lines, each of which is composed of a set of large protein machines. . . . Why do we call the large protein assemblies that underlie cell function protein *machines*? Precisely because, like machines invented by humans to deal efficiently with the macroscopic world, these protein assemblies contain highly coordinated moving parts.[7]

Although Alberts notes the strong resemblance between molecular machines and machines designed by human engineers, he nonetheless sides with the majority of biologists in regarding the cell's marvelous complexity as only apparently designed. Lehigh University biochemist Michael Behe disagrees. In *Darwin's Black Box* Behe argues for actual design in the cell. Central to his argument is his notion of *irreducible complexity*. As Behe defines it, a system is irreducibly complex if it consists of several mutually adapted and interrelated parts such that removing even a single part completely destroys the system's function.[8] According to Behe, irreducibly complex molecular machines pose a decisive obstacle to the Darwinian mechanism.

As an everyday example of an irreducibly complex system, Behe offers the mousetrap. A mousetrap consists of a platform, a hammer, a spring, a catch, and a holding bar. Remove any one of these five components, and the remaining components no longer suffice to build a functional mousetrap (see figure 5.1). This is not to say one cannot omit components and obtain a functional mousetrap if one suitably modifies the remaining components.

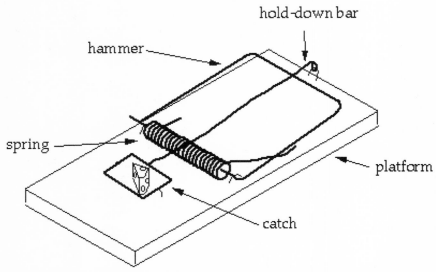

hold-down bar

hammer

spring

platform

catch

**Figure 5.1.** **The Standard Mousetrap. (Copyright © 2001 John H. McDonald. Reprinted with permission. All rights reserved.)**

But the point of irreducible complexity as Behe defines it is that any reduced set of components—so long as they remain unmodified—surrenders functionality. This raises the question whether eliminating components and suitably modifying others can restore functionality and thus whether the Darwinian mechanism may have a way to circumvent irreducible complexity after all. I take up this concern in sections 5.6 and 5.7.

Irreducible complexity may be contrasted with *cumulative complexity*. A system can be defined as cumulatively complex if the components of the system can be arranged sequentially so that the successive removal of components never leads to the complete loss of function. An example of a cumulatively complex system is a city. It is possible successively to remove people and services from a city until one is down to a tiny village—all without losing the sense of community, which in this case constitutes function. From this characterization of cumulative complexity, it is clear that the Darwinian selection mechanism can readily account for cumulative complexity. Indeed, the gradual accrual of complexity via selection mirrors the retention of function as components are successively removed from a cumulatively complex system.

But what about irreducible complexity? Can the Darwinian selection mechanism account for irreducible complexity? If selection acts with reference to a goal, then there is no difficulty for selection to produce irreducible complexity. Take Behe's mousetrap. Given the goal of constructing a mousetrap, one can specify a goal-directed selection process that in turn selects a platform, a hammer, a spring, a catch, and a holding bar, and at the end puts all these components together to form a functional mousetrap. In the case of an organism forming a new structure over the course of several generations, we can think of the organism as fashioning certain components and selection as setting aside those components and then, once all the components are in place, putting them together to form that new structure. Given a prespecified goal, selection has no difficulty producing irreducibly complex systems.

But the selection operating in biology is Darwinian natural selection. And this form of selection operates without goals, has neither plan nor purpose, and is wholly undirected. The great appeal of Darwin's selection mechanism was that it would eliminate teleology (i.e., goal-directed processes and therefore design) from biology. Yet by making selection an undirected process, Darwin unduly restricted the type of complexity that biological systems could manifest. Behe argues that according to Darwin's theory biological systems should fail to exhibit irreducible complexity:

> An irreducibly complex system cannot be produced . . . by slight, successive modifications of a precursor system, because any precursor to an irreducibly complex system that is missing a part is by definition nonfunctional. . . . Since natural selection can only choose systems that are already working, then if a biological system cannot be produced gradually it would have to arise as an integrated unit, in one fell swoop, for natural selection to have anything to act on.[9]

Behe is saying that for an irreducibly complex system, function is attained only when all components of the system are in place simultaneously. It follows that natural selection, if it is going to produce an irreducibly complex system, has to produce it all at once or not at all. This would not be a problem if the systems in question were simple. But they are not. The irreducibly complex biochemical systems Behe considers are protein machines consisting of numerous distinct proteins, each indispensable for function, and together beyond what natural selection can muster in a single generation.

One such irreducibly complex biochemical system that Behe considers is the bacterial flagellum (see figure 5.2). The flagellum is an acid-powered rotary motor with a whip-like tail whose rotating motion enables a bacterium to navigate through its watery environment. Behe shows that the intricate machinery in this molecular motor—including a rotor, a stator, O-rings,

**Figure 5.2.   The Bacterial Flagellum. (Copyright © 2001 Discovery Media Productions. Reprinted with permission. All rights reserved.)**

bushings, and a drive shaft—requires the coordinated interaction of about thirty proteins and another twenty or so proteins to assist in their assembly. Yet the absence of any one of these proteins would result in the complete loss of motor function.[10] On a Darwinian view, a bacterium with a flagellum evolved via the Darwinian selection mechanism from a bacterium without a flagellum. For this mechanism to produce the flagellum, chance modifications have to generate the various proteins that constitute the flagellum and then selection must preserve them, gather them to the right location in the bacterium, and then properly assemble them.

But how is selection to accomplish this? Selection is nonteleological, so it cannot cumulate proteins, holding them in reserve until with the passing of many generations they are finally available to form a complete flagellum. The environment contains no blueprint of the flagellum that selection can extract and then transmit to an organism to form a flagellum. Selection can only build on partial function, gradually improving function that already exists. But a flagellum without its full complement of protein parts does not function at all. Behe therefore concludes that if the Darwinian mechanism is going to produce the flagellum, it will have to do so in one generation. But that is to place an impossible demand on the Darwinian mechanism. The Darwinian mechanism works by cumulating and sifting small changes, not by introducing sudden massive changes.

For Behe, the irreducible complexity of biochemical systems like the bacterial flagellum counts decisively against the Darwinian mechanism, and indeed against any naturalistic evolutionary mechanism proposed to date. Moreover, because irreducible complexity occurs at the biochemical level, he argues that there is no more fundamental level of biological analysis to which the irreducible complexity of biochemical systems can be referred and at which a Darwinian analysis in terms of selection and mutation can still hope for success. What undergirds biochemistry is ordinary chemistry and physics, neither of which can explain biological complexity. Also, whether a biochemical system is irreducibly complex is an empirical question: Individually knock out each protein constituting a biochemical system to determine whether function is lost. If so, we are dealing with an irreducibly complex system. Protein knock-out experiments of this sort are routine in biology.[11]

The irreducibly complex systems Behe considers require numerous components specifically adapted to each other and each necessary for function. Such systems are not only highly improbable (see section 5.10), but also specified in their function. Biological specification always denotes function. An organism is a functional system comprising many functional subsystems. The functionality of organisms can be cashed out in any number of ways. Behe cashes it out in terms of the *minimal function* of biochemical systems.[12] Arno Wouters cashes it out globally in terms of the *viability* of whole organisms.[13] Even the staunch Darwinist Richard Dawkins will admit that life is specified functionally, cashing out functionality in terms of the *reproduction* of genes. Thus, in *The Blind Watchmaker* Dawkins writes, "Complicated things have some quality, specifiable in advance, that is highly unlikely to have been acquired by random chance alone. In the case of living things, the quality that is specified in advance is . . . the ability to propagate genes in reproduction."[14]

Behe concludes that the Darwinian mechanism cannot account for the origin of irreducibly complex biomolecular machines. But he goes further. According to Behe, the irreducible complexity of biochemical systems provides not just negative evidence against Darwinism, but also positive evidence for design. The irreducibly complex systems Behe considers require numerous components specifically adapted to each other and each necessary for function. According to Behe, such systems are not only specified in virtue of their function, but also highly improbable or complex in the sense required by the complexity-specification criterion. But highly improbable specified structures are the key trademark of intelligence—they exhibit specified complexity. Behe therefore takes irreducible complexity as a reliable empirical marker of design in biology.

## 5.3   Scaffolding and Roman Arches

Behe's book was published in 1996. Since then it has been widely reviewed, both in the popular press and in scientific journals.[15] It has also been widely discussed over the Internet, with entire websites devoted specifically to refuting the connection Behe draws between irreducible complexity and design.[16] By and large critics have conceded the scientific accuracy of Behe's claims (including his literature-search demonstrating the absence of detailed neo-Darwinian accounts of how the irreducibly complex systems he examines could have come about). Nonetheless, they have objected to his argument on theoretical and methodological grounds. Behe argues that the irreducible complexity of biochemical machines is inaccessible to the Darwinian evolutionary mechanism and that only design can properly account for such complex systems. I want in this and the next four sections to enumerate the main objections to Behe's argument, assess their merit, and then draw a general conclusion about the significance of his argument.

Many of the objections to Behe's argument try to show that an irreducibly complex system could upon closer examination have been produced by gradual increments apart from design after all. According to the scaffolding objection, to generate an irreducibly complex system, first some nonirreducibly complex system arises by mutation and selection incrementally adding components. Then at some point a subsystem arises that is able to function autonomously (i.e., without the rest of the system). Since it can function autonomously, the other components are now vestigial and drop away. When all have dropped away, we have a system that is irreducibly complex. In short, what appears to be a qualitative difference is really only the result of a lot of small quantitative changes.[17]

The scaffolding objection claims that eliminating functional redundancy is a plausible route to irreducible complexity. But is it? According to Thomas Schneider, there are situations in which "a functional species can survive without a particular genetic control system but . . . would do better to gain control *ab initio*."[18] In such situations, Schneider continues,

> Any new function must have this property until the species comes to depend on it, at which point it can become essential if the earlier means of survival is lost by

atrophy or no longer available. I call such a situation a "Roman arch" because once such a structure has been constructed on top of scaffolding, the scaffold may be removed, and will disappear from biological systems when it is no longer needed. Roman arches are common in biology, and they are a natural consequence of evolutionary processes.[19]

R. H. Thornhill and D. W. Ussery elaborate further on this analogy of the arch.[20] To build an arch requires a scaffold. So long as the scaffold is in place, pieces of the arch can be shifted in and out of position. But once all the pieces of the arch are in position and the scaffold is removed (i.e., redundancy is eliminated), each of the pieces of the arch become indispensable and the arch forms an irreducibly complex system. Thornhill and Ussery next consider biological examples. They focus most of their attention on the origin of the mammalian jaw. They also consider the origin of feathers from scales, shared domains in different proteins, and two cases where a protein gets co-opted to perform a different function. But they fail to address the type of biological system that prompted Behe's challenge in the first place, namely, irreducibly complex biochemical machines.

A fundamental theoretical difficulty confronts the scaffolding objection. With the Darwinian mechanism, selection has to work on function. We know that we have function with an irreducibly complex system like the bacterial flagellum. Let us concede that we have function with the irreducibly complex system plus its scaffold (which eventually gets eliminated as it becomes redundant). In building up to the aggregate system of irreducibly complex system plus scaffold, when did function begin? With a bacterial flagellum plus scaffold, for instance, when did we get outboard bi-directional rotary motion for propelling the bacterium through its watery environment? Indeed, even with the assistance of a scaffold, there is no reason to think that function was attained *until* all the pieces of the final irreducibly complex system were in place. Given an irreducibly complex system to be explained by scaffolding, the challenge for the Darwinist is to identify a sequence of gradual *functional* intermediaries that starts from some initial simple system and eventually leads to an irreducibly complex system plus scaffold. Even though the scaffold can help build the irreducibly complex system, the scaffold's function is parasitic on the function of the thing it is helping to construct (in this case the flagellum), and the only evidence of that function is from the irreducibly complex system itself.

There is also a practical problem with the scaffolding objection, and this problem recurs with the objections listed in the next sections. Human imagination is notoriously hard to discipline. Indeed, there is no way to argue

against a putative transmutation that seems plausible enough to our imagina-
tions but has yet to be concretely specified—as if we could demonstrate with
complete certainty that Dr. Jekyll really could not transform into Mr. Hyde
by some unspecified process. Unless a concrete model is put forward that is
detailed enough to be seriously criticized, then it is not going to be possible
to determine the adequacy of that model. This is of course another way of
saying that the scaffolding objection has yet to demonstrate causal specificity
when applied to actual irreducibly complex biochemical systems. The absence of detailed models in the biological literature that employ scaffoldings
to generate irreducibly complex biochemical systems is therefore reason to
be skeptical of such models. If they were the answer, then one would expect
to see them in the relevant literature, or to run across them in laboratory.
But we do not. That, Behe argues, is good reason to think they are not the
answer.[21]

## 5.4 Co-optation, Patchwork, and Bricolage

Another common objection to Behe's claim that irreducibly complex biochemical systems lie beyond the remit of the Darwinian mechanism is the
co-optation objection. According to this objection, proteins previously
targeted for various cellular systems sometimes break free and are co-opted
into novel systems. It is as though pieces from a car, bicycle, motorboat, and
train can be suitably recombined to form an airplane. This objection is also
sometimes called the patchwork or bricolage objection. Thus any such airplane would be a patchwork or bricolage of preexisting materials originally
targeted for different uses.

Certainly there is no logical impossibility that prevents such patchworks
from forming irreducibly complex systems. But a patchwork, if sufficiently
intricate and elegant, does beg a precise causal account of how it arose. The
bacterial flagellum, for instance, is an engineering marvel of miniaturization
and performance (e.g., it spins at close to 20,000 rpm and can change direction in a quarter turn). Simply to call such a system a patchwork of co-opted
preexisting materials is hardly illuminating and does nothing to answer the
causal specificity problem.

The idea of one structure originally serving one purpose being co-opted
for another purpose is a theme in evolutionary biology. Stephen Jay Gould
and Elisabeth Vrba, for instance, refer to such structures as *exaptations* to
distinguish them from adaptations. Douglas Futuyma elaborates on their use
of this distinction: "If an adaptation is a feature evolved by natural selection
for its current function, a different term is required for features that, like the

hollow bones of birds or the sutures of a young mammal's skull, did not evolve because of the use to which they are now put."[22] Futuyma continues: "[Gould and Vrba] suggest that such characters that evolved for other functions, or for no function at all, but which have been co-opted for a new use be called *exaptations*."[23]

The problem with trying to explain an irreducibly complex system like the bacterial flagellum as a patchwork is that it requires multiple coordinated exaptations. It is not just that one thing evolves for one function (or no function at all), and then through some quick and dirty modification gets used for some completely different function. It is that multiple protein parts from different functional systems have to break free and then coalesce to form a new integrated system. Sheer possibilities being what they are, multiple coordinated exaptations cannot be excluded on a priori grounds. But short of some concrete, causally specific example where such a system emerged via the "bricolage method," there is no reason to view such sheer possibilities as live possibilities.

Think of it this way: Even if all the pieces (i.e., proteins) for a bacterial flagellum are in place within a cell but serving other functions, there is no reason to think that those pieces can come together spontaneously to form a tightly integrated system like the flagellum. In addition to those proteins that go into a flagellum, the cell in question will have many other proteins that could play no conceivable role in a flagellum. The majority of proteins in the cell will be of this sort. How then can those and only those proteins that go into a functional flagellum be brought together and guided to their proper locations in the cell without interfering cross-reactions from the other proteins in the cell. It is like going through a grocery store, randomly taking items off the racks, and hoping that what ends up in the shopping cart when all mixed together will make a cake. It does not happen that way (cf. the localization probability in section 5.10).

But do not take my word for it. Allen Orr, who is no fan of Behe or of his notion of irreducible complexity, defends him against the co-optation objection:

First it will do no good to suggest that all the required parts of some biochemical pathway popped up simultaneously by mutation. Although this "solution" yields a functioning system in one fell swoop, it's so hopelessly unlikely that no Darwinian takes it seriously. As Behe rightly says, we gain nothing by replacing a problem with a miracle. Second, we might think that some of the parts of an irreducibly complex system evolved step by step for some other purpose and were then recruited wholesale to a new function. But this is also unlikely. You may as well hope that half your car's transmission will suddenly help out in the airbag department. Such things might happen very, very rarely, but they surely do not offer a general solution to irreducible complexity.[24]

Actually, prominent Darwinians do take the co-optation objection seriously—notably Kenneth Miller. One of his main objections against Behe in *Finding Darwin's God* is the co-optation objection.[25] According to Miller, the parts of an irreducibly complex system are never totally functionless. Rather, those parts have some function and thus are grist for selection's mill. Accordingly, selection can work on those parts and thereby form irreducibly complex systems. Two years after the publication of *Finding Darwin's God*, this seems to have become Miller's main argument against Behe.[26] Moreover, Miller has publicly stated that Allen Orr is wrong in rejecting co-optation as a way of forming irreducibly complex systems.[27]

Yes, the Darwinian mechanism requires a selectable function if that mechanism is going to work at all. And yes, functional pieces cobbled together from various systems via the bricolage method are selectable by the Darwinian mechanism. But what is selectable here is the individual functions of the individual pieces and not the function of the yet-to-be-produced cobbled-together system. The Darwinian mechanism selects for preexisting function. It does not select for future function. Once that function is realized, the Darwinian mechanism can select for it as well. But getting over the hump is the hard part. How does one get from functional pieces that are selectable in terms of their individual functions to a system that consists of those pieces and exhibits a novel function? The Darwinian mechanism is no help here.

As with the scaffolding objection, the co-optation objection founders on the problem of causal specificity. As a sheer possibility, it could happen that irreducibly complex biochemical systems form by proteins being co-opted from other functional systems. But the scientific literature shows a complete absence of concrete, causally detailed proposals for how this might happen with actual irreducibly complex biochemical systems (Miller would disagree, but see section 5.7).

## 5.5   Incremental Indispensability

Although Allen Orr rejects the co-optation objection, he nonetheless thinks there can be gradual routes to irreducibly complex systems. Accordingly, he argues that an irreducibly complex system may arise by gradually enfolding parts that initially were dispensable but eventually become indispensable (as required with an irreducibly complex system, for which every part is indispensable). I refer to this objection as the incremental indispensability objection. Orr states this objection as follows:

> An irreducibly complex system can be built gradually by adding parts that, while initially just advantageous, become—because of later changes—essential [i.e., indispensable]. The logic is very simple. Some part (A) initially does some job (and not very well, perhaps). Another part (B) later gets added because it helps A. This new part isn't essential, it merely improves things. But later on, A (or something else) may change in such a way that B now becomes indispensable. This process continues as further parts get folded into the system. And at the end of the day, many parts may all be required. . . . I'm afraid there's no room for compromise here: Behe's key claim that all the components of an irreducibly complex system "have to be there from the beginning" is dead wrong.[28]

Orr refers to Behe's claim that irreducibly complex biochemical systems are inaccessible to Darwinian evolution as "Behe's colossal mistake."[29] But in what sense is Behe committing a mistake? Orr's incremental indispensability objection is similar to the scaffolding and co-optation objections in offering a narrative schema for how an irreducibly complex system might conceivably have evolved by Darwinian means despite prima facie indications to the contrary. A cumulatively complex system in which each piece is dispensable but nonetheless enhances function is the easiest to account for in Darwinian terms. An irreducibly complex system, on the other hand, by its very integration and interrelatedness of parts does not admit so straightforward an incremental analysis. Consequently, any Darwinian narrative that accounts for such systems requires additional steps besides those that simply add novel function-enhancing components. As with the scaffolding and co-optation objections, the incremental indispensability objection adds such further steps to account for irreducibly complex systems, in this case structural changes in components over time that render all the components indispensable.

There are two difficulties with the incremental indispensability objection, one empirical and the other logical. The empirical difficulty is that there is no evidence that the irreducibly complex biochemical systems Behe considers came about by this method of add a component, make it indispensable, add another component, now make it indispensable, etc. As with the scaffolding and co-optation objections, the incremental indispensability objection requires additional steps besides those that simply add novel function-enhancing components. But there is no evidence for those additional steps. All we have evidence for is the actual components in an irreducibly complex system, all of which are by definition indispensable.

Orr offers several counterexamples to Behe's claim that irreducibly complex biochemical systems cannot arise by gradual Darwinian means, but none of them succeeds. Here are the counterexamples Orr considers:

1. **The evolution of the lung.** "The transformation of air bladders into lungs that allowed animals to breathe atmospheric oxygen was initially just advantageous: such beasts could explore open niches—like dry land—that were unavailable to their lung-less peers. But as evolution built on this adaptation (modifying limbs for walking, for instance), we grew thoroughly terrestrial and lungs, consequently, are no longer luxuries—they are essential. The punch-line is, I think, obvious: although this process is thoroughly Darwinian, we are often left with a system that is irreducibly complex."[30]

2. **Gene duplication.** "Molecular evolutionists have shown that some genes are duplications of others. In other words, at some point in time an extra copy of a gene got made. The copy wasn't essential—the organism obviously got along fine without it. But through time this copy changed, picking up a new, and often related, function. After further evolution, this duplicate gene will have become essential. (We're loaded with duplicate genes that are required: myoglobin, for instance, which carries oxygen in muscles, is related to hemoglobin, which carries oxygen in blood. Both are now necessary.)"[31]

3. **Computer programming.** "Anyone who programs knows how easy it is to write yourself into a corner: a change one makes because it improves efficiency may become, after further changes, indispensable. Improvements might be made one line of code at a time and, at all stages, the program does its job. But, by the end, all the lines may be required. This programming analogy captures another important point: If I were to hand you the final program, it's entirely possible that you would not be able to reconstruct its history—that this line was added last and that, in a previous version, some other line sat between these two. Indeed, because the very act of revising a program has a way of wiping out clues to its history, it may be impossible to reconstruct the path taken. Similarly, we have no guarantee that we can reconstruct the history of a biochemical pathway. But even if we can't, its irreducible complexity cannot count against its gradual evolution any more than the irreducible complexity of a program does—which is to say, not at all."[32]

Orr's first example looks at a complex organ system—the lung—and not at an irreducibly complex biochemical machine. Moreover, it is not clear what in this example is supposed to be irreducibly complex. The lung treated as a collection of cells and tissues is not irreducibly complex—patients with lung cancer can have portions of their lungs removed and recover. Is Orr ascribing irreducible complexity to the whole organism and identifying the lung as one of its indispensable parts? But clearly, the organism also includes organ systems that are dispensable, like the appendix in humans. In what sense, then, is the organism irreducibly complex (irreducible complexity according to Behe requires that all components of a system be indispensable)? Orr does not elaborate. Finally, the evidence is hardly compelling that lungs evolved from air bladders, much less that Darwinian evolution was the driving force here.[33]

Orr's second example is about gene duplication and hence is at least in the right biochemical ballpark. As in the last example, we are dealing here with structures that evolve and become indispensable. More specifically, Orr is here looking at proteins that evolved by gene duplication where the initial protein maps loosely onto the later evolved protein, which then becomes indispensable for the organism in its own right. But indispensability is not the same as irreducible complexity (though there is a reverse implication: all the parts of an irreducibly complex system are indispensable for the functioning of that system[34]). What exactly is the irreducibly complex system here? Individual amino acids in a protein usually allow some degree of substitution and sometimes can be dropped entirely without destroying the protein's function. As a consequence, it is not evident in general whether an individual protein taken as a precise sequence of amino acids is going to be irreducibly complex. Nor are entire organisms going to be irreducibly complex since viability can usually be maintained despite removing parts of the organism. This second example therefore misses the mark as well.

Orr's third example is even more problematic than the last two. The last two at least focused on biology, though even here Orr avoided addressing any of Behe's actual examples. But with computer programming we have an example that is chock-full of design and for which any irreducible complexity must properly be attributed to design. Orr is of course correct that a computer program can be irreducibly complex, that it can acquire this feature by repeated revisions, and that it may be impossible to reconstruct precisely how a program took its final form. Even so, computer programming does follow certain broad design constraints that are anything but Darwinian: The identification of a problem that needs to be solved, the theoretical determination that a solution exists or can be approximated, the breaking down of the problem into manageable algorithmic modules, and finally the assimilation of these modules into a fully functioning program that solves the original problem. All of this is done by a designing intelligence.

Orr's examples illustrate the empirical difficulty facing the incremental indispensability objection. As with the other objections considered so far, the incremental indispensability objection lacks causal specificity. How exactly did (or could) an irreducibly complex biochemical machine like the bacterial flagellum evolve by Darwinian means? In his computer programming example Orr seems willing to concede ignorance: "Because the very act of revising a program has a way of wiping out clues to its history, it may be impossible to reconstruct the path taken. Similarly, we have no guarantee that we can reconstruct the history of a biochemical pathway."[35] That may be, but Behe's challenge is hardly answered by offering the incremental indispensability objection as a narrative schema for how irreducibly complex biochemical machines might be formed historically, and then turning around and claiming that the actual historical pathways to such systems are invariably occluded. Nor does it help that Orr cannot offer any causally specific possible pathways to such systems (independent of actual history). Even if Orr is right that tracing actual historical pathways constitutes an intractable problem, the utter absence of causally specific possible pathways to irreducibly complex biochemical systems constitutes a major conceptual void within Darwinism. For the incremental indispensability objection to have any force, there have to be causally specific examples of how irreducibly complex biochemical machines can be formed by successively adding components and alternately rendering them indispensable. Orr fails to provide such examples.

Another difficulty with Orr's incremental indispensability objection is logical rather than empirical. Behe identifies this difficulty as follows:

> Orr questions the concept of irreducible complexity on logical grounds. He agrees with me that "You cannot . . . gradually improve a mousetrap by adding one part and then the next. A trap having half its parts doesn't function half as well as a real trap; it doesn't function at all." So Orr understands the point of my mousetrap analogy—but then mysteriously forgets it. He later writes, "Some part (A) initially does some job (and not very well, perhaps). Another part (B) later gets added, because it helps A." Some part initially does some job? Which part of the mousetrap is he talking about? A mouse has nothing to fear from a "trap" that consists of just an unattached holding bar, or spring, or platform, with no other parts.[36]

The incremental indispensability objection posits a gradual increase in complexity such that novel parts that enhance function are added and alternately rendered indispensable. But which function or job are we talking about? If it is a function different from the one that the system ultimately attains, then we are really talking about a system formed by co-optation (i.e., a system that comes together from parts previously deployed in other systems and for different functions). But then the incremental indispensability objection becomes a special case of the co-optation objection, which we have already seen to be defective—indeed, Alan Orr himself rejected it, and for compelling reasons. Clearly, then, the incremental indispensability objection can stand only if the function or job that the system is doing remains unchanged as components are added to the system and alternately rendered indispensable. But for that to happen, it must always be possible for simpler systems to do the same job (though perhaps not quite as well) as an irreducibly complex system. This is itself a problematic assumption and brings us to the next objection.

## 5.6   Reducible Complexity

In replying to Orr's incremental indispensability objection, Behe points to the mousetrap. Orr claims that successively adding components and rendering them indispensable is how irreducibly complex systems can be formed: "Some part (A) initially does some job (and not very well, perhaps). Another part (B) later gets added, because it helps A."[37] Behe counters by asking how this might work with a mousetrap: "Some part initially does some job? Which part of the mousetrap is he talking about? A mouse has nothing to fear from a 'trap' that consists of just an unattached holding bar, or spring, or platform, with no other parts."[38] Behe's mousetrap consists of five components: a catch, a hammer, a holding bar, a platform, and a spring (see figure 5.3). Although such a mousetrap is irreducibly complex in the sense that all five components as given are indispensable, might it not be possible by suitably modifying these components to reduce the number needed for a functional mousetrap?

Enter John H. McDonald. On his website titled "A Reducibly Complex Mousetrap" McDonald presents a striking progression of increasingly simpler mousetraps, each irreducibly complex, but each derivable from its more complicated predecessor by the removal of some part and the readjustment of another part.[39]

Figure 5.3 illustrates the standard "snap mousetrap" that one buys in stores. As we have already seen, it has five main parts. McDonald describes these parts as follows: "A hammer, which kills the mouse; a spring, which snaps the hammer down on to the mouse; a hold-down bar, which holds the hammer in the cocked position; a catch, which holds the end of the hold-

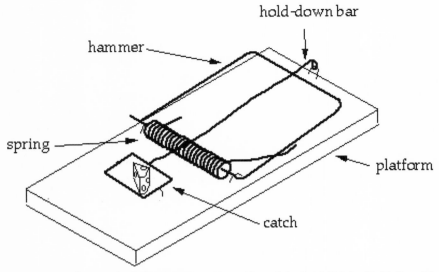

**Figure 5.3.  The Standard Five-Part Mousetrap. (Copyright © 2001 John H. McDonald. Reprinted with permission. All rights reserved.)**

down bar and releases it when the mouse jiggles the catch; and a platform, to which everything else is attached."[40] As McDonald observes, the bait illustrated in figure 5.3 (i.e., a chunk of cheese) is not one of the irreducible parts of the mousetrap since unbaited traps will occasionally catch mice that stumble upon them.

McDonald next describes the progression of increasingly simpler mouse-traps as follows:

> The first step in reducing the complexity of a mousetrap is to remove the catch [see figure 5.4]. The hold-down bar is then bent a little so that it will catch on the end of the hammer that protrudes out from the spring; this end of the hammer might need a little filing to make the action nice and delicate. Clearly this is not as good a mousetrap as the five-part mousetrap, but it will catch mice: the unlucky mouse that is standing on the platform when it jiggles the bait will be just as dead as if it were killed by the more complex trap.
>
> The next step is to remove the hold-down bar and bend the hammer so that one end is resting right at the edge of the platform, holding the hammer up in the cocked position [see figure 5.5]. This is not as good a mousetrap as the four-part mousetrap; for one thing, it needs to be put at the edge of a stair or shelf so that the hammer doesn't hit the floor and send the platform flying, thereby tossing the mouse to safety. But properly positioned, it will catch mice.
>
> The next step is to remove the hammer and bend the straight part of the spring

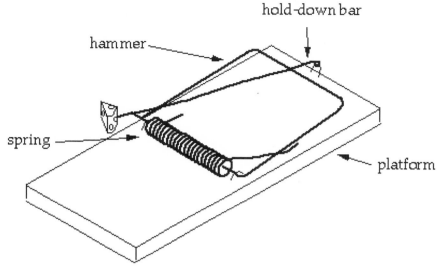

**Figure 5.4.   A Four-Part Mousetrap. (Copyright © 2001 John H. McDonald. Reprinted with permission. All rights reserved.)**

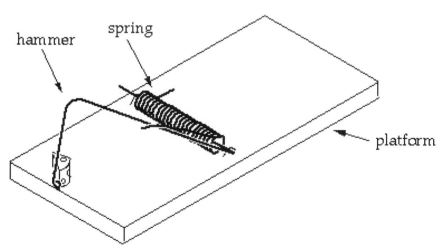

**Figure 5.5.   A Three-Part Mousetrap. (Copyright © 2001 John H. McDonald. Reprinted with permission. All rights reserved.)**

to resemble the hammer of the three-part mousetrap [see figure 5.6]. Without straightening any coils, the gap is just big enough for a mouse's paw or tail, so you'll only catch a few, very unlucky mice. If you could straighten out a few coils of the spring (which is easier said than done—mousetrap springs are pretty tough), you could make a two-part trap that was basically the same as the three-part trap. In either case, the two-part trap will catch mice.

[The final step] is to straighten out a few coils of each end of the spring [see figure 5.7]. One straight piece of the wire is then bent so the end points up; the other piece of wire comes across and rests delicately on the upraised point. The unlucky mouse that is standing under the top wire when it jiggles the trap will be just as dead as if it were killed by the much more complex five-part mousetrap.[41]

What does this progression of mousetraps establish? For the sake of argument, let us assume that each of these mousetraps is functional in the sense that it can, with varying degrees of success, actually catch mice. Let us also assume that the more complicated mousetraps (i.e., those with more components) are more successful at catching mice. Has McDonald therefore established that the standard five-part mousetrap is "reducibly complex"? There is an equivocation here. It is certainly true that the complexity has been reduced if one means that one can build functional mousetraps with fewer than the five parts in a standard mousetrap (complexity here being measured strictly in terms of number of parts). But to say that the complexity of the mousetrap has been reduced suggests that the original mousetrap was not in

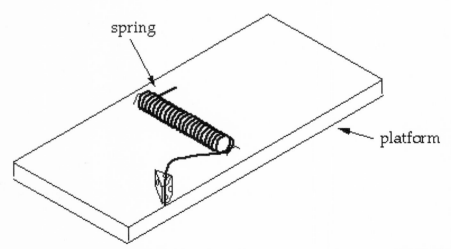

**Figure 5.6. A Two-Part Mousetrap. (Copyright © 2001 John H. McDonald. Reprinted with permission. All rights reserved.)**

spring

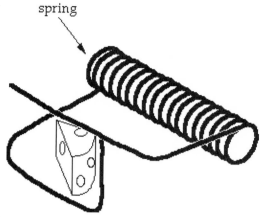

**Figure 5.7. A One-Part Mousetrap. (Copyright © 2001 John H. McDonald. Reprinted with permission. All rights reserved.)**

fact irreducibly complex. That is not the case, and that is where the equivocation comes in. Irreducible complexity as laid out by Behe is a well-defined notion—remove a part and the system in question can no longer perform its original function. That is the key idea behind irreducible complexity, and the standard five-part mousetrap is irreducibly complex in just that sense.

Consequently, a system does not fail to be irreducibly complex because the joint action of removing a part and then modifying other parts retains the system's original function. Irreducible complexity fails only if removal of one part *without* modification of other parts retains the system's original function. As McDonald's mousetraps proceed from more complex to increasingly simple (complexity here being measured strictly in terms of number of parts), parts get not only removed but also modified. McDonald therefore has not demonstrated that the standard five-part mousetrap is not irreducibly complex. Rather, he has shown that there are simpler systems with parts from more complicated systems such that when those parts are suitably modified, the simpler system can perform the same function as the original system (though perhaps not as well). What McDonald has therefore actually accomplished is to give some teeth to the incremental indispensability objection. Indeed, the process of removing and modifying parts to obtain simpler systems can be reversed and corresponds to systems that grow increasingly complex by adding and modifying parts. The reducible complexity objection is therefore a beefed-up version of the incremental indispensability objection, with some details filled in.

But in beefing-up the incremental indispensability objection, has McDonald shown that irreducible complexity does not in fact pose a critical obstacle to the Darwinian mechanism? McDonald himself admits that the progression of mousetraps in figures 5.3 to 5.7 is "not intended as an analogy of how evolution works."[42] He does not elaborate on why the analogy breaks down, but let me offer three reasons. First, consider his successive modifications of mousetrap parts. These modifications counteract the removal of some given part so that what remains of the previous mousetrap can continue to function. But in each case these modifications are designed. Indeed, there is nothing Darwinian about the modifications. They are carefully designed to ensure that what is left of the previous mousetrap can still catch mice. The progression of reducibly complex mousetraps is therefore chock-full of the very teleology that Darwin sought to eliminate.

Another reason to be skeptical of McDonald's mousetraps is that we are still waiting for a Darwinian account of an irreducibly complex biochemical machine like a bacterial flagellum. To his credit, McDonald is approaching the problem correctly and providing the detailed progression of intermediate forms that is needed to show that the Darwinian mechanism can account for an irreducibly complex system. McDonald is therefore presenting the right sort of counterexample to irreducible complexity. The problem is that his progression of mousetraps has little connection to biological reality. Indeed, no such progression that is causally specific and faithful to biological reality has yet to be proposed for the irreducibly complex biochemical systems considered by Behe in *Darwin's Black Box*—notably the flagellum.[43] McDonald claims that his progression of mousetraps shows that the inability of a person to "imagine something doesn't mean it is impossible; it may just mean that the person has a limited imagination."[44] Yet if there is a failure of imagination, it is not simply on Behe's part. Rather, it is on the part of the entire biological community, which to date has offered no detailed Darwinian causal pathways to account for systems like the bacterial flagellum. McDonald has answered the causal specificity problem for mousetraps (and then only by introducing design at crucial points), not for irreducibly complex biochemical machines.

The final reason to be skeptical of McDonald's mousetraps is that for such a counterexample to irreducible complexity to work, it must be possible to build a simpler system that performs the same function as a more complicated system. But there is no reason to think that all irreducibly complex systems can be simplified in this way. Some irreducibly complex systems performing a given function will be *minimally complex*, admitting no simpler system that performs the same function.[45] Thus, even if we grant that a functional mousetrap can be built out of a single spring (see figure 5.7), it does not follow that the bacterial flagellum admits so drastic a simplification. There is in fact not just one bacterial flagellum but many. Different bacteria have different flagella. Some have only one flagellum. Some have tufts of flagella. The flagellum that Behe focuses on is perhaps the best known—the one attached to *E. coli*. It is known that this is not the simplest type of flagellum. As John Postgate remarks: "*E. coli*'s flagellar structure is also not the simplest, in that its hook terminates in a rod with four rather than two discs."[46]

Even so, all known flagella exhibit considerable complexity and give no indication of admitting the vast simplification that we see in McDonald's mousetraps. John Postgate indicates some of the complexity that is involved in all known flagella:

> A typical bacterial flagellum, we now know, is a long, tubular filament of protein. It is indeed loosely coiled, like a pulled-out, left-handed spring, or perhaps a corkscrew, and it terminates, close to the cell wall, as a thickened, flexible zone, called a hook because it is usually bent. . . . One can imagine a bacterial cell as having a tough outer envelope within which is a softer more flexible one, and inside that the jelly-like protoplasm resides. The flagellum and its hook are attached to the cell just at, or just inside, these skins, and the remarkable feature is the way in which they are anchored. In a bacterium called *Bacillus subtilis*, which has a fairly simple structure, the hook extends, as a rod, through the outer wall, and at the end of the rod, separated by its last few nanometres, are two discs. There is one at the very end which seems to be set in the inner membrane, the one which covers the cell's protoplasm, and the near-terminal disc is set just inside the cell wall. In effect, the long flagellum seems to be held in place by its hook, with two discs acting as a double bolt, or perhaps a bolt and washer.[47]

Of course this is just scratching the surface of the complexities involved with even the simplest flagella. The above quote merely describes what amounts to a propeller and its attachment to the cell wall. Additionally there needs to be a motor that runs the propeller. This motor needs to be mounted and stabilized. Moreover, it must be capable of bidirectional rotation. The complexities quickly rise, and we have every reason to suspect that the very simplest flagellum is substantially complex and does not admit anything like the vast simplification of McDonald's mousetraps.

## 5.7 Miscellaneous Objections

In concluding this discussion of objections to Behe's notion of irreducible complexity, I want to consider a few remaining objections. Unlike the previous objections, none of these is substantive. Though initially persuasive, these objections quickly collapse when probed. In every instance they sidestep the actual challenge to the Darwinian mechanism raised by irreducible complexity and instead substitute some simpler or irrelevant problem that was never at issue in the first place. Here are the objections.

**Redundant Complexity**. Philosopher Niall Shanks and biologist Karl Joplin argue that "real biochemical systems . . . manifest *redundant complexity*—a characteristic result of evolutionary processes."[48] Redundant systems allow parts to be removed and their role to be assumed by other parts that thereby preserve the function of the system. As Behe notes in his response, "The observation that some biochemical systems are redundant, however, does not entail that all are. And, in fact, some are not redundant."[49] All Behe's examples of irreducibly complex systems given in *Darwin's Black Box* continue to apply. In responding specifically to Shanks and Joplin, Behe spotlights "inborn errors of metabolism,"[50] which are fatal and for which no redundant backup is available.

**Redefining Irreducible Complexity**. Kenneth Miller conveniently redefines irreducible complexity to discredit it. At a conference in the summer of 2000 he contended that the bacterial flagellum is in fact not irreducibly complex because a subsystem of the flagellum is still functional. The subsystem of the flagellum that Miller identified was a pump—it was therefore no longer an outboard rotary motor.[51] For irreducible complexity to fail, one must be able to remove parts of a system and retain the *same function* as the *original system*. It is no argument against the irreducible complexity of the standard five-part mousetrap to say that it is not irreducibly complex because the base can be used as a doorstop. The issue is not whether fewer parts can be used for some other function, but whether they can accomplish the same function as the original system.

**Argument from Ignorance**. Miller also invokes an *argument from ignorance* against irreducible complexity. Consider again the bacterial flagellum. Like the rest of the biological community, Miller does not know how the bacterial flagellum originated. The biological community's ignorance about the flagellum, however, does not end with its origin but extends to the very functioning of the flagellum. For instance, according to David DeRosier, "The mechanism of the flagellar motor remains a mystery."[52] Miller takes this admission of ignorance by DeRosier and uses it to advantage: "Before [Darwinian] evolution is excoriated for failing to explain the evolution of the flagellum, I'd request that the scientific community at least be allowed to figure out how its various parts work."[53]

But in the article by DeRosier that Miller cites, Miller conveniently omits the following quote: "More so than other motors, the flagellum resembles a machine designed by a human."[54] So apparently we know enough about the bacterial flagellum to know that it is designed or at least design-like. Indeed, we know what most of its individual parts do. Moreover, we know that the flagellum is irreducibly complex. Far from being a weakness of irreducible complexity as Miller suggests, it is a strength of the concept that one can determine whether a system is irreducibly complex without knowing the precise role that each part in the system plays (one need only knock out individual parts and see if function is preserved; knowing exactly what the individual parts do is not necessary). Miller's appeal to ignorance obscures just how much we know about the flagellum and how compelling the case is for its design.

**The Red Herring.** In *Darwin's Black Box* Behe challenges the Darwinian community to exhibit even one published paper in the peer-reviewed literature that provides a causally specific Darwinian account of an irreducibly complex biochemical system. Miller purports to take up Behe's challenge but then conveniently mischaracterizes it as the inability "to find a single published Darwinian explanation for the origin of a biochemical machine."[55] Not surprisingly, Miller next claims to find "four glittering examples of what Behe claimed would never be found."[56] The mathematician George Polya used to quip that if you cannot solve a problem, find an easier problem and solve it. That is in fact what Miller has done. Behe's challenge was not simply to find a Darwinian explanation for the origin of a biochemical machine, but to find a *detailed* Darwinian explanation for the origin of an *irreducibly complex* biochemical machine. For lack of space, let us leave aside the question of whether the glittering examples Miller cites are sufficiently detailed to be causally specific. The fact is that none of the papers Miller cites deals with irreducibly complex systems.[57]

**Painting a Rosy Picture.** Miller is a master at *painting a rosy picture* about the power of Darwinian evolution. According to him "a true acid test" of the power of Darwinian evolution would be to "[use] the tools of molecular genetics to wipe out an existing multipart system and then see if evolution can come to the rescue with a system to replace it."[58] Fair enough. Miller then claims that Barry Hall's experiments with the *lac* operon in *E. coli*, in which this operon was "wiped out" and then "reevolved," spectacularly meets the acid test. Miller exults, "Think for a moment—if we were to happen upon the interlocking biochemical complexity of the reevolved lactose system, wouldn't we be impressed by the intelligence of its design? . . . Except we *know* that it was *not* designed. We know it evolved because we watched it happen right in the laboratory! No doubt about it—the evolution of biochemical systems, even complex multipart ones, is explicable in terms of evolution. Behe is wrong."[59]

Miller's rosy picture disintegrates quickly when subjected to scrutiny. Even more ironic, the actual facts vindicate not Darwinism but design. Consider the actual facts: only one gene was deleted from the *lac* operon, the gene for galactosidase; a variant of galactosidase in the cell was left intact; and an artificial inducer (IPTG) was added to permit the variant of galactosidase to perform that job of the original galactosidase. This is like having two microwave ovens to fix food, one that works (the original galactosidase) and one that does not (the variant). Get rid of the microwave that works and fix the microwave that was broken, and there is no surprise that you can still prepare food. What's more, in the case of the *lac* operon both its disruption and its restoration were the result of design. Once the gene for galactosidase was removed, Barry Hall's *E. coli* were on life support and required the *artificial* inducer IPTG to stay alive. IPTG was not picked out of a hat but was carefully chosen by an intelligent agent—in this case Barry Hall—with full knowledge of what the modified *E. coli* needed to survive. Thus, as Behe rightly notes, "*The system was being artificially supported by intelligent intervention.*"[60]

**Ignored in the Scientific Literature**. Finally, the charge is frequently made that irreducible complexity can be dismissed because it is ignored in the scientific literature. Behe published his work on irreducible complexity in the popular press (the concept was introduced in *Darwin's Black Box*, which appeared with Free Press). For the scientific community to take this concept seriously, the peer-reviewed scientific literature needs to address this concept explicitly and acknowledge it as a genuine problem for biology. That has happened. In 2000 Thornhill and Ussery published an article in the *Journal of Theoretical Biology* addressing "the accessibility by Darwinian evolution of irreducibly complex structures of functionally indivisible components."[61]

The challenge of irreducible complexity to Darwinian evolution is real, and to claim that Behe's ideas have been refuted is false. I close this section with two observations, one by David Griffin, the other by James Shapiro. Griffin is a philosopher interested in the relation between science and religion. Shapiro is a well-respected molecular biologist at the University of Chicago. Neither are fans of intelligent design. Griffin's observation is useful because as an outsider to the biological community with no stake in intelligent design, he has nonetheless concluded that biologists are stonewalling when they purport that Behe's ideas have been thoroughly discredited. Griffin writes:

> The response I have received from repeating Behe's claim about the evolutionary literature—which simply brings out the point being made implicitly by many others, such as Crick, Denton, [Robert] Shapiro, Stanley, Taylor, Wesson—is that I obviously have not read the right books. There are, I am assured, evolutionists who have described how the transitions in question could have occurred. When I ask in which books I can find these discussions, however, I either get no answer or else some titles that, upon examination, do not in fact contain the promised accounts. That such accounts exist seems to be something that is widely known, but I have yet to encounter someone who knows where they exist.[62]

Shapiro explains why:

> There are no detailed Darwinian accounts for the evolution of any fundamental biochemical or cellular system, only a variety of wishful speculations. It is remarkable that Darwinism is accepted as a satisfactory explanation for such a vast subject—evolution—with so little rigorous examination of how well its basic theses work in illuminating specific instances of biological adaptation or diversity.[63]

## 5.8  The Logic of Invariants

In the *Origin of Species* Darwin issued the following challenge: "If it could be demonstrated that any complex organ existed, which *could not possibly* have been formed by numerous, successive, slight modifications, my theory would absolutely break down. But I can find out no such case."[64] Darwin's challenge is an invitation to falsify his theory rather than merely to confirm it. Darwin's theory is confirmed every time the Darwinian mechanism of natural selection and random variation is shown sufficient to account for some transformation between organisms. At issue, however, is the all-sufficiency of the Darwinian mechanism to bridge all differences between organisms. Are there differences between organisms that we should rightly think are beyond the power of the Darwinian mechanism to bridge? In his challenge Darwin indicated that he thought there were none. Behe, on the other hand, thinks there are and that irreducibly complex biochemical systems constitute one such barrier to the Darwinian mechanism.

Is Darwinism impervious to falsification? Perhaps not in principle, though in practice Darwin thought it was. But let us probe deeper. Whatever else we might want to say about intelligent design, it and Darwinism differ significantly in what it would take to refute them. With regard to a particular biochemical system like the bacterial flagellum, intelligent design asserts that "*No* undirected natural process could produce this system." By contrast, Darwinism asserts that "*Some* undirected natural process could produce this system." To falsify the first claim it is enough to exhibit but one causally specific Darwinian pathway capable of producing the system. But how does one falsify the second claim? The Darwinian community seems to assume that to falsify the second claim would require showing that the system could not have been formed by any of a potentially infinite number of Darwinian pathways. Moreover, it places the burden of proof on Darwinism's detractors, who are then called to run through all these possibilities item by item—clearly an impossible task.[65]

Hence a frequent charge made against intelligent design is that its proponents trade in arguments from ignorance. Indeed, the Darwinist literature is filled with reproaches like the following: "Just because you with your recently evolved intellect can't imagine how a system like the bacterial flagellum might have evolved by Darwinian means, don't go thinking that nature didn't find a Darwinian solution to this problem. Give nature more credit. Nature is a lot smarter than we can imagine or may ever be able to imagine. To invoke design because you haven't figured out how nature did it hardly constitutes a profound insight but rather signifies incredulity and laziness on your part. Instead of short-circuiting science by invoking design, get to work and do the honest labor of figuring out how nature—unassisted by mysterious or occult powers—did what it had to do to produce the flagellum. You may not succeed, but at least you'll have given it your best shot instead of capitulating to a long-outdated and discredited view of biological design. All you're offering is *Paley Redux*. William Paley is dead, killed at the hands of David Hume and Charles Darwin. Don't go resurrecting his corpse."[66]

If this seems a bit dramatic, it nonetheless captures the attitude of many Darwinists toward intelligent design. Indeed, it has come to the point where merely entertaining the possibility that Darwin's theory might provide a less than complete account of biological diversity is regarded as barbaric. We have already seen (section 1.8) Daniel Dennett tout Darwin's theory as the greatest idea ever thought—not much quibble-room for Darwin's theory here.[67] Skeptic Michael Shermer writes, "No one, and I mean *no one*, working in the field is debating whether natural selection is the driving force behind evolution, much less whether evolution happened or not."[68] Not much skepticism here. Biologist Paul Ewald even accords Darwin's theory the same status as arithmetic: "You have heritable variation, and you've got differences in survival and reproduction among the variants. That's the beauty of it. It has to be true—it's like arithmetic. And if there is life on other planets, natural selection has to be the fundamental organizing principle there, too."[69]

What, then, are we to make of Darwin's challenge? In issuing his challenge to falsify his theory, was Darwin in fact sticking his neck out, or was he merely going through rhetorical motions to convince his readers that he did not hold his theory dogmatically and that he was willing to take objections against it seriously even though privately he saw his theory as impervious to falsification? Let us give Darwin the benefit of the doubt that he meant his challenge seriously, but also that he did not see his challenge as adequately met. The question then before us is how the irreducibly complex biochemical machines, to which Michael Behe has called our attention, might meet Darwin's challenge. In other words, how can one show that such systems "*could not possibly* have been formed by numerous, successive, slight modifications"?

I have emphasized the phrase "could not possibly" because how we interpret it will make all the difference between whether Darwin's challenge constitutes a real challenge or merely a rhetorical flourish. Let us therefore inquire how in general we determine that something could not possibly have happened. First off, let us be clear that humans are in the habit of making proscriptive generalizations (i.e., claims asserting that something cannot happen) all the time, and in many instances do so quite reliably. *You can't square a circle. You can't win that election. You can't turn lead into gold.* Such claims are common coin. The crucial issue with regard to irreducible complexity is how we justify such claims.

How, then, do we justify a proscriptive generalization? Alternatively, how do we show that something could not possibly happen? There are two ways and only two ways to show that something could not possibly have happened:

1. Conduct an exhaustive search.
2. Find an invariant.

Conducting an exhaustive search is clear enough—simply run through all the possibilities and determine that in each instance the thing being sought does not obtain. If an exhaustive search fails to locate the thing being sought, then the thing being sought can be decisively ruled out.

Frequently, however, the search space is too large to accommodate an exhaustive search. In that case we need to find an invariant. Invariants are the standard way to establish a proscriptive generalization when an exhaustive search is not possible. Often, to say that something could not possibly happen is to say that some process is unable to transform one thing into another. An invariant is some property that does not change under the action of such a process. Most proscriptive generalizations, in ruling out some occurrence, take the form of denying that some process $\varphi$ is capable of transforming X into Y. Typically X and $\varphi$ are presupposed, and the point at issue is whether $\varphi$ is able to transform X into Y. Now, if there is an invariant that does not change (i.e., remains constant) under the action of $\varphi$, it becomes possible to rule out conclusively the transformation of X into Y. To do so, the invariant simply needs to assign a different value to X and Y. Since the action of $\varphi$ does not change the value of the invariant, if X could be transformed into Y, whatever value the invariant assigns to X should remain unchanged for Y. The fact that it does not is enough to rule out the transformation of X into Y.

More formally, the logic of invariants can be described as follows. Suppose we are given a reference class of possibilities $\Omega$ (also known as a phase space) and a process $\varphi$ on $\Omega$ (i.e., $\varphi$ is a function that maps $\Omega$ to itself). Let **Init** and **Term** denote nonempty subsets of $\Omega$ (**Init** for initial states and **Term** for terminal states). The crucial question now is whether by repeated applications of $\varphi$ (i.e., by running the process $\varphi$ indefinitely), it is possible to get from **Init** to **Term**. To answer this question define a function **Invar** on $\Omega$ (for definiteness assume **Invar** is real-valued, i.e., **Invar** takes values in the real numbers **R**). Let $A = \{\, r \in \mathbf{R} \mid$ there exists some natural number n and some X in **Init** such that $\mathbf{Invar}(\varphi^n(X)) = r \,\}$ and $B = \{\, r \in \mathbf{R} \mid$ there exists some Y in **Term** such that $\mathbf{Invar}(Y) = r \,\}$ (note that $\varphi^n(X)$ denotes the application of $\varphi$ to X n times—in case n equals zero, $\varphi^n(X)$ is by definition just X, i.e., $\varphi^0$ is the identity function on $\Omega$). The set A comprises all the values that **Invar** assumes on **Init** together with all the values **Invar** assumes on all possibilities in $\Omega$ accessible from **Init** via $\varphi$. The set B comprises all the values that **Invar** assumes on **Term**. If A and B are disjoint (i.e., do not share any members), then we say that **Invar** is an *invariant function* with respect to **Init**, **Term**, and $\varphi$, and that it is impossible to transform **Init** into **Term** via $\varphi$. This logic of invariants generalizes straightforwardly to continuous dynamical systems in which $\varphi$ maps $\mathbf{R} \times \Omega$ (i.e., the Cartesian product of the reals with $\Omega$) into $\Omega$ such that $\varphi(s+t,X) = \varphi(s,\varphi(t,X))$ for s and t in **R** and X in $\Omega$.[70]

To illustrate this use of invariants, consider the modified chessboard in figure 5.8. This figure depicts a standard chessboard, but with lower left and upper right squares removed. Below the chessboard are a number of tiles. Each of the tiles can cover exactly two adjacent squares of the chessboard. I have shown only three such tiles, but we imagine that there are an unlimited number. Suppose your task is to cover this modified chessboard with such

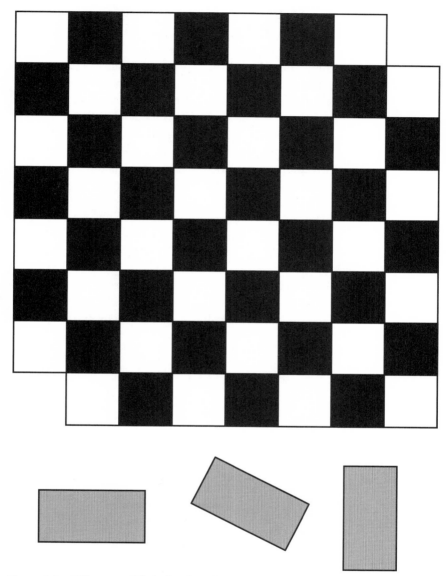

**Figure 5.8.  Tiling a Modified Chessboard.**

tiles so that each tile covers exactly two adjacent squares and so that the entire chessboard ends up covered. Can it be done? With current computational power it might just be possible to program this task and run through all possibilities of tiling the chessboard. Nevertheless, such an exhaustive search is neither necessary nor efficient. Instead, it is enough to note that whenever a tile covers a pair of adjacent squares, it covers both a white and a black square. Thus, whatever tiling procedure is used, the number of white minus the number of black squares covered remains constant, namely, zero. This difference between the number of white and black squares covered by tiles is therefore an invariant. Now notice that the two squares deleted from the standard chessboard were both black. Thus, if it were possible to cover the entire chessboard with tiles, two more white squares would be covered than black squares. This would violate the invariant, which guarantees that the difference between white and black squares always remains zero and never rises to two. It follows that this modified chessboard cannot be completely covered with tiles.

Sometimes an exhaustive search and an invariant are both capable of settling a proscriptive generalization. Consider the famous Königsberg bridge problem (see figure 5.9). Running through Königsberg (today known as Kali-

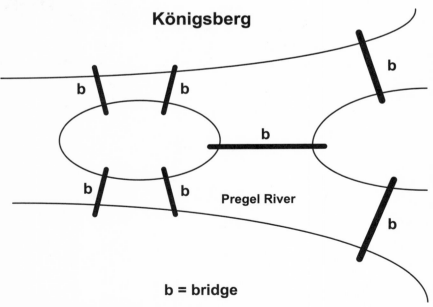

**Figure 5.9. The Königsberg Bridge Problem.**

ningrad) is the Pregel river. Within the city limits of Königsberg, the Pregel has two islands and seven bridges connecting those islands. The citizens of Königsberg used to amuse themselves by attempting to cross all seven bridges in a continuous circuit without recrossing any bridge.[71] Because the complexities of this problem are so minimal, the citizens simply by trial and error (i.e., exhaustive search) managed to convince themselves that no such circuit was possible. Even so, the famous mathematician Leonard Euler, while in St. Petersburg in the 1730s, was able to prove this result mathematically. He simplified the representation of this problem, converting it to what nowadays is called a graph (see figure 5.10). Euler then proved the following theorem: For a graph as in figure 5.10 to have a continuous circuit that traverses all the edges and does not recross any edges, the number of vertices with an odd number of edges must be either zero or two.[72] These numbers constitute

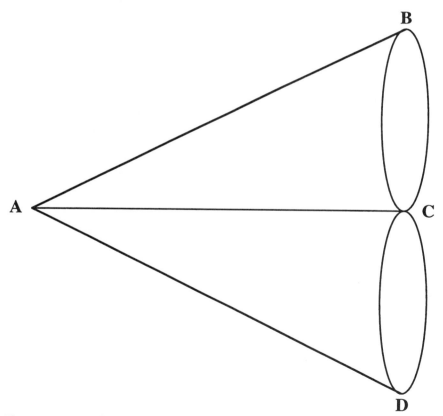

**Figure 5.10.   Graph Corresponding to the Königsberg Bridge Problem.**

an invariant of the problem. What's more, the graph in figure 5.10 violates this invariant and therefore does not permit a continuous nonrepeating circuit.

Both the modified chessboard and the Königsberg bridge problem involve strictly mathematical invariants. Mathematical invariants are exceedingly common. Indeed, mathematics abounds with invariants. Classic problems of geometry like squaring a circle or trisecting an angle were finally shown to have no solution in the nineteenth century when Galois developed his theory of groups, which attached invariants to the operations of ruler and compass.[73] Until then the claim that one cannot square a circle would simply have reflected the failure of the mathematical community to perform that feat (squaring a circle means with ruler and compass deriving a square having the same area as a given circle). After that, its status changed: though previously unattained, the feat was now recognized as unattainable. Or consider algebraic topology. Algebraic topology consists in attaching algebraic invariants to geometric objects and therewith establishing that certain geometric objects are not topologically equivalent to others (topologists refer to topologically equivalent spaces as homeomorphic).[74] So too, consistency theorems in mathematical logic depend on establishing an invariant that holds for all provable sentences and then showing that the invariant is violated for some sentences.[75] Invariants are the standard way to establish proscriptive generalizations within mathematics.

But what about outside mathematics? Invariants also apply outside mathematics, but in that case theoretical coherence and empirical adequacy rather than strict mathematical proof govern the applicability of invariants. The conservation laws of physics provide the most notable example of invariants that are empirically rather than strictly mathematically based. For instance, the law of conservation of energy states that in an isolated system total energy neither increases nor decreases. Total energy thus serves as an invariant for a system whose operation is self-contained. Consequently, should the total energy fluctuate, this would indicate that the system was not self-contained but was exchanging energy with the outside. The justification of the law of conservation of energy derives not from a strict mathematical proof but from long experience with energetic systems as well as from fine-tuning our notion of energy over time so that the law can continue to be maintained (there is a bit of circularity here, but it is virtuous rather than vicious).

Perhaps the best known cases of proscriptive generalizations derive from the second law of thermodynamics. The second law asserts that the entropy of a closed system is extremely likely statistically to increase. Since increased entropy corresponds to the diffusion of energy so that it can no longer be exploited to perform work, in practice the second law amounts to the claim that the effective work capable of being extracted from a system remains constant or diminishes. And since all real physical systems diffuse energy, it follows that without the addition of outside energy the actual work capable of being performed by such systems continually diminishes. Entropy thus becomes an invariant in the sense that for an isolated system its value can never fall below its starting value. Thus for a system to increase in its capacity to perform work or even for that capacity to remain constant is statistically so unlikely as to be effectively impossible. This is how the second law is used to preclude perpetual motion machines (see chapter 3 and especially section 3.10).

The refutation of alchemy also looks to an invariant. Though fission and fusion can change the atomic number of elements, for many elements not subjected to these or other highly energetic processes, their atomic number is quite stable. In particular, for many elements their atomic number is immune to change from low-energy interventions like subjecting them to potions and furnaces, as was the case with alchemy. Lead and gold are both elements of this sort. Lead has atomic number 82 and gold has atomic number 79. These numbers serve as invariants with respect to the processes that alchemists used to try to convert lead into gold. It follows that alchemy, by limiting itself to low-energy interventions, cannot succeed in converting lead into gold.

What about irreducible complexity? According to Behe irreducible complexity is likewise an invariant for the Darwinian process of random variation and natural selection. Thus for an organism that possesses an irreducibly complex biochemical system (e.g., a bacterial flagellum) and one that does not, Behe argues that it is effectively impossible for an organism lacking such a system to evolve by purely Darwinian processes into one possessing such a system. Whether Behe's invariant can legitimately be used to place such a restriction on Darwinian processes is the subject of the next section.

## 5.9  Fine-Tuning Irreducible Complexity

Few ideas in science when first posed or even when well developed are exactly correct. Most are either salvageable or wrong. Wrong ideas can be hopelessly and irretrievably wrong. They can be helpfully wrong in the sense of providing a useful foil for better ideas. Or they can be, in the words of Wolfgang Pauli, "not even false"—in other words they are so misconceived as not to merit the appellation "false" or "wrong." Ideas that are not even false are like the question "Have you stopped beating your wife lately?" Better not to entertain them at all. In its more charitable moments the Darwinian community regards Behe's idea of irreducible complexity as helpfully wrong in the sense of identifying a problem to which Darwinists need to devote more attention (though their confidence remains unshaken that irreducibly complex biochemical systems will eventually yield to a Darwinian analysis). In its less charitable moments, however, the Darwinian community regards irreducible complexity as hopelessly and irretrievably wrong in the sense that not only does it not pose a serious challenge to Darwinism, but also it confuses nonbiologists into thinking there is a problem with Darwinism when in fact there is not. And then there are those, like Pauli, who sneer that irreducible complexity is not even false.

Behe's idea of irreducible complexity is neither exactly correct nor wrong in any of the three senses described (i.e., helpfully, hopelessly, or abysmally). Instead it is salvageable. Salvageable ideas need some fixing and fall on a broad continuum. The precise point on the continuum where a salvageable idea lands depends on how much fixing it needs. The mathematician Gauss, for instance, used to complain about the amount of fixing he had to do to Laplace's mathematical work (German *Genauigkeit* putting the brakes on French *éclat*). Presumably Gauss thought the amount of fixing was inordinate. On the other hand salvaging an idea may not require much effort at all. Godfrey Hardy's tidying of Ramanujan's work in the theory of numbers is a case in point. Salvageable ideas can be substantially correct, though requiring a bit of touch-up. They can be sloppy mess, requiring a complete overhaul. And they can be everything in between. My own view is that Michael Behe's ideas about irreducibly complex biochemical systems are substantially correct and require merely a bit of fine-tuning. The aim of this section, then, is to fine-tune irreducible complexity and show how irreducible complexity, suitably fine-tuned, decisively challenges Darwinism.

Let us start by reviewing Behe's original definition of irreducible complexity. Here it is:

> **Definition IC$_{init}$**—A system is *irreducibly complex* if it is "composed of several well-matched, interacting parts that contribute to the basic function, wherein the removal of any one of the parts causes the system to effectively cease functioning."[76]

Implicit in this definition is of course that the *original* function of the system cannot be recovered once a part is removed. This is probably best made explicit since critics like Kenneth Miller have argued that finding a different function for a proper subsystem is enough to refute irreducible complexity for the whole system (see section 5.7). If we require that the original function be preserved and make this requirement explicit, the definition of irreducible complexity looks as follows:

> **Definition IC$_2$**—A system is *irreducibly complex* if it is composed of several well-matched, interacting parts that contribute to the basic function, wherein the removal of any one of the parts effectively prevents the system from maintaining its basic, and therefore original, function.

So redefined, irreducible complexity still requires further clarification. Russell Doolittle put his finger on one conceptual difficulty when he proposed a putative counterexample to the irreducible complexity of the blood-clotting cascade. Even though the counterexample itself failed, the logical point it raised was well taken. In *Darwin's Black Box* Behe had considered the irreducible complexity of the blood-clotting cascade. Doolittle subsequently cited an article in *Cell*, which he interpreted as arguing that mice deficient in both plasminogen and fibrinogen (each components of the blood-clotting cascade) were viable whereas those lacking either but not both were not.[77] As it turned out, mice deficient in both plasminogen and fibrinogen had the same problems with lack of clotting that mice only deficient in fibrinogen had.[78] Even though Doolittle's counterexample failed, it was nonetheless instructive: Though knocking out any one component of an irreducibly complex system might destroy the original function of the system, knocking out more than one might in principle allow it to be recovered. Let us imagine, for instance, two parts of an irreducibly complex system that are joined together and that when both removed allow the system to continue functioning normally. Jointly, then, the two parts are useless to the system. Nevertheless, we can imagine that when only one of these parts is removed, the other that is retained acts as a wrench that gums up the system and destroys its original function. To forestall this possibility the definition of irreducible complexity needs to be revised as follows:

> **Definition IC$_3$**—A system is *irreducibly complex* if it is composed of several well-matched, interacting parts that contribute to the basic function, wherein the removal of any one or more parts effectively prevents the system from maintaining its basic, and therefore original, function.

Although this definition is logically tighter than the previous ones, it has the disadvantage of not being as experimentally tractable. With the old definition, it is enough to verify irreducible complexity for an N component system by knocking out the N components individually and checking in each case whether the original function is preserved. With the new definition, to verify irreducible complexity it is now necessary to knock out all possible subsets of these N components. But the number of these subsets is exponential in N. Whereas it took only N knock-outs to determine irreducible complexity before, it now takes $2^N$ (i.e., the number of subsets of a set with N elements).[79]

Even so, tractability can often be recovered by combining an experimental with a theoretical analysis. In determining whether a system is irreducibly complex, the whole point to removing parts from the system, both individually as well as in combination, is to ascertain whether each of the parts is *indispensable* to the system. For a part to be indispensable to a system, not only must knocking it out destroy the system's function, but also knocking it out in combination with other parts must destroy the system's function. Indispensability, however, can be assessed apart from such knock-out experiments through theoretical analyses. A bacterial flagellum, for instance, is an outboard rotary motor that propels a bacterium through its watery environment. It therefore requires—on theoretical or conceptual grounds—some part outside the cell membrane that rotates. For the flagellum a long, tubular filament accomplishes this task. What's more, there are no backup components that accomplish the same task. A theoretical analysis therefore demonstrates that this component is indispensable to the flagellum and cannot be eliminated from it. It follows that indispensability can be assessed in two ways: either experimentally by removing parts or theoretically by determining the precise role that the parts play in the system as a whole. Let us therefore revise the definition of irreducible complexity as follows:

> **Definition IC₄**—A system performing a given basic function is *irreducibly complex* if it is composed of several well-matched, interacting parts that are each indispensable to maintaining its basic, and therefore original, function.

Several aspects of this definition, which were also included in Behe's original definition, now require clarification. A system's *basic function* is the main function it performs. The basic function of the heart is to pump blood. The basic function of a flagellum is to propel a bacterium through solution. The basic function of a mousetrap is to catch mice. Basic function needs to be distinguished from peripheral function. It is a peripheral function of a mouse-trap to act as a doorstop. A mousetrap may function as a superb doorstop, but that is not its basic function. Two other aspects of this definition require clarification: that the parts of an irreducibly complex system *interact* and be *well-matched*.

The requirement that parts of an irreducibly complex system interact is a continuity requirement ensuring that the system hangs together and does not decompose into multiple subsystems that have nothing do to with each other. For instance, the lava lamp and boom box on my desk together con-duce toward making my work experience more pleasant, which is their joint function. Although they interact in the sense of having this joint effect on me, they are not properly regarded as interacting systems in the sense of causally influencing each other in any significant way (which is the sense relevant to the definition of irreducible complexity). The lava lamp and boom box are discrete systems, each having a basic function of its own and each capable of operating on its own. The joint system of lava lamp plus boom box is therefore not properly regarded as consisting of interacting parts because the parts that constitute the lava lamp do not causally influence in any significant way the parts that constitute the boom box (or vice versa). This is perhaps more clearly stated by saying that the parts of an irreducibly complex system have to be not just interacting but *mutually* interacting. The parts of the joint lava lamp plus boom box system are not mutually inter-acting.

The requirement that parts of an irreducibly complex system be well-matched is a fittedness requirement ensuring that the system's parts are adapted or adjusted to each other and not just thrown together in a puréed soup. Shanks and Joplin, for instance, offer the Belousov-Zhabotinsky reac-tion as a counterexample to show that irreducible complexity can arise by natural processes after all.[80] In the Belousov-Zhabotinsky reaction, chemicals are mixed together and give rise to circular shapes.[81] All the chemicals are required and must appear in certain proportions for the reaction to take place. Even so, they are not well-matched in the sense of being specifically adapted to each other. Compare this to a car in which, for instance, the fan belt is specifically adapted to the cooling fan. Or compare this to a bacterial flagellum in which the tubular filament is specifically adapted to the hook joint.

Even with these clarifications, the definition of irreducible complexity can use some further fine-tuning. One problem is that even with parts well-matched and mutually interacting, the way the parts are individuated may be wholly arbitrary and thus fail to articulate the system at its functional joints. Consider, for instance, a jigsaw puzzle. The pieces are well-matched and mutually interacting in virtue of their shapes, but those shapes are purely arbitrary and bear no relation to the picture signified by the puzzle. Or consider a car engine whose parts are individuated simply in terms of spatial coordinates—for instance, consider the cubic foot of engine at the right rear portion on the engine block. No doubt, removing a cubic foot anywhere from the engine will destroy its function. But individuating the engine into disjoint cubes, even if the cubes are well-matched, interacting, and indispensable, is hardly going to make for a useful definition of irreducible complexity. If we allow ourselves to individuate parts willy-nilly, we can divide up just about any system into indispensable parts and thus render it irreducibly complex. To avoid this problem it is necessary to require that the parts of an irreducibly complex system be *nonarbitrarily individuated*. For instance, the parts of a protein machine like the bacterial flagellum must not be individuated by picking and choosing a few amino acids from each of the proteins that together constitute the machine. Instead, the proteins themselves or higher order structures built out of those proteins must comprise the parts that are individuated. Fortunately, in practice the individuation of a system's parts proceeds as a matter of course and is nonarbitrary.

While the requirement that the individuation of a system's parts be nonarbitrary may seem like a restriction on the concept of irreducible complexity, it can also be seen as a way of liberating the concept. Consider, for instance, a parachute. A parachute consists of a canopy, a harness, and suspension lines that connect the canopy to the harness. Is the parachute irreducibly complex? Certainly the canopy and the harness are both indispensable to the parachute's function of slowing a skydiver's fall. But what about the suspension lines? There are multiple suspension lines, and any one of them is dispensable. Nonetheless, taken jointly they are indispensable. Now it seems that there is nothing arbitrary about taking the suspension lines jointly and treating them as a single part of the parachute. If we do this, the suspension lines, taken jointly, become a nonarbitrarily individuated part of the parachute. This freedom, to individuate parts of a system in ways that are nonarbitrary and yet not immediately obvious, can render a part indispensable and thus a system that contains the part irreducibly complex even if on another individuation it would not be irreducibly complex. This freedom greatly expands the scope of irreducible complexity. For instance, redundant systems with parts that serve as backups for some primary part can all be folded into one grand part that includes the original primary part as well as the backups. In this way redundant systems can become irreducibly complex.[82]

There is still another way the definition of irreducible complexity can use some fine-tuning. Irreducible complexity as defined thus far requires that *all* components of a system be indispensable. This seems an unnecessary limitation. Consider an old-fashioned pocket watch with a winding mechanism. The basic function of the watch is to tell time. What's more, many parts of the watch are indispensable to that basic function, for instance, the spring, the face, the hour hand, and the minute hand. On the other hand, other parts of the watch are dispensable, for instance, the crystal, the metal cover holding the crystal, and the chain. By focusing purely on the indispensable parts of the pocket watch one obtains what can be called an *irreducible core* that has all the crucial properties of irreducibly complex systems considered so far. It therefore makes sense to define an irreducibly complex system as one that contains an irreducible core whose parts are each indispensable, but where the system itself is permitted to retain certain unnecessary or redundant elements. This redefinition greatly expands the scope of irreducible complexity. It allows the definition to encompass many designed objects that would otherwise be omitted. It also broadens the range of biological systems that are irreducibly complex and potentially inaccessible via the Darwinian mechanism.

We are now in a position to offer the following final definition of irreducible complexity:

**Definition IC$_{final}$**—A system performing a given basic function is *irreducibly complex* if it includes a set of well-matched, mutually interacting, nonarbitrarily individuated parts such that each part in the set is indispensable to maintaining the system's basic, and therefore original, function. The set of these indispensable parts is known as the *irreducible core* of the system.

Given this definition, does it immediately follow that if we find a biochemical system that satisfies it, that system is inaccessible via the Darwinian mechanism? The answer is obviously no. Extremely simple systems can be irreducibly complex according to this definition (e.g., a system consisting of a single indispensable part—like a rock that is used as a doorstop). What will prove crucial in applying this definition are the auxiliary conditions that supplement it and turn it into an effective invariant for the Darwinian mechanism (think of an effective invariant here as an insurmountable obstacle for the Darwinian mechanism—see section 5.8). What are these auxiliary conditions?

To answer this question let us review why Behe thinks irreducible complexity poses such a significant challenge to Darwinism in the first place:

> An irreducibly complex system cannot be produced directly (that is, by continuously improving the initial function, which continues to work by the same mechanism) by slight, successive modifications of a precursor system, because any precursor to an irreducibly complex system that is missing a part is by definition nonfunctional. An irreducibly complex biological system, if there is such a thing, would be a powerful challenge to Darwinian evolution. Since natural selection can only choose systems that are already working, then if a biological system cannot be produced gradually it would have to arise as an integrated unit, in one fell swoop, for natural selection to have anything to act on.[83]

Irreducible complexity challenges Darwinism by attempting to exploit the Achilles' heel of the Darwinian mechanism, namely, its compulsion to track selective advantage. I call this the *tracking problem*. The Darwinian mechanism consists of random variation, which provides the raw material for Darwinian evolution, and natural selection, which sifts that material. For natural selection to accomplish anything, it must latch onto random variations that are advantageous to the organism. Now the problem is that irreducibly complex systems are discrete integrated combinatorial objects whose parts fit together in highly circumscribed ways. Organisms, on the other hand, if they evolved by Darwinian evolution, must trace a virtually continuous figure through space and time. The problem, then, is to coordinate the gradual Darwinian evolution of an organism with the emergence of an irreducibly complex system that the organism now houses but did not always possess. This is the tracking problem, and the Darwinian community has been utterly stymied in explaining the emergence of irreducibly complex systems once the complexity of these systems becomes palpable.[84]

What makes the complexity of an irreducibly complex system palpable and thus turns irreducible complexity into an effective invariant for assessing the power and limits of the Darwinian mechanism? Crucial in this regard is the addition of two auxiliary conditions that effectively rule out the functional intermediates that the Darwinian mechanism needs if it is to produce an irreducibly complex system gradually (which is the only way it can produce it). The fundamental intuition underlying irreducible complexity is that *irreducibly complex systems cannot be substantially simplified and yet preserve function*. The Darwinian mechanism requires such simplification if step by gradual step it is to succeed in generating an irreducibly complex system. The two auxiliary conditions added to the definition of irreducible complexity, on the other hand, are meant to forestall precisely such simplification. By being jointly satisfied, these conditions rule out any substantial simplification of an irreducibly complex system and thus effectively prevent the Darwinian mechanism from exploiting the functional intermediates it would need to account for the system. Here are the two conditions:

- **Numerous and Diverse Parts**. If the irreducible core of an irreducibly complex system consists of one or only a few parts, there may be no insuperable obstacle to the Darwinian mechanism explaining how that system arose in one fell swoop. But as the number of indispensable well-fitted, mutually interacting, nonarbitrarily individuated parts increases in number and diversity, there is no possibility of the Darwinian mechanism achieving that system in one fell swoop.
- **Minimal Complexity and Function**. Given an irreducibly complex system with numerous and diverse parts in its core, the Darwinian mechanism must produce it gradually. But if the system needs to operate at a certain minimal level of function before it can be of any use to the organism and if to achieve that level of function it requires a certain minimal level of complexity already possessed by the irreducible core, the Darwinian mechanism has no functional intermediates to exploit.

These two conditions transform the definition of irreducible complexity into a vise that allows the Darwinian mechanism no room to maneuver. To achieve an irreducibly complex system the Darwinian mechanism has but two options. First, it can try to achieve the system in one fell swoop. But if an irreducibly complex system's core consists of numerous and diverse parts, that option is decisively precluded. The only other option for the Darwinian mechanism then is to try to achieve the system gradually by exploiting functional intermediates. But this option can only work so long as the system admits substantial simplifications. The second condition blocks this other option. Let me stress that there is no false dilemma here—it is not as though there are other options that I have conveniently ignored but that the Darwinian mechanism actually has at its disposal. Ideas like coordinated macromutations, lateral gene transfer, set-aside cells, and punctuated saltational events are thoroughly non-Darwinian.

In closing this section, let me briefly describe how these two auxiliary conditions apply to the bacterial flagellum. By requiring around thirty distinct proteins in even its simplest known forms, the bacterial flagellum clearly possesses numerous and distinct parts. It therefore satisfies the first auxiliary condition and therefore cannot be achieved by the Darwinian mechanism in one fell swoop. To see that the bacterial flagellum also satisfies the second condition, consider that in propelling a bacterium through its watery environment, the flagellum must overcome Brownian motion. The main reason flagella need to rotate bidirectionally is because Brownian motion sets bacteria off their course as they try to wend their way up a nutrition gradient. Reversing direction of the rotating filament causes the bacterium to tumble, reset itself, and try again to get to the food it needs. The minimal functional requirements of a flagellum, if it is going to do a bacterium any good at all in propelling it through its watery environment, is that the filament rotate bidirectionally and extremely fast. Flagella of known bacteria spin at rates well above 10,000 rpm (actually, closer to 20,000 rpm). Anything substantially less than this is not going to overcome the disorienting effects of Brownian motion.[85]

But how simple can a flagellum be and still attain this minimal level of function? It will need a bidirectional motor. Moreover, because it spins so fast, the motor will need to be attached to the cell wall and stabilized with stators, rings, and bushings. It will also need a propeller unit outside the cell wall. What's more, the entire flagellum needs to be self-assembling. Thus it will require various additional proteins that facilitate and regulate its construction even though these proteins do not appear in the actual flagellum. Now while it is true that various known flagella differ in complexity, the differences are in no way drastic. Moreover, a theoretical analysis of the sort just sketched, where one considers what is required for a flagellum to achieve a certain minimal level of function, indicates that the complexity of known flagella is not very different from the minimal complexity that such systems might in principle require.

It follows that the bacterial flagellum satisfies both auxiliary conditions and thus constitutes an irreducibly complex system that is unattainable by the Darwinian mechanism. How firm is this conclusion? Very firm. Does this conclusion have the status of a mathematical proof? No. Irreducible complexity, even when combined with the two auxiliary conditions, is not irrefutable complexity. As throughout science, claims are always subject to further testing and thus to refutation. Nonetheless, with respect to the bacterial flagellum the burden of proof is on the Darwinist to show how this system can be substantially simplified and still retain the minimal level of function needed to overcome Brownian motion. Combined with the two auxiliary conditions, irreducible complexity provides an effective invariant for the Darwinian mechanism. This invariant distinguishes organisms with and without such a system and thus provides strong evidence against the power of the Darwinian mechanism to merge such organisms. Granted, in the case of the bacterial flagellum this argument does not constitute an absolute proof. Even so, it is reason enough to treat Darwinism as a provisional theory of biological diversity and to open the discussion to other theoretical possibilities.

## 5.10   Doing the Calculation

The Darwinian mechanism is powerless to produce irreducibly complex systems for which the irreducible core consists of numerous diverse parts that are minimally complex relative to the minimal level of function they need to maintain. This was the conclusion of the last section. It is also a conclusion that for Behe does not go far enough. Not only does he regard such systems as beyond the power of the Darwinian mechanism, but he also regards them as clearly signaling design. In *Darwin's Black Box* Behe's argument for the design of irreducibly complex systems was informal, arguing that our only well-supported accounts of such systems, once they attain a certain level of complexity, are those that appeal to a designing intelligence. My own more theoretical work on detecting design appeared subsequently in *The Design Inference* and identified specified complexity as the key marker of design. The connection between these two forms of complexity could thus be established by showing that irreducible complexity is a special case of specified complexity.

I want therefore in this section to show how irreducible complexity is a special case of specified complexity, and in particular I want to sketch how one calculates the relevant probabilities needed to eliminate chance and infer design for such systems. Determining whether an irreducibly complex system exhibits specified complexity involves two things: showing that the system is specified and calculating its probability (recall that probability and complexity are correlative notions—see section 1.3). Specification is never a problem. The irreducibly complex systems we consider, particularly those in biology, always satisfy independently given functional requirements (see section 3.7). For instance, in the case of the bacterial flagellum, humans developed outboard rotary motors well before they figured out that the flagellum was such a machine. This is not to say that for the biological function of a system to constitute a specification humans must have independently invented a system that performs the same function. Nevertheless, independent invention makes the detachability of a pattern from an event or object all the more stark (see section 2.5). At any rate, no biologist I know questions whether the functional systems that arise in biology are specified. At issue always is whether the Darwinian mechanism, by enlisting natural selection, can overcome the vast improbabilities that at first blush seem to arise with such systems and therewith break a vast improbability into a sequence of more manageable probabilities.[86]

To illustrate what's at stake in breaking vast improbabilities into more manageable probabilities, suppose a hundred pennies are tossed. What is the probability of getting all one hundred pennies to exhibit heads? The probability depends on the chance process by which the pennies are tossed. If, for instance, the chance process operates by tossing all the pennies simultaneously and does not stop until all the pennies simultaneously exhibit heads, it will require on average about $10^{30}$ such simultaneous tosses for all the pennies to exhibit heads. If, on the other hand, the chance process tosses each penny individually and keeps tossing until it lands heads, then after about a hundred tosses all the pennies will on average exhibit heads (and after a thousand tosses it will be almost certain that all the pennies exhibit heads). Darwinists tacitly assume that all instances of biological complexity are like the second case, in which a seemingly vast improbability can be broken into a sequence of reasonably probable events by gradually improving on an existing function (in the case of our pennies improved function corresponds to exhibiting more heads). The challenge of irreducible complexity is to show that there are instances of biological complexity that must be attained all at once (as when the pennies are tossed simultaneously) and thus for which gradual Darwinian improvement offers no help in overcoming the improbability.

Consider therefore an irreducibly complex system whose irreducible core contains numerous diverse parts that are minimally complex relative to the minimal level of function they need to maintain. Such a system clearly resists the divide-and-conquer approach typical of Darwinian gradualism. Richard Dawkins has memorably described this gradualistic approach to achieving biological complexity as "climbing Mount Improbable."[87] Climbing Mount Improbable requires taking a slow serpentine route up the backside of the mountain and avoiding precipices. For irreducibly complex systems that have numerous diverse parts and that exhibit the minimal level of complexity needed to retain a minimal level of function, such a gradual ascent up Mount Improbable is no longer possible. The mountain is, as it were, all one big precipice. Certainly this seems intuitively right. It is now time to put some numbers with this intuition.

An irreducibly complex system is a discrete combinatorial object. Probabilities therefore naturally arise and attach to such objects. Such objects are invariably composed of building blocks. Moreover, these building blocks need to be converge on some location. Finally, once at this location the building blocks need to be configured to form the object. It follows that the probability of obtaining an irreducibly complex system is the probability of originating the building blocks required for the system, multiplied times the probability of locating them in one place once the building blocks are given, multiplied times the probability of configuring them once the building blocks are given and located in one place. We therefore have three probabilities that figure into the probability of a discrete combinatorial object: $p_{orig}$ = the probability of originating the building blocks for that object; $p_{local}$ = the probability of locating the building blocks in one place once they are given; $p_{config}$ = the probability of configuring the building blocks once they are given and in one place. Let us call the first of these the *origination probability*, the second the *localization probability*, and the third the *configuration probability*. It follows that the probability of a discrete combinatorial object equals the product of these three probabilities: $p_{dco}$ = $p_{orig}$ × $p_{local}$ × $p_{config}$.[88]

To illustrate the probability of a discrete combinatorial object, imagine you want to bake a cake that requires several distinct ingredients. The probability of baking the cake entirely by chance is then the product of three probabilities. First there is $p_{orig}$, the probability that the ingredients (i.e., building blocks) for your cake will arise by chance and show up at your supermarket. Next there is $p_{local}$, the probability that by going through a supermarket and randomly picking items off the shelf, you just happen to select the right ingredients for your cake and put them in your shopping cart. Finally there is $p_{config}$, the probability that randomly configuring the ingredients in your shopping cart—even if they are the right ones for your cake—will produce the desired cake. Since $p_{dco}$ = $p_{orig}$ × $p_{local}$ × $p_{config}$ and since probabilities never exceed 1, it follows that if even one of the three probabilities multiplied on the right side of this equation is small, then $p_{dco}$, which is the probability of getting the cake by chance, will also be small.

Mere inspection of the equation $p_{dco}$ = $p_{orig}$ × $p_{local}$ × $p_{config}$ shows why the Miller-Urey experiment and experiments like it are virtually irrelevant to the origin of life.[89] All such experiments can show is that $p_{orig}$, the origination probability for building blocks like amino acids, may not be too small. Indeed, the sole concern of such experiments is to determine whether stochastic chemistry is sufficient to account for the building blocks of life like amino acids and nucleotide bases. But that still leaves $p_{local}$ and $p_{config}$. Even if we take $p_{orig}$ equal to 1 and thus assume that there is no probabilistic obstacle to generating the building blocks for some discrete combinatorial object, $p_{local}$ and $p_{config}$ can pose huge obstacles to the chance formation of such objects.

Consider again the example of making a cake. Suppose you have a well-stocked supermarket, so that all the ingredients (building blocks) you could possibly want for your cake are there. The problem is that your supermarket also has lots of things that have no conceivable relevance to your cake. Detergents, pet food, diapers, and indeed most of the items on the supermarket shelves not only will not have any relevance to your cake but would in fact destroy it if they got mingled with the ingredients during the configuration phase (the problem here is that of interfering cross-reactions). Let us say your cake requires 50 separate ingredients and that your supermarket stocks about 4,000 separate items. Let us be generous and assume that the 50 ingredients you need for your cake permit various alternates so that there are 10 interchangeable items on the supermarket shelf for each ingredient (in particular, the alternates are able to avoid interfering cross-reactions whereas other items on the shelves entail cross-reactions that inevitably would destroy the cake during the configuration phase—like for instance Draino™). Then by going through the supermarket and randomly filling your shopping cart with items, the probability that each of the 50 required ingredients (allowing alternates) makes it into your cart is on the order of $10^{-66}$.[90] Note that you need to get the right 50 ingredients and only those ingredients into the shopping cart: fewer ingredients means not being able to make the cake at all whereas the wrong ingredients, even with all the right ingredients present, entail cross-reactions that prevent the right ingredients from coming together and forming the cake. Although $10^{-66}$ does not quite attain to the level of improbability set by the universal probability bound (i.e., $10^{-150}$), it is getting uncomfortably small.

What relevance does baking a cake have to irreducibly complex biochemical systems? Consider the bacterial flagellum in E. coli. E. coli's genome has 4,639,221 basepairs and codes for 4,289 proteins.[91] These 4,289 proteins are the items on the supermarket shelves and the 50 proteins needed to make a bacterial flagellum are the 50 ingredients for the cake. Even if we are generous and assume that each of the proteins required for the flagellum's construction permits 10 interchangeable proteins (there is no evidence that the flagellar proteins of E. coli permit this much interchangeability), the probability of getting a single representative of each of the right proteins for a flagellum together in one spot would be the probability of going through the supermarket and getting each of the right ingredients for the cake into your shopping cart. Again the probability is $10^{-66}$. Actually, it is even smaller than this because the "supermarket" of proteins in E. coli contains even more items (i.e., 4,289) than the 4,000 in the cake example, thus permitting even more interfering cross-reactions and lowering the probability even more.

But the localization probability for the bacterial flagellum gets even worse.

The problem is not just getting one instance of each of the 50 proteins required for the bacterial flagellum in *E. coli* to a given location. For instance, the filament that serves as the propeller for the flagellum makes up over 90 percent of the flagellum's mass and is comprised of more than 20,000 subunits of flagellin protein (FliC). That means going to *E. coli*'s "protein supermarket" and picking 20,000 items of flagellin off the shelf. The three ring proteins (Flgh, I, and F) are present in about 26 subunits each. The proximal rod requires 6 subunits, FliE 9 subunits, and FliP about 5 subunits. The distal rod consists of about 25 subunits. The hook (or U-joint) consists of about 130 subunits of FlgE.[92]

Let us therefore assume that 5 copies of each of the 50 proteins required to construct *E. coli*'s flagellum are required for a functioning flagellum (this is extremely conservative—all the numbers above were at least that and some far exceeded it, for example, the 20,000 subunits of flagellin protein in the filament). We have already assumed that each of these proteins permits 10 interchangeable proteins. That corresponds to 500 proteins in *E. coli*'s "protein supermarket" that could legitimately go into a flagellum. By randomly selecting proteins from *E. coli*'s "protein supermarket," we need to get 5 copies of each of the 50 required proteins or a total of 250 proteins. Moreover, since each of the 50 required proteins has by assumption 10 interchangeable alternates, there are 500 proteins in *E. coli* from which these 250 can be drawn. But those 500 reside within a "protein supermarket" of 4,289 proteins. Randomly picking 250 proteins and having them all fall among those 500 therefore has probability $(500/4,289)^{250}$, which has order of magnitude $10^{-234}$ and falls considerably below the universal probability bound of $10^{-150}$.

Granted, I have oversimplified. But every simplifying assumption I have made favors more manageable probabilities and thus more conservative estimates of the actual improbabilities involved. The improbability of $10^{-234}$ will become even more extreme the more of the right proteins have to be located in the same place. What's more, by allowing alternates to the 50 required proteins, I have factored in the possibility of extraneous proteins that might be present but innocuous in the construction of the bacterial flagellum. But that still leaves most of the proteins in *E. coli*, which will not be innocuous but ruin the flagellum's construction through interfering cross-reactions. Indeed, what makes these localization probabilities so effective at blocking the chance formation of irreducibly complex biochemical machines is the constant threat of interfering cross-reactions. Moreover, the reason for these cross-reactions is the presence, whether in the wild or in the supermarket, of other building blocks besides those needed for the irreducibly complex structure to be formed. In fact, irrelevant building blocks targeted for other structures will inevitably preponderate over those actually required in a given irreducibly complex structure. That very preponderance will make it highly improbable to localize the building blocks needed to construct a given irreducibly complex structure.

The only way to mitigate the localization probability is by reducing the number of building blocks required for the flagellum or by reducing the genome of the bacterium housing the flagellum and thus reducing the total number of building blocks available within the bacterium. In our cake analogy, either cut down the number of ingredients in the cake or cut down the number of items on the supermarket shelves. Neither avenue holds any promise for the bacterial flagellum. It is true that there exist variant flagella in other bacteria. But these bacteria do not have substantially simpler genomes than *E. Coli* (at least not for the purpose of radically mitigating a localization probability). What's more, simplifications in flagellar structure result primarily from other bacteria having outer surfaces that do not require an additional L-ring. Otherwise, however, the structures are very similar as are the genes encoding the structures.[93]

Instead of merely mitigating the localization probability, one can also try to circumvent it entirely. Imagine the odds of getting water to localize in your neighborhood pond or stream. Out of all the stuff in the world, all those water molecules just happen to end up there! But clearly there is no vast improbability involved with localizing those water molecules—natural processes like evaporation and condensation account for the water localizing just fine. The appeal here is to necessities of nature to circumvent chance. This appeal, however, does not work with discrete combinatorial objects since it is not just one type of element that needs to be localized in one location but a diversity of types of elements. Moreover, in this case natural processes have no way of singling out one collection of diverse types of elements among the vast combinatorial possibilities for collecting diverse types of elements. Consequently, there is no way to eliminate localization probabilities by collapsing chance to necessity.

Although the localization probability seems enough to secure specified complexity for the bacterial flagellum, the probabilistic hurdles facing the flagellum do not end there. Severe improbabilities also arise at the configuration phase, where typically most of the important design work gets done. I want therefore next to turn to the configuration probability—the probability of putting all the building blocks together in the right way once they are in the right location. Strictly speaking, the configuration probability for a discrete combinatorial object that exhibits some function is the ratio of all the ways of arranging its building blocks that preserve the function divided by all the possible ways whatsoever of arranging the building blocks. Note that it is not enough, as has often been done in the creationist literature, to take some unique arrangement of building blocks and divide by the total number possible arrangements.[94] In biology and outside, what is of interest is not a unique arrangement but all the arrangements that preserve some function as well as how these function-preserving arrangements relate to the totality of possible arrangements.

So defined, the configuration probability seems thoroughly beyond our ken. Indeed, how can we ever figure out all the possible arrangements of building blocks that fulfill some function? This problem seems completely intractable. There is, however, a way to get a handle on configuration probabilities—through what I call *perturbation probabilities*. To see what is at stake, suppose you are given an English text of 1,000 characters. Let us say each character admits 30 possibilities—26 for capital Roman letters and 4 for spaces and punctuation. For definiteness let us say the text coincides with Lincoln's Gettysburg Address (the actual text is about 1,500 characters,[95] but let us say it is 1,000). Suppose now that you want to figure out what is the probability of coming up with a text that conveys the same meaning as the Gettysburg Address (let us identify the text's meaning with its function). Clearly we cannot enumerate all the possible English texts that convey the same meaning. Even so, if we are convinced that there is not much redundancy in the Gettysburg Address and that texts that convey the same meaning will require around 1,000 characters (cf. the minimal complexity requirement for irreducible complexity), then we can still get some handle on the improbability of texts conveying the same meaning as the given text.

What we do is consider a *perturbation tolerance factor* and a *perturbation identity factor*. The perturbation tolerance factor asks what proportion of characters in a text can be randomly altered and still preserve the meaning of the text. A single typo or two never prevented anybody from extracting the meaning of a long text. Multiple typos, if not too numerous, tend also not to be a problem. True, if the typos all bunch up in one location and eliminate, for instance, the word "not," then the meaning of the text will be fundamentally altered. But that just lowers the perturbation tolerance factor and makes it more restrictive. The perturbation tolerance factor can be defined as the percentage of random changes, in this case changes in the characters of the Gettysburg Address, that on average still allow the text (or discrete combinatorial object in general) to convey its meaning (i.e., preserve its function). For texts 1,000 characters in length the perturbation tolerance factor is surely less than 10 percent (with such a long text and this percentage of errors, errors are likely to bunch and destroy meaning). Note that this factor will vary with medium. For computer source codes with non-object oriented programming languages like Fortran, the tolerance factor is going to be less than 1 percent (often there is zero tolerance).

Whereas the perturbation tolerance factor asks what percentage of random changes permit continued functioning (or preserve meaning in the case of the Gettysburg Address), the perturbation identity factor asks what percentage of random changes allow unique identification of function. Think of it this way: Even though the function of a perturbed object may be lost, if it had a function, could it be other than the function of the object that was perturbed? If not, then we are still within the perturbation identity factor. This is one instance where American popular culture helps clarify what is at stake. Anyone who has watched the television game show *Wheel of Fortune* knows that the object of the game is to identify some target phrase as increasingly many letters in the phrase become known. Initially none of the letters is known. Eventually all the letters are known. Somewhere in the middle, however, enough letters are known to uniquely identify the phrase but not enough to make its meaning evident to the contestants. Thus somewhere in the middle, deviation from the target phrase falls within the perturbation identity factor (the partial phrase displayed could not be other than the target phrase) but outside the perturbation tolerance factor (the contestants cannot fathom its meaning). The perturbation identity factor can be defined as the percentage of random changes—in our running example changes in the characters of the Gettysburg Address—that on average still allow the meaning of the text (or function of the discrete combinatorial object in general) to be uniquely identified. The perturbation identity factor will always be at least as large as the perturbation tolerance factor since the perturbation tolerance characterizes function that is actually preserved whereas the perturbation identity also includes function that can only be recovered. The whole point of the perturbation probability is to exploit the disparity between perturbation tolerance and identity factors.

Suppose now that for the Gettysburg Address we set the perturbation tolerance factor at 10 percent and the perturbation identity factor at 20 percent. Both numbers seem conservative (the perturbation tolerance factor is probably less than 10 percent and the perturbation identity factor is probably more than 20 percent). With 1,000 characters in the Gettysburg address and 30 possibilities at each character location, this means a total of $T = 30^{1,000} = 10^{1,477}$ possible sequences. Of these sequences, there are $M = \Sigma_{1 \leq i \leq 100} C(1000,i)29^i$ sequences that differ from the target sequence in at most 100 places (i.e., a 10 percent or less difference corresponding to the perturbation tolerance factor) and $N = \Sigma_{1 \leqslant i \leqslant 200} C(1000,i)29^i$ sequences that differ from the target sequence in at most 200 places (i.e., a 20 percent or less difference corresponding to the perturbation identity factor). Here $C(n,k) = n!/k!(n-k)!$ where $n! = 1 \cdot 2 \cdot 3 \cdots n$. The perturbation probability is then defined as the proportion of objects, or in this case sequences, within the perturbation tolerance factor (here M/T) divided by the proportion of objects, or in this case sequences, within the perturbation identity factor (here N/T). It follows that T cancels out, and what we have is the ratio M/N. Since M and N are sums and since the last terms of the sums dominate (i.e., are considerably larger than the rest), we can approximate the order of magnitude of M/N by the ratio

$$\frac{C(1000,100) \cdot 29^{100}}{C(1000,200) \cdot 29^{200}}.$$

Using Stirling's formula, this ratio comes out to on the order of $10^{-288}$, again well below the universal probability bound.

It is important to understand the rationale behind perturbation probabilities. Imagine a bullet is shot at a wall and hits a fly sitting on the wall. How can we preclude that the bullet happened just by chance to hit the fly? Perhaps most of the wall is just covered with flies, and a random bullet was bound to hit a fly. But what if the local area surrounding the fly that was hit was empty of other flies and what if hitting that particular fly by chance within that local area was highly improbable? In that case it does not matter if the wall in question is the Wall of China and if all but the local area surrounding the fly that was hit is carpeted with flies. Indeed, it does not matter what the global density distribution of flies is on the wall but only the density distribution in the local area surrounding the fly that was hit.[96] This, then, is the point of the perturbation probability: to fix a local area around some discrete combinatorial object, and then show that the target of functional variants gotten by perturbing that object is nonetheless minuscule in relation to the local area. In terms of the information-theoretic apparatus developed in chapter 3, to compute a perturbation probability means fixing a reference class of possibilities by means of a perturbation identity factor, specifying a target by means of the perturbation tolerance factor, and then computing the probability of the target within that reference class. The probability of this target is the perturbation probability.

More formally, the rationale behind the perturbation probability is as follows. Suppose a particular discrete combinatorial object X within a reference class of possible combinatorial objects $\Omega$ performs a particular function

F. Let $\Omega(F)$ denote the collection of objects in $\Omega$ that performs the function F. Ideally we would like to know the probability of $\Omega(F)$ within $\Omega$ (i.e., $P(\Omega(F)|\Omega)$, which is the overall proportion of objects that perform the function F). Since there may be many other objects in $\Omega$ that perform the same function as X despite being structured quite differently from X, estimating the probability of $\Omega(F)$ within $\Omega$ is usually not possible. We therefore attempt to obtain an upper bound estimate of this probability by restricting our analysis to a subset of the whole combinatorial space. This subset contains X and all objects that are clearly variants of X. Let us denote this subset by $\Omega(X)$. Any discrete combinatorial object Y that performs F but is not a variant of X will therefore not be in $\Omega(X)$. Since all objects in $\Omega(X)$ are X-like and no object in this subset is Y-like, all functional objects in $\Omega(X)$ must function in a distinctly X-like manner. Call the functional objects in this subset $\Omega(X,F)$. By measuring the tolerance of X to perturbation, we can estimate the number of these functional X-like objects. If we can also estimate the number of objects that are clearly variants of X, then the ratio of these numbers is the prevalence of functional objects among X-like objects.

This ratio is the perturbation probability, which we denote by $p_{perturb}$. The perturbation probability can thus be represented alternately as $P(\Omega(X,F)|\Omega(X))$ or $P(\Omega(X,F)|\Omega)/P(\Omega(X)|\Omega)$. By definition $\Omega(X,F)$ is then the set of discrete combinatorial objects within the perturbation tolerance factor and $\Omega(X)$ is the set of discrete combinatorial objects within the perturbation identity factor. Since X is known to be functional, and in the absence of any non-X-like objects known to perform F, we can safely assume that the prevalence of functional objects among X-like objects is higher than the prevalence of functional objects in the whole space (i.e., $P(\Omega(X,F)|\Omega(X)) \geq P(\Omega(F)|\Omega)$). But note that even if at some point some non-X-like object Y is found that performs F and for which $P(\Omega(Y,F)|\Omega(Y))$ is much bigger than $P(\Omega(X,F)|\Omega(X))$, $\Omega(X,F)$ will nonetheless constitute a specification of X within the reference class $\Omega(X)$. Moreover, if the probability $P(\Omega(X,F)|\Omega(X))$ is small enough, the ordered pair $(\Omega(X,F),X)$ will constitute an instance of complex specified information (see section 3.5). Thus for the perturbation probabilities to effectively rule out chance and implicate design it is not necessary to use perturbation probabilities to estimate overall proportion of functionality. Provided $P(\Omega(X,F)|\Omega(X))$ is small enough, X can exhibit specified complexity even if $P(\Omega(F)|\Omega)$ is large (for instance, just because model airplanes powered by propeller and rubber band might be reasonably probable and thus not too implausibly referred to chance does not mean that model airplanes powered by propeller and gasoline engine can likewise be referred to chance—both airplanes exhibit the same function, namely flight, but the mechanism for the latter cannot reasonably be referred to chance even if the former can).

How does a perturbation analysis apply to irreducibly complex biochemical systems like the flagellum of *E. coli*? In fact, the perturbation analysis of the Gettysburg Address example is relevant to the bacterial flagellum. Although the flagellum requires about 50 separate proteins for its construction, in its final form it utilizes only about 30 proteins (corresponding to the 30 character types—alphabetic and punctuation—required to express the Gettysburg Address). We can think of the remaining 20 proteins as scaffolding that falls away. Also, many more individual protein subunits constitute a flagellum than the 1,000 or so sequenced characters that make up the Gettysburg Address. Finally, a perturbation tolerance factor of 10 percent and a perturbation identity factor of 20 percent seem quite conservative for the flagellum (i.e., the perturbation tolerance factor is likely to be much smaller and the perturbation identity factor is likely to be much bigger).

Why? Most discrete combinatorial objects that perform a function are far more sensitive to perturbation than human texts. In the case of texts like the Gettysburg Address, function corresponds to meaning that humans are able to extract from it. What's more, humans have the ability to track meaning despite substantial perturbation of texts. Unlike humans, who can forgive typographical mistakes, most contexts allow no or very little forgiveness for such mistakes. Computer source codes, mechanical devices, and architectural edifices in general have much lower perturbation tolerance factors than apply to texts intended for human interpreters. On the other hand, perturbation identity factors tend to be large across contexts—we can reconstruct the function of an object that through perturbation has lost its function even if the perturbation is substantial. It follows that $C(1000,100)29^{100}/C(1000,200)29^{200}$, or approximately $10^{-288}$, can reasonably be taken as an upper bound on the perturbation probability for the bacterial flagellum.

In fact a much smaller perturbation probability is indicated. To see this, here is a good rule of thumb for approximating the perturbation probability. For a discrete combinatorial object with N subunits, k different types of subunits, perturbation tolerance factor q, and perturbation identity factor r ($0 \leq q, r \leq 1$), we saw in the Gettysburg Address example that the perturbation probability can be approximated as follows (we assume qN and rN are integers or else the integers closest to these real numbers):

$$P_{perturb} \approx \frac{C(N,qN)}{C(N,rN)}(k - 1)^{qN - rN}.$$

The expression on the right side of this equation can be simplified further. This expression becomes incredibly small very quickly as N increases and as the disparity between q and r increases. Since C(N,rN) will usually completely dominate C(N,qN), the first factor C(N,qN)/C(N,rN) will usually be close to zero. For k substantially bigger than 2 (say 10 or more), a quick and dirty approximation for the perturbation probability is therefore

$$P_{perturb} \approx\approx k^{(q-r)N}.$$

Since the bacterial flagellum has at least N = 20,000 subunits, for k = 30, q = .1, and r = .2, a quick and dirty approximation of $P_{perturb}$ is therefore $30^{-2,000}$ or on the order of $10^{-2954}$. Again, there is no problem here satisfying the universal probability bound.

An objection may now be raised against this analysis. By identifying specified complexity and thus issuing in a design inference, my analysis seems to imply that flagellar assembly requires direct intelligent intervention, which clearly I do not intend. Thus according to this objection my analysis focuses exclusively on chance whereas in reality law plays a crucial role. Imagine the odds of a snowflake ever forming if $10^{19}$ water molecules had to simultaneously converge in the appropriate orientations. Yet it happens all the time. Likewise, the parts of a flagellum do not have to simultaneously converge by chance—they self-assemble in order when chance collisions allow specific, cooperative, local electrostatic interactions to lock the structure together, one piece at a time.

The problem with this objection is two-fold. First, unlike the snowflake, which consists entirely of water molecules, the flagellum requires multiple diverse parts. The analogy is therefore weak. Second, a flagellum has to self-assemble within the right cellular context. Just to have all the right protein components for a bacterial flagellum localized in one spot is not going to produce a flagellum. The right ambient conditions have to obtain (including the right cell-membrane for the flagellum to attach to) for the flagellar components to self-assemble into a flagellum. My analysis in terms of perturbation probabilities captures the contingencies that remain even when all the right components for a flagellum are in the same place (just because the right components are in the same place does not mean they are in the right context to self-assemble). More particularly, my analysis assesses the improbability of configuring a bacterial flagellum if the right protein subunits in sufficient quantities to produce a flagellum are localized in one place. Chance rather than law now comes to dominate the configuration probability of the flagellum (and therewith the perturbation probability), moving it toward zero rather than one, because the exact number of protein subunits of different types, their precise delivery schedule to a particular location in the cell, and the cellular context itself are all contingent and all condition that probability. Granted, once all these factors are in place, probabilities collapse to one. But these factors need themselves to be accounted for and come with probabilities.

The perturbation probability approximates the configuration probability. Interestingly, the perturbation probability can also be used to approximate the origination probability for the building blocks that then need to be local-ized and configured. The reason for this is that the very building blocks for a discrete combinatorial object may themselves be discrete combinatorial ob-jects and thus can be assigned configuration probabilities and approximated with perturbation probabilities. Consider, for instance, the proteins that make up a bacterial flagellum. A modestly sized protein has N = 300 amino acid subunits strung together with peptide bonds. There are k = 20 different types of amino acids. Preliminary indications are that proteins permit a per-turbation tolerance factor of no more than 10 percent (thermodynamic con-siderations seem to preclude proper protein folding for more than this percentage of random substitutions).[97] On the other hand, proteins fall into only a limited number of taxonomic groups performing certain types of func-tions. Thus proteins of a given length and sharing only a few key sites will often be uniquely determined functionally. It follows that the perturbation identity factor will be significantly larger than the perturbation tolerance factor. For enzymes that number appears to be larger than 30 percent.[98] Structural proteins tend to be more tolerant of substitution than enzymes, but it seems safe to say that protein sequences differing in no more than 20 percent of their positions can be presumed to be variants of the same design.

In that case we may take the perturbation tolerance factor q equal to .1 and the perturbation tolerance factor r equal to .2. Then for a protein com-prising 300 amino acid subunits, our quick and dirty approximation of its perturbation probability is $20^{(.1)300-(.2)300} = 20^{-30}$ or on the order of $10^{-39}$. Thus if four distinct proteins comprising 300 amino acid subunits were re-quired to perform a function, the estimated improbability of getting all four would fall below the universal probability bound (the probabilities would multiply). In the case of the bacterial flagellum we require at least thirty such distinct proteins. Since $10^{-39}$ is just the improbability of getting one of many building blocks that goes into the construction of the flagellum, the origination probability $p_{orig}$ will be much smaller than this. Indeed, each building block formed by chance will have a probability like $10^{-39}$ associated with it, and these probabilities will all need to be multiplied to form the origination probability.

Although it may seem as though I have cooked these numbers, in fact I have tried to be conservative with all my estimates. To be sure, there is plenty of biological work here to be done. The big challenge is to firm up these numbers and make sure they do not cheat in anybody's favor. Getting solid, well-confirmed estimates for perturbation tolerance and perturbation identity factors will require careful scientific investigation. Such estimates, however, are not intractable. Perturbation tolerance factors can be assessed empirically by random substitution experiments where one, two, or a few substitutions are made. If early on, when only a few substitutions are made, the discrete combinatorial object breaks down (say it reliably fails after four substitutions pretty much regardless where those substitutions are made), then there is a reasonable presumption that it will fail with still more substitutions. For discrete combinatorial objects that are more robust under perturbation and resist failure for just a few random substitutions, such an approach will quickly become intractable, and perturbation tolerance will have to be assessed using different techniques—and perhaps purely on theoretical grounds. Similarly, perturbation identity factors will need to be assessed both empirically and theoretically. Note that perturbation identity is already part of molecular biology. For instance, perturbation identity is how we know we are dealing with a pseudogene. A pseudogene looks like a functional gene, but has been sufficiently perturbed so that it is no longer functional. Nonetheless, *if it were functional*, it could have no other function than the functional gene for which it is a pseudogene.[99]

Origination, localization, and configuration probabilities as well as the perturbation probabilities needed to approximate configuration probabilities pervade biology at all levels of complexity and organization. I predict that at all these levels of complexity and organization, save at the lowest level for the very simplest building blocks (i.e., amino acids and nucleotide bases), these probabilities will be extremely small and regularly fall below the universal probability bound of $10^{-150}$.

# Notes

1. Franz Bardon, *Initiation into Hermetics: A Course of Instruction of Magic Theory and Practice*, 4th ed., trans. A. Radspieler (Wuppertal, Germany: D. Rüggeberg, 1981). Carl G. Jung, *Mysterium Coniunctionis: An Inquiry into the Separation and Synthesis of Psychic Opposites in Alchemy*, in *Collected Works of C. G. Jung*, vol. 14 (Princeton: Princeton University Press, 1963). On page 114 Jung writes: "The alchemists sought for that effect which would heal not only the disharmonies of the physical world but inner psychic conflict as well, the 'affliction of the soul,' and they called this effect the *lapis philosophorum* [i.e., the philosopher's stone]. In order to obtain it, they had to loosen the age-old attachment of the soul to the body and thus make conscious the conflict between the purely natural and the spiritual man."

2. Not only has alchemy failed as a scientific project, but also alchemy as a metaphysical project seems not to be in much better a state. Consider the following admission by Carl Jung toward the end of his life (apparently alchemy had not enabled him to resolve the connection between body and soul—see previous note): "I observe myself in the stillness of Bollingen, with the experience of almost eight decades now, and I have to admit that I have found no plain answer to myself. I am in doubt about myself as ever, the more I try to say something definite. It is even as though through familiarity with oneself one became still more alienated." Quoted in Gerhard Wehr, *Jung: A Biography*, trans. D. M. Weeks (Boston: Shambhala, 1987), 416. According to Jung's biographer (407), Jung regarded it as speaking well for the honesty of alchemists that "after years of continuing toil they were able to produce neither gold nor the highly praised philosopher's stone and openly admitted this. To these men, failures in the popular sense, Jung compared himself. He too had in the end been unable to solve the riddle of the *mysterium coniunctionis*."

3. Even so, it is worth remembering that Isaac Newton devoted a full half of his writings to theology and alchemy. See the introduction by Brad Gregory to Baruch Spinoza, *Tractatus Theologico-Politicus*, trans. S. Shirley, intro. B. S. Gregory (1670; reprint, Leiden: Brill, 1989), 9.

4. Terry Bossomaier and David Green, *Patterns in the Sand* (Reading, Mass.: Perseus Books, 1998), 39.

5. Darwinists hold that this extrapolation follows as a matter of course. Ernst Mayr, for instance, writes, "All evolution is due to the accumulation of small genetic changes guided by natural selection and . . . transpecific evolution is nothing but an extrapolation and magnification of the events which take place within populations and species." Ernst Mayr, *Animal Species and Evolution* (Cambridge, Mass.: Harvard University Press, 1963), 586. Indeed, it is often hard to get Darwinists even to admit that there is a distinction between small-scale evolutionary changes (i.e., microevolution) and the extrapolated large-scale evolutionary changes (i.e., macroevolution), so tightly are the two fused within Darwinian theory. Nevertheless, evidence and extrapolations from evidence are not rightly conflated.

6. See Charles Thaxton, Walter Bradley, and Roger Olsen, *The Mystery of Life's Origin* (New York: Philosophical Library, 1984); Michael Denton, *Evolution: A Theory in Crisis* (Bethesda, Md.: Adler & Adler, 1985); Percival Davis and Dean Kenyon, *Of Pandas and People*, 2nd ed. (Dallas, Tex.: Haughton, 1993); and Michael Behe, *Darwin's Black Box* (New York: Free Press, 1996).

7. Bruce Alberts, "The Cell as a Collection of Protein Machines: Preparing the Next Generation of Molecular Biologists," *Cell* 92 (8 February 1998): 291.

8. See Behe, *Darwin's Black Box*, 39–45. Behe's exact definition reads, "An irreducibly complex system is one that requires several closely matched parts in order to function and where removal of one of the components effectively causes the system to cease functioning" (39).

9. Ibid., 39.

10. Ibid., 69–73.

11. See, for example, Nicholas Gaiano, Adam Amsterdam, Koichi Kawakami, Migeul Allende, Thomas Becker, and Nancy Hopkins, "Insertional Mutagenesis and Rapid Cloning of Essential Genes in Zebrafish," *Nature* 383 (1996): 829–832; Carolyn K. Suzuki, Kitaru Suda, Nan Wang, and Gottfried Schatz, "Requirement for the Yeast Gene *LON* in Intramitochondrial Proteolysis and Maintenance of Respiration," *Science* 264 (1994): 273–276; Qun-Yong Zhou, Carol J. Qualfe, and Richard D. Palmiter, "Targeted Disruption of the Tyrosine Hydroxylase Gene Reveals that Catecholamines are Required for Mouse Fetal Development," *Nature* 374 (1995): 640–643.

12. Behe, *Darwin's Black Box*, 45–46.

13. Arno Wouters, "Viability Explanation," *Biology and Philosophy* 10 (1995): 435–457.

14. Richard Dawkins, *The Blind Watchmaker* (New York: Norton, 1987), 9.

15. For reviews in the popular press see James Shreeve, "Design for Living," *New York Times*, Book Review Section (4 August 1996): 8; Paul R. Gross, "The Dissent of Man," *Wall Street Journal* (30 July 1996): A12; and Boyce Rensberger, "How Science Responds When Creationists Criticize Evolution," *Washington Post* (8 January 1997): H01. For reviews in the scientific journals see Jerry A. Coyne, "God in the Details," *Nature* 383 (19 September 1996): 227–228; Neil W. Blackstone, "Argumentum Ad Ignorantiam," *Quarterly Review of Biology* 72(4) (December 1997): 445–447; and Thomas Cavalier-Smith, "The Blind Biochemist," *Trends in Ecology and Evolution* 12 (1997): 162–163.

16. See John Catalano's web page titled "Behe's Empty Box": http://www.world-of-dawkins.com/box/behe.htm (last accessed 11 June 2001).

17. I am indebted to James Bradley for this particularly apt formulation of the scaffolding objection.

18. Thomas D. Schneider, "Evolution of Biological Information," *Nucleic Acids Research* 28(14) (2000): 2794.

19. Ibid.

20. R. H. Thornhill and D. W. Ussery, "A Classification of Possible Routes of Darwinian Evolution," *Journal of Theoretical Biology* 203 (2000): 111–116.

21. Personal communication with author, 3 June 1999. For Behe's literature search of the relevant technical journals, a search that revealed the overwhelming absence of gradualistic Darwinian explanations for irreducibly complex biomolecular systems, see Behe, *Darwin's Black Box*, 165–186.

22. Douglas J. Futuyma, *Evolutionary Biology*, 3rd ed. (Sunderland, Mass.: Sinauer, 1998), 355.

23. Ibid.

24. H. Allen Orr, "Darwin v. Intelligent Design (Again)," *Boston Review* (December/January 1996–1997): 29.

25. See, for instance, Kenneth Miller, *Finding Darwin's God* (New York: HarperCollins, 1999), 152–158, in which Miller offers a co-optation story for the evolution of blood clotting.

26. In fact, co-optation was Miller's only substantive argument at a debate between Behe and Miller at a summer workshop sponsored by the American Association for the

Advancement of Science, the Center for Theology and the Natural Sciences, and the Philadelphia Center for Religion and Science. The workshop was titled *Interpreting Evolution* and took place at Haverford College (14–19 June 2001). The actual encounter between Behe and Miller occurred 17 June 2001. For the ongoing Internet debate between Behe and Miller, see respectively http://www.discovery.org/crsc/fellows/MichaelBehe/index.html (click on "Articles by Michael J. Behe") and http://biocrs.biomed.brown.edu/Darwin/DI/Design.html (both last accessed 26 June 2001).

27. Miller made this statement during the question and answer session of a talk that I gave: William A. Dembski, "Detecting Design in the Sciences," talk presented at conference titled *Design and Its Critics* (Mequon, Wis.: Concordia University, 22–24 June 2000).

28. Orr, "Darwin v. Intelligent Design (Again)," 29.

29. Ibid.

30. Ibid., 30.

31. Ibid.

32. Ibid.

33. See Denton, *Evolution: A Theory in Crisis*, 109–110.

34. A further clarification is worth adding: A part may be indispensable for maintaining an irreducibly complex system's function but not be indispensable for the life of the organism, which is the sense of indispensability that Orr seems to have in mind. A bacterium missing a part of the flagellum will not have a functioning flagellum, but may be able to grow, reproduce, etc., so that the part is dispensable in regard to the bacterium's life. I am indebted to Michael Behe for this clarification.

35. Orr, "Darwin v. Intelligent Design (Again)," 30.

36. Michael J. Behe, "The Sterility of Darwinism," *Boston Review* (February/March 1997): 24.

37. Orr, "Darwin v. Intelligent Design (Again)," 29.

38. Behe, "The Sterility of Darwinism," 24.

39. I am grateful to John H. McDonald for permission to reprint these diagrams. They may be found on his website at http://udel.edu/~mcdonald/mousetrap.html (last accessed 11 June 2001).

40. Ibid.

41. Ibid.

42. Ibid.

43. See Behe's responses to critics at http://www.discovery.org/crsc/fellows/Michael-Behe/index.html (last accessed 11 June 2001).

44. Quoted from http://udel.edu/~mcdonald/mousetrap.html.

45. See Harold Morowitz, *Beginnings of Cellular Life: Metabolism Recapitulates Biogenesis* (New Haven, Conn.: Yale University Press, 1992), ch. 5, titled "The Minimal Cell," in which Morowitz illustrates the concept of minimal complexity at the level of a complete cell.

46. John Postgate, *The Outer Reaches of Life* (Cambridge: Cambridge University Press, 1994), 161.

47. Ibid., 160.

48. Niall Shanks and Karl H. Joplin, "Redundant Complexity: A Critical Analysis of Intelligent Design in Biochemistry," *Philosophy of Science* 66(2) (June 1999): 268.

49. Michael J. Behe, "Self-Organization and Irreducibly Complex Systems: A Reply to Shanks and Joplin," *Philosophy of Science* 67(1) (March 2000): 160.

50. Ibid.

51. Miller made this claim during the question and answer session of my talk at this conference: William A. Dembski, "Detecting Design in the Sciences," talk presented at conference titled *Design and Its Critics* (Mequon, Wis.: Concordia University, 22–24 June 2000).

52. David J. DeRosier, "The Turn of the Screw: The Bacterial Flagellar Motor," *Cell* 93 (1998): 17–20.

53. Miller, *Finding Darwin's God*, 148.

54. DeRosier, "The Turn of the Screw."

55. Miller, *Finding Darwin's God*, 148.

56. Ibid.

57. The first two articles that Miller cites (by Dean and Golding and by Logsdon and Doolittle) focus on individual proteins whose specificity increases under selection pressure. These studies show that selection can increase the specificity of already functional proteins, but do not describe irreducibly complex systems of multiple coordinated parts each indispensable for function. The article by Musser and Chan that Miller cites focuses on a proton pump. Here we are at least in the right ball park, focusing on biochemical machines of the sort that Behe focused on in *Darwin's Black Box*. Even so, this system is not known in enough detail to determine whether it is irreducibly complex. Moreover, the article by Musser and Chan is utterly lacking in causal specificity. Consider the following remark in the article: "It makes evolutionary sense that the cytochrome bc1 and cytochrome c oxidase complexes arose from a primitive quinol terminal oxidase complex via a series of beneficial mutations." What exactly is this series of beneficial mutations? Behe's challenge to the biological community was to exhibit causally specific accounts of how the Darwinian mechanism could produce irreducibly complex biochemical machines. Musser and Chan are offering nothing like this. Finally, the article by Melendez-Hevia et al. works out a scheme for how the organic-chemical components of a certain metabolic pathway may have arisen gradually. But as Behe emphasized in *Darwin's Black Box* (141–142, 150–151), metabolic pathways are not irreducibly complex because components can gradually be added to a previous pathway.

See A. M. Dean and G. B. Golding, "Protein Engineering Reveals Ancient Adaptive Replacements in Isocitrate Dehydrogenase," *Proceedings of the National Academy of Sciences* 94 (1997): 3104–3109; J. M. Logsdon Jr. and W. F. Doolittle, "Origin of Antifreeze Protein Genes: A Cool Tale in Molecular Evolution," *Proceedings of the National Academy of Sciences* 94 (1997): 3485–3487; S. M. Musser and S. I. Chan, "Evolution of the Cytochrome C Oxidase Proton Pump," *Journal of Molecular Evolution* 46 (1998): 508–520; and E. Melendez-Hevia, T. G. Waddell, and M. Cascante, "The Puzzle of the Krebs Citric Acid Cycle: Assembling the Pieces of Chemically Feasible Reactions, and Opportunism in the Design of Metabolic Pathways during Evolution," *Journal of Molecular Evolution* 43 (1996): 293–303. For Behe's full response to Miller's "four glittering examples" consult http://www.discovery.org/crsc/fellows/MichaelBehe/index.html (last accessed 26 June 2001) and specifically Behe's article titled "Irreducible Complexity and the Evolutionary Literature: Response to Critics."

58. Miller, *Finding Darwin's God*, 145.

59. Ibid., 146–147.

60. Michael J. Behe, "Answering Scientific Criticisms of Intelligent Design," 133–149 in *Science and Evidence for Design in the Universe*, eds. M. J. Behe, W. A. Dembski, and S. C. Meyer (San Francisco: Ignatius Press, 2000), 141. Emphasis in the original.

61. R. H. Thornhill and D. W. Ussery, "A Classification of Possible Routes of Darwinian Evolution," *Journal of Theoretical Biology* 203 (2000): 111–116.

62. David Griffin, *Religion and Scientific Naturalism: Overcoming the Conflicts* (Albany, N.Y.: State University of New York Press, 2000), 287, n. 23.

63. James Shapiro, "In the Details . . . What?" *National Review* (16 September 1996): 62–65. This statement is from a popular publication. Shapiro also makes the same point in his scholarly work. See James Shapiro, "Genome System Architecture and Natural Genetic Engineering in Evolution," *Annals of the New York Academy of Sciences* 870 (18 May 1999): 23–35.

64. Charles Darwin, *On the Origin of Species*, facsimile 1st ed. (1859; reprinted Cambridge, Mass.: Harvard University Press, 1964), 189. Emphasis added.

65. See Behe, "Answering Scientific Criticisms of Intelligent Design," 144–147.

66. Though not a direct quote, this passage captures some of the main concerns of Darwinists regarding intelligent design.

67. Daniel Dennett, *Darwin's Dangerous Idea* (New York: Simon & Schuster, 1995), 21.

68. Michael Shermer, *Why People Believe Weird Things* (New York: W. H. Freeman, 1997), 148. Shermer is the editor of *Skeptic Magazine*. On two occasions I have offered to join the editorial advisory board of *Skeptic Magazine* to be its resident skeptic regarding Darwinism. Shermer has yet to respond to either offer.

69. Quoted in Jill Cooper, "A New Germ Theory," *The Atlantic* (February 1999).

70. This can be generalized even further so that in place of the real numbers we substitute a group. See I. P. Cornfeld, S. V. Fomin, and Ya. G. Sinai, *Ergodic Theory* (New York: Springer-Verlag, 1982), 10–11.

71. Morris Kline, *Mathematical Thought from Ancient to Modern Times*, vol. 3 (Oxford: Oxford University Press, 1972), 1164.

72. See Alan Gibbons, *Algorithmic Graph Theory* (Cambridge: Cambridge University Press, 1985), 155.

73. Irving Kaplansky, *Fields and Rings*, 2nd ed. (Chicago: University of Chicago Press, 1972), 8–9, section titled "Ruler and Compass Constructions."

74. Cf. the homotopy groups and homology classes that attach to topological spaces. See Edwin H. Spanier, *Algebraic Topology* (New York: Springer-Verlag, 1966), 43–44 and 157 respectively.

75. For a readily accessible example of such a consistency theorem, see Ernest Nagel and James R. Newman, *Gödel's Proof* (New York: New York University Press, 1958), 45–56. Here Nagel and Newman prove the consistency of sentential logic (i.e., logic without quantification).

76. Behe, *Darwin's Black Box*, 39.

77. Doolittle's criticism of Behe appeared in Russell F. Doolittle, "A Delicate Balance," *Boston Review* (February/March 1997): 28–29. Doolittle was citing T. H. Bugge, K. W. Kombrinck, M. J. Flick, C. C. Daugherty, M. J. Danton, and J. L. Degen, "Loss of Fibrinogen Rescues Mice from the Pleiotropic Effects of Plasminogen Deficiency," *Cell* 87 (1996): 709–719.

78. Behe, "Answering Scientific Criticisms of Intelligent Design," 143.

79. Actually, it is unnecessary to consider the case where all N components are knocked out (that would leave no system at all) or where none of the components is knocked out (that would leave the original system intact). Thus the number of knockouts to determine irreducible complexity in this stronger sense is actually $2^N-2$.

80. Shanks and Joplin, "Redundant Complexity," 271–275.

81. See Peter Coveney and Roger Highfield, *Frontiers of Complexity: The Search for Order in a Chaotic World* (New York: Fawcett Columbine, 1995), 175–178.

82. See Shanks and Joplin, "Redundant Complexity."

83. Behe, *Darwin's Black Box*, 39.

84. See David Berlinski, "The Deniable Darwin," *Commentary* (June 1996): 24.

85. See Howard C. Berg, *Random Walks in Biology*, exp. ed. (Princeton: Princeton University Press, 1993), 134. Berg writes: "*E. coli* has receptors for oxygen and other electron acceptors, sugars, amino acids, and dipeptides. It monitors the occupancy of these receptors as a function of time. The probability that a cell will run (rotate its flagella counterclockwise) rather than tumble (rotate its flagella clockwise) depends on the time rate of change of receptor occupancy. We know from responses of cells to short pulses of chemicals delivered by micropipettes that this measurement spans some 4 sec. A cell compares the occupancy of a given receptor measured over the past second—the aspartate receptor is the only receptor that has been studied in detail—with that measured over the previous 3 sec and responds to the difference. Now given rotational diffusion, *E. coli* wanders off course about 60 degrees in 4 sec. If measurements of differences in concentration took much longer than this, they would not be relevant, because cells would change course before the results could be applied. On the other hand, if these measurements were made on a much shorter time scale, their precision would not be adequate. *E. coli* counts molecules as they diffuse to its receptors, and this takes time. The relative error (the standard deviation divided by the mean) decreases with the square root of the count. Thus in deciding whether life is getting better or worse, *E. coli* uses as much time as it can, given the limit set by rotational Brownian movement."

86. Consider, for instance, Dawkins's metaphor of climbing Mount Improbable in Richard Dawkins, *Climbing Mount Improbable* (New York: Norton, 1996). Consider as well Dawkins's account of "cumulative selection" in *The Blind Watchmaker*, 45.

87. Dawkins, *Climbing Mount Improbable*.

88. These probabilities do in fact multiply because the equality is of the form $P(A\&B\&C) = P(A) \times P(B|A) \times P(C|A\&B)$. It is therefore not the case that some unwarranted assumption about probabilistic independence is being slipped in. The rationale behind the equation $p_{dco} = p_{orig} \times p_{local} \times p_{config}$ parallels that behind the Drake equation—see Carl Sagan, *Cosmos* (New York: Random House, 1980), 299. Indeed, $p_{dco} = p_{orig} \times p_{local} \times p_{config}$ plays much the same role in the study of intelligent design in biology that the Drake equation plays in the search for extraterrestrial intelligence.

89. For the original papers by Miller and Urey see Harold Urey, "On the Early Chemical History of the Earth and the Origin of Life," *Proceedings of the National Academy of Sciences* 38 (1952): 351–363 and Stanley Miller, "A Production of Amino Acids under Possible Primitive Earth Conditions," *Science* 117 (1953): 528–529. For critiques of their work as well as subsequent work that attempts to produce the building blocks of life from stochastic chemistry see Charles Thaxton, Walter Bradley, and Roger Olsen, *The Mystery of Life's Origin: Reassessing Current Theories* (New York: Philosophical Library, 1984), ch. 4 and Jonathan Wells, *Icons of Evolution* (Washington, D.C.: Regnery, 2000), ch. 2.

90. To calculate this probability let $X_1$ be the first item selected off the supermarket shelves, $X_2$ the second, and so on up to $X_{50}$. Next let $A_1$ be the first ingredient required for the cake, let $A_2$ be the second, and so on up to $A_{50}$. Then the probability of $X_i \in A_j$ for any i and j is 10/4000, or 1/400 (there are 10 alternates permitted for each ingredient and there 4000 items total on the supermarket shelves). Since the $X_i$s are probabilistically independent, the probability of $P(X_{i1} \in A_1, X_{i2} \in A_2, ..., X_{i50} \in A_{50})$ for any permutation of the $X_i$s is the product of the probabilities $P(X_{i1} \in A_1) \times P(X_{i2} \in A_2) \times \cdots \times P(X_{i50} \in A_{50}) = (1/400)^{50}$. The probability of getting all the right ingredients into the shopping cart must therefore sum across all 50! permutations of the $X_i$s. Multiplying 50! times $(1/400)^{50}$ yields the probability we are after, i.e., $50! \times (1/400)^{50} = 2.4 \times 10^{-66}$.

91. This genome information is from the National Center for Biotechnology Information and can be obtained at the following URL: http://www.ncbi.nlm.nih.gov/PMGifs/Genomes/bact.html (last accessed 11 June 2001).

92. Robert McNab, "Flagella and Motility," in *Escherichia Coli and Salmonella: Cellular and Molecular Biology*, 2 vols., eds. F. C. Neidhardt et al. (Washington, D.C.: ASM Press, 1996), 123–145.

93. Ibid. See also Postgate, *Outer Reaches of Life*, 158–167.

94. For instance, consider the following argument by Frank Salisbury: "A medium [sized] protein might include about 300 amino acids. The DNA gene controlling this would have about 1,000 nucleotides in its chain. Since there are four kinds of nucleotides in a DNA chain, one consisting of 1,000 links could exist in $4^{1000}$ different forms. Using a little algebra (logarithms) we can see that $4^{1000} = 10^{600}$. Ten multiplied by itself 600 times gives the figure 1 followed by 600 zeros! This number is completely beyond our comprehension." Quoted from Frank B Salisbury, "Doubts about the Modern Synthetic Theory of Evolution," *American Biology Teacher* (September 1971): 336. Right after quoting this passage, a popular creationist text remarks: "It seems beyond all question that such complex systems as the DNA molecule could never arise by chance, no matter how big the universe nor how long is time." Quoted from Henry Morris, ed., *Scientific Creation-*

*ism* (San Diego: Creation-Life Publishers, 1975), 62. The problem here is not to compute the probability of a given DNA or protein sequence, but to compute the probability of all such sequences that perform the same function as the given sequence.

95. Abraham Lincoln, Gettysburg Address, in *Selected Writings of Abraham Lincoln*, ed. H. Mitgang (New York: Bantam, 1992), 279–280.

96. John Leslie describes this fly on the wall example in *Universes* (London: Routledge, 1989), 156–162.

97. See Douglas Axe, "Extreme Functional Sensitivity to Conservative Amino Acid Changes on Enzyme Exteriors," *Journal of Molecular Biology* 301 (2000): 585–595.

98. Ibid. See particularly Axe's discussion of the relation between TEM-1 beta-lactamase and *Proteus mirabilis* beta-lactamase, which are 50 percent identical in terms of sequence comparisons but clearly distinct as functional entities. Thus *Proteus mirabilis* beta-lactamase falls beyond the perturbation identity factor for TEM-1 beta-lactamase.

99. See Benjamin Lewin, *Genes*, 6th ed. (Oxford: Oxford University Press, 1997), 694.

~

# CHAPTER SIX

# Design as a Scientific Research Program

## 6.1   Outline of a Positive Research Program

Logic does not require that a scientific theory be rejected only after a better alternative is found. It does seem to be a fact about the sociology of science, however, that scientific theories give way not to criticism but to new, improved theories. Informed critiques of Darwinism have consistently appeared ever since Darwin published his *Origin of Species* (consider the work of Louis Agassiz, St. George Mivart, Richard Goldschmidt, Pierre Grassé, Gerald Kerkut, Michael Polanyi, Marcel Schützenberger, and Michael Denton). Yet all these critiques never succeeded in transforming design into a viable scientific alternative to Darwinism. For intelligent design to succeed as an intellectual project, the crucial next step is therefore to develop a design-theoretic research program as a positive alternative to Darwinism and other naturalistic approaches to the origin and history of life. In broad strokes, such a positive research program is now in place and looks as follows (here I am going to offer a conceptual rather than a historical reconstruction):

1. Much as Darwin began with the commonsense recognition that artificial selection in animal and plant breeding experiments is capable of directing organismal variation (which he then bootstrapped into a general mechanism to account for all organismal variation),[1] so too a design-theoretic research program begins with the commonsense recognition that humans draw design inferences routinely in ordinary life, explaining some things in terms of purely natural causes and other things in terms of intelligence or design (cf. archeologists attributing rock formations in one case to erosion and in another to design—as with the megaliths at Stonehenge).

2. Just as Darwin formalized and extended our commonsense understanding of artificial selection to natural selection, a design-theoretic research program next attempts to formalize and extend our commonsense understanding of design inferences so that they can be rigorously applied in scientific investigation. At present, my codification of design inferences as an extension of Fisherian hypothesis testing has attracted the most attention. It is now being vigorously debated whether my approach is valid and sustainable. The only alternative on the table at this point is a likelihood approach, which I argued in chapter 2 is inadequate.[2] Yet regardless how things fall out with my codification of design inferences, the question whether design is discernible in nature is now squarely on the table for discussion. This in itself is significant progress.

3. At the heart of my codification of design inferences is the notion of specified complexity, which is a statistical and complexity-theoretic concept. Provided this concept is well-defined (see chapter 2) and can effectively be applied in practice (see chapter 5), the next question is whether specified complexity is exhibited in actual physical systems where no evolved, reified, or embodied intelligence was involved. In other words, the next step is to apply the codification of design inferences in step 2 to natural systems and see whether it properly leads us to infer design. The most exciting area of application is of course biology, with Michael Behe's irreducibly complex biochemical systems, like the bacterial flagellum, having thus far attracted the most attention (see chapter 5). In my view, however, the most promising research in this area is now being done at the level of individual proteins (i.e., certain enzymes) to determine just how sparsely populated island(s) of a given functional enzyme type are within the greater sea of nonfunctional polypeptides. Preliminary indications are that they are very sparsely populated indeed, making them an instance of specified complexity (see section 5.10).

4. Once it is settled that certain biological systems are designed, the door is open to a new set of research problems. Here are some of the key problems:

- Detectability Problem—Is an object designed? An affirmative answer to this question is needed before we can answer the remaining questions. The whole point of steps 2 and 3 is to make an affirmative answer possible.
- Functionality Problem—What is the designed object's function? This problem is separate from the detectability problem. For instance, archeologists have discovered tools that they recognize as tools but do not understand what their function is.

- Transmission Problem—What is the causal history of a designed object? Just as with Darwinism, intelligent design seeks historical narratives (though not the just-so stories of Darwinists).
- Construction Problem—How was the designed object constructed? Given enough information about the causal history of an object, this question may admit an answer.
- Reverse-Engineering Problem—In the absence of a reasonably detailed causal history, how could the object have come about?
- Constraints Problem—What are the constraints within which the designed object functions optimally?
- Perturbation Problem—How has the original design been modified and what factors have modified it? This requires an account of both the natural and the intelligent causes that have modified the object over its causal history.
- Variability Problem—What degree of perturbation allows continued functioning? Alternatively, what is the range of variability within which the designed object functions and outside of which it breaks down?
- Restoration Problem—Once perturbed, how can the original design be recovered? Art restorers, textual critics, and archeologists know all about this.
- Optimality Problem—In what sense is the designed object optimal?
- Separation of Causes Problem—How does one tease apart the effects of intelligent causes from natural causes, both of which could have affected the object in question? For instance, a rusted old Cadillac exhibits the effects of both design and weathering.
- Ethical Problem—Is the design morally right?
- Aesthetics Problem—Is the design beautiful?
- Intentionality Problem—What was the intention of the designer in producing a given designed object?
- Identity Problem—Who or what is the designer?

To be sure, the last four questions are not questions of science, but they arise very quickly once design is back on the table for serious discussion. As for the other questions, they are strictly scientific (indeed, many special sciences, like archeology or SETI, already raise them). Now, it is true that some of these questions make perfect sense within a naturalistic framework (e.g., the functionality problem). But others clearly do not. For instance, the separation of causes problem (i.e., teasing apart the effects of intelligent causes from natural causes) and the restoration problem (i.e., recovering the original design) properly belong to a design-theoretic framework.

## 6.2   The Pattern of Evolution[3]

Intelligent design is a scientific research program that examines the role of specified complexity in nature. Since many special sciences already employ specified complexity as a criterion for detecting design (e.g., SETI and archeology), there can be no principled objection to teaching intelligent design within a science curriculum, and particularly whenever the origin and history of life comes up in grades K–12. To affirm the legitimacy of intelligent design as a proper subject for study within a science curriculum, however, raises two practical questions: (1) How is intelligent design to be taught? and (2) How will its teaching affect the teaching of other scientific subjects, notably biological evolution? One of the worries about intelligent design is that it will jettison much that is accepted in science, and that an "ID-based curriculum" will look very different from current science curricula. Although intelligent design has radical implications for science, I submit that it does not have nearly as radical implications for science education.

First off, intelligent design is not a form of anti-evolutionism. Intelligent design does not claim that living things came together suddenly in their present form through the efforts of a supernatural creator. Intelligent design is not and never will be a doctrine of creation. A doctrine of creation presupposes not only a designer that in some manner is responsible for organizing the structure of the universe and its various parts but also a creator that is the source of being of the universe.[4] A doctrine of creation thus invariably entails metaphysical and theological claims about a creator and the creation. Intelligent design, on the other hand, merely concerns itself with features of natural objects that reliably signal the action of an intelligence, whatever that intelligence might be. More significantly for the educational curriculum, however, is that intelligent design has no stake in living things coming together suddenly in their present form. To be sure, intelligent design leaves that as a possibility. But intelligent design is also fully compatible with large-scale evolution over the course of natural history, all the way up to what biologists refer to as "common descent" (i.e., the full genealogical interconnectedness of all organisms). If our best science tells us that living things came together gradually over a long evolutionary history and that all living things are related by common descent, then so be it. Intelligent design can live with that result and indeed live with it cheerfully.

But—and this is the crucial place where an ID-based curriculum will differ from how biological evolution is currently taught—intelligent design is not willing to accept common descent as a consequence of the Darwinian mechanism. The Darwinian mechanism claims the power to transform a single organism (known as the last common ancestor) into the full diversity of life that we see both around us and in the fossil record. If intelligent design is correct, then the Darwinian mechanism of natural selection and random variation lacks that power (see chapter 4). What's more, in that case the justification for common descent cannot be that it follows as a logical deduction from Darwinism. Darwinism is not identical with evolution understood merely as common descent. Darwinism comprises a historical claim (common descent) and a naturalistic mechanism (natural selection operating on random variations), with the latter being used to justify the former. According to intelligent design, the Darwinian mechanism cannot bear the weight of common descent. Intelligent design therefore throws common descent into question but at the same time leaves open as a very live possibility that common descent is the case, albeit for reasons other than the Darwinian mechanism.

What, then, are teachers who are persuaded of intelligent design to teach their students? Certainly they should teach Darwinian theory and the evidence that supports it. At the same time, however, they should candidly report problems with the theory, notably that its mechanism of transformation cannot account for the specified complexity we observe in biology. But that still leaves the question, "What happened when?" There is a lot of persuasive evidence for large-scale evolution that does not invoke the Darwinian mechanism, notably from biogeography and molecular sequence comparisons involving DNA and proteins. At the same time, discontinuities in the fossil record (preeminently in the Cambrian explosion) are more difficult to square with common descent.[5]

To establish evolutionary interrelatedness invariably requires exhibiting similarities between organisms. Within Darwinism, there is only one way to connect such similarities, and that is through descent with modification as driven by the Darwinian mechanism. But within a design-theoretic framework, this possibility, though not precluded, is also not the only game in town. It is possible for descent with modification instead to be driven by telic processes inherent in nature (and thus by a form of design). Alternatively, it is possible that the similarities are not due to descent at all but result from a similarity of conception, just as designed objects like your TV, radio, and computer share common components because designers frequently recycle ideas and parts. Teasing apart the effects of intelligent and natural causation, as noted in the last section, is one of the key questions confronting a design-theoretic research program. Unlike Darwinism, therefore, intelligent design has no immediate and easy answer to the question of common descent.

Darwinists necessarily see this as a bad thing and as a regression to ignorance. From the design theorists' perspective, however, frank admissions of ignorance are much to be preferred to overconfident claims to knowledge that cannot in the end be adequately justified. Or as David Berlinski put it, "It is not necessary to choose between doctrines. The rational alternative to Darwin's theory is intelligent uncertainty."[6] Despite advertisements to the contrary, science is not a juggernaut that relentlessly pushes back the frontiers of knowledge. Rather, science is an interconnected web of theoretical and factual claims about the world that are constantly being revised and for which changes in one portion of the web can induce radical changes in another. In particular, science regularly confronts the problem of having to retract claims that it once confidently asserted.

Consider the following example from geology. In the nineteenth century the geosynclinal theory was proposed to account for how mountain ranges originate. This theory hypothesized that large troughlike depressions, known as geosynclines, filled with sediment, gradually became unstable, and then, when crushed and heated by the earth, elevated to form mountain ranges. To the question "What happened when?" geologists as late as 1960 confidently asserted that the geosynclinal theory provided the answer. Thus in the 1960 edition of Clark and Stearn's *Geological Evolution of North America*, the status of the geosynclinal theory was compared favorably with Darwin's theory of natural selection:

> The geosynclinal theory is one of the great unifying principles in geology. In many ways its role in geology is similar to that of the theory of evolution, which serves to integrate the many branches of the biological sciences. . . . Just as the doctrine of evolution is universally accepted among biologists, so also the geosynclinal origin of the major mountain systems is an established principle in geology.[7]

Whatever became of the geosynclinal theory? Within ten years of this statement, the theory of plate tectonics, which explained mountain formation through continental drift and seafloor spreading, had decisively replaced the geosynclinal theory.[8] The history of science is filled with such turnabouts in which confident claims to knowledge suddenly vanish from the scientific literature. Often they are replaced with more accurate claims. At times no suitable replacement can be found.

But that still leaves the question, What does an ID-based curriculum teach actually happened in the course of biological evolution? As I already indicated, an ID-based curriculum will teach Darwinian theory, both the evidence that supports it as well as the countervailing evidence.[9] Such a curriculum will also teach progress to date on the research problems specific to a design-theoretic research program (see section 6.1). In particular, as regards the shape of natural history, it will teach what at the time is the best scientific account of the pattern of evolution consistent with specified complexity not being a free lunch. What I mean here is that evolutionary relationships cannot be drawn simply because some naturalistic mechanism is posited as capable of generating specified complexity. Naturalistic mechanisms, notably the Darwinian mechanism, are in principle incapable of generating specified complexity. Consequently, whenever evolution exhibits a net increase in specified complexity, that net increase must be sought in factors other than naturalistic mechanisms (e.g., reshuffling of preexisting specified complexity or its deliberate insertion from outside).

Darwinism takes a top-down approach to evolution. Darwinian theory posits a great tree of life that connects all organisms by descent from a last common ancestor and accounts for that tree in terms of the Darwinian mechanism of natural selection and random variation. Once Darwinian theory is presupposed, reconstructing natural history becomes a matter fitting the data of nature to Darwin's great tree of life. Some data are consistent with that tree, other data are not. In place of a top-down approach that requires in advance that all organisms be evolutionarily interconnected, intelligent design proposes a bottom-up approach in which evolution is confirmed within increasingly wider envelopes of variability. Whether an envelope can be expanded to include all living forms (thus implying common descent) is for now an open question facing intelligent design.

When it comes to integrating intelligent design with current science curricula, it is important to understand that intelligent design departs from these curricula principally only over the origin of biological complexity. True, intelligent design also takes up design in cosmology.[10] But arguing for design at the level of cosmology does not contradict any of the theories currently held by cosmologists (for instance, Big Bang and inflationary cosmologies can be interpreted as consistent with intelligent design). Arguing for design in biology, on the other hand, does squarely challenge Darwinian theory and more generally all purely naturalistic accounts of biological complexity. But that is about all intelligent design challenges. Thus one can be quite conservative in adapting intelligent design to a science curriculum. There is no need, for instance, to alter our understanding of cosmology or geology regarding the formation of the universe, galaxies, our solar system, or the Earth. Nor for that matter is there any need to challenge the standard chronologies scientists have assigned to these events (e.g., 12 or so billion years for the age of the universe and 4.5 billion years for the age of the earth).

How, then, would a design-theoretic research program play out in relation to evolutionary biology? I want to focus on this question for the remainder of this section. Let us start by noting that intelligent design leaves plenty of room for purely natural processes to play a significant role in biological evolution. True, specified complexity is not a free lunch in the sense that natural causes cannot generate it. Nevertheless, natural causes can take already existing specified complexity and shift it around; and since there is nothing to prevent specified complexity from being abundant in the universe, there is nothing to prevent natural causes from taking already existing specified complexity and expressing it in biological systems. We could therefore simply treat specified complexity as a raw datum.[11]

Although this move can be justified philosophically, it is unsatisfying scientifically. As scientists we want to know how specified complexity gets expressed in actual physical systems. In reference to the origin of life, we want to know the informational pathway that takes whatever specified complexity is inherent in a lifeless universe and translates it into the first organism. In reference to the evolution of life, we want to know the informational pathway that takes whatever specified complexity is inherent in an already existing organism together with its environment and translates that specified complexity into a new organism of still greater complexity. Even if the origin of specified complexity admits no naturalistic explanation, its flow surely does. How, then, do natural processes control the flow of specified complexity into and out of biological systems?

The answer to this question, at least in broad terms, is clear: The specified complexity inherent in an organism consists of the specified complexity it acquired at birth together with whatever specified complexity it acquires during the course of its life. The specified complexity acquired at birth derives from *inheritance with modification* (i.e., the specified complexity inherent in the parent(s) as well as any modifications of this specified complexity by chance). The specified complexity acquired after birth consists of *selection* (i.e., the environmental pressure that selects some organisms to reproduce and eliminates others before they can reproduce) along with *infusion* (i.e., the direct introduction of novel information from outside the organism).

Modification, as used here, is more general than mutation. Mutations are random genetic errors that get passed from one generation to the next. Not all randomly induced changes between generations, however, are errors. In sexual reproduction, for instance, genetic information from both parents combines according to a well-defined stochastic process that is functionally specified by the organism. Modification signifies not only chance errors but also chance processes specifically under the direction of an organism.

The Darwinian mechanism admits selection and inheritance with modification, but historically has minimized infusion. Not all evolutionary mechanisms, however, are limited in this way. The Lamarckian mechanism, for instance, focuses mainly on infusion. For Lamarck, characteristics acquired by an organism in the course of its life could be passed on to its offspring. These acquired characteristics arise from interaction with the environment. In passing these characteristics to their offspring, organisms therefore transmit information from the environment to the next generation.

Infusion as Lamarck conceived it has largely been discredited. Except for the direct introduction of genetic information into an organism's germ cells, organisms that acquire characteristics in one generation do not appear to pass them to the next. What evidence is there for the infusion of genetic information that once introduced into an organism gets transferred via reproduction to succeeding generations? It is well-established that bacteria exchange plasmids (i.e., circular pieces of genetic information) as a way of developing antibiotic resistance.[12] More speculative is the lateral gene transfer that is said to have occurred within the cellular community ancestral to the Archaea, Bacteria, and Eukarya.[13] Also speculative is Lynn Margulis's idea of symbiosis, where one organism assimilates another to form a more complex organism.[14] In all these instances, one organism co-opts specified complexity from another.

Inheritance with modification, selection, and infusion account for the specified complexity inherent in biological systems. Together they comprise all the sources of specified complexity in biology. I want therefore to examine more closely how these three sources contribute to the specified complexity of an organism. Consider first inheritance with modification. According to Franklin Harold, "There are many generalizations in biology but precious few universal laws; and of these, the least controversial may well be that like begets like. Offspring resemble their parents in form as well as function: roses and rabbits, yeast and *Escherichia coli* display the same forms, generation after generation, within a narrow range of variations."[15] From fertilization through to the adult phenotype, organisms follow well-defined developmental pathways.[16] These pathways are organism-specific and largely invariant, and supply organisms with all the structures and functions they inherit from their parents.[17] Inheritance is the developmental pathway by which already existing information is transferred from parents to offspring. Inheritance is thus merely a conduit for already existing specified complexity.

Because organisms do not merely repeat the information inherent in their parents, but also modify it through chance, the specified complexity organisms acquire at birth derives not just from inheritance, but also from modification. By modification I mean all the instances where chance enters an organism's developmental pathway and modifies its specified complexity. For instance, modification of genetic material includes point mutations, base deletions, genetic crossover, transpositions, and recombination generally.[18] Thus, while inheritance is merely a conduit for already existing information, modification is the operation of chance on the information passing through that conduit. Given the Law of Conservation of Information (see section 3.9), it follows that inheritance with modification by itself is incapable of explaining the increases of specified complexity that organisms have exhibited in natural history. Inheritance with modification needs therefore to be supplemented.

The most obvious candidate here is, of course, selection. Selection presupposes inheritance with modification, but instead of merely shifting around already existing information, selection also introduces new information. By exploiting advantageous modifications, selection is able to introduce new information into a population. The majority view in biology—known as the neo-Darwinian synthesis—is that selection and inheritance with modification together are adequate to account for all the specified complexity inherent in organisms. As a parsimonious account of the origin and history of life, this view has much to commend it. Nonetheless, as we have seen in previous chapters, this view places undue restrictions on the flow of biological information, restrictions that biological systems routinely violate. Michael Behe's irreducibly complex biochemical systems are a case in point (see chapter 5).

If the joint action of selection and inheritance with modification is unable to account for the specified complexity in biological systems (and specifically for the irreducible complexity of certain biochemical systems like the bacterial flagellum), there remains but one source for it: infusion, that is, the direct introduction of novel information from outside the biological system. In principle there is nothing problematic or controversial about infusion. To innovate a given informational structure an organism has informational needs, and these needs can be supplied from outside the organism in but one of two ways. Either it can be supplied indirectly through selection (whose efficacy, however, is limited to enhancing existing function—hence its inability to produce irreducibly complex systems, which require all components in place at the same time for any function at all). On the other hand, an organism's informational needs can be supplied directly by the insertion of ready-to-go information into the organism. The latter is infusion.

Although at this level of generality infusion is unproblematic, it quickly becomes problematic once we start tracing backwards informational pathways of infused information. Consider, for instance, what is perhaps the best confirmed instance of infusion in biology, namely, plasmid exchange among bacteria to develop antibiotic resistance.[19] Plasmids are small circular pieces of DNA that can easily be exchanged among bacteria and are capable of conferring antibiotic resistance. When one bacterium releases a plasmid and another absorbs it, information is infused from one into the other. By itself this is unproblematic. Problems begin, however, when we ask, Where did the bacterium that released the plasmid in turn derive it? There is a regress here, and this regress always terminates in something nonorganismal. We cannot just keep explaining plasmid infusion into a bacterium by plasmid release from another bacterium. Eventually, as we trace the informational pathway back, we must tell a different kind of story. If, for instance, the plasmid is cumulatively complex, then it could have arisen through selection and inheritance with modification (see section 5.2). But if it is irreducibly complex, whence could it have arisen?

It will be helpful here to distinguish between *biotic* and *abiotic* infusion, and correspondingly between *endogenous* and *exogenous* information. Biotic infusion is the infusion of information from one organism to another; abiotic infusion is the infusion of information from something other than an organism to an organism. Correspondingly, endogenous information comprises biotically infused information (and thus information already present within biological systems); exogenous information comprises abiotically infused information (and thus information external to biological systems). Now, regardless whether plasmids are irreducibly complex (the relevant analysis has yet to be performed), the fact remains that there exist irreducibly complex biochemical systems (see chapter 5). What's more, even though biotic infusion may explain how a particular instance of an irreducibly complex biochemical system came to exist in a given organism, it cannot explain how such a system arose in the first place. Because organisms have a finite trajectory back in time, biotic infusion must ultimately give way to abiotic infusion, and endogenous information must ultimately derive from exogenous information. And as always, the type of information of interest here is specified complexity.

The abiotic infusion of exogenous information is the great mystery confronting modern evolutionary biology. It is the mystery posed by Manfred Eigen in section 3.7. Why is it a mystery? Not because the abiotic infusion of exogenous information is inherently incomprehensible, but because evolutionary biology is only now beginning to appreciate the relevance and centrality of information to its task. The task of evolutionary biology is to explain the origin and history of life. But the key feature of life is specified complexity. Caught up in the Darwinian mechanism of selection and inheritance with modification, evolutionary biology has tended to neglect the informational hurdles organisms need to jump in the course of natural history.[20] To jump those hurdles, organisms require information. What's more, a significant part of that information is exogenous and must originally have been infused abiotically.

What, then, would evolutionary biology look like if information were taken as its central and unifying concept? First off, let us be clear that the Darwinian mechanism of selection and inheritance with modification will continue to occupy a significant place in evolutionary theory. Nevertheless, its complete and utter dominance in evolutionary theory—the inflated view that this mechanism can account for the full diversity of life—will have to be relinquished. As a mechanism for conserving, adapting, and honing already existing biological structures, the Darwinian mechanism is ideally suited. But as a mechanism for innovating irreducibly complex biological structures, it utterly lacks the informational resources. As for biotic infusion, its role within an information-theoretic framework must always remain limited, for even though it can account for how organisms trade already existing biological information, it can never get at the root question of how that information came to exist in the first place.

Not surprisingly, therefore, the key task an information-theoretic approach to evolutionary biology faces is to make sense of abiotically infused specified complexity. Abiotically infused specified complexity is information exogenous to an organism, but which nonetheless gets transmitted to and assimilated by the organism. Two obvious questions now arise: (1) How is abiotically infused specified complexity transmitted to an organism? and (2) Where does this information reside prior to being transmitted? If this information is clearly represented in some empirically accessible nonbiological physical system, and if there is a clear informational pathway from that system to the organism, and if that informational pathway can be shown suitable for transmitting this information to the organism so that the organism properly assimilates it, only then will these two questions receive an empirically adequate naturalistic answer.

But note that this naturalistic answer, far from eliminating the information question, simply pushes it one step further back, for how did the specified complexity that was abiotically infused into an organism first get into a prior nonbiological physical system? Because of the Law of Conservation of Information, whenever we inquire into the source of specified complexity, we never resolve the information problem until we trace it back to a designing intelligence. Short of that, we only intensify the information problem. This is not to say that inquiries that stop short of a designing intelligence are unilluminating. We learn an important fact about a pencil when we learn that a certain pencil-making machine made it. Nonetheless, the information in the pencil-making machine exceeds the information in the pencil. The Law of Conservation of Information guarantees that as we trace informational pathways backwards, we have more information to explain than we started with—until, that is, we locate the intelligence responsible for the informational pathway we are tracing back. For instance, my copy of *King Lear* has a convoluted causal history including printers, marketers, distribution networks, etc. But ultimately that causal history terminates in a designing intelligence, namely, the mind of William Shakespeare. (Note that it is a red herring to ask Who designed Shakespeare? Even if this question admits an answer—e.g., Shakespeare's parents, God, or aliens from Alpha Centauri—Shakespeare's mind remains the informational bottleneck through which all the specified complexity in *King Lear* gets expressed. See section 6.8.)

Where, then, do the informational pathways of life terminate as we trace them backwards? The possibilities are limited. One possibility is that we get nowhere, unable even to begin tracing backwards the information in a biological system. Thus we may discover an irreducibly complex biological system, but be unable to trace it back to any abiotic source of exogenous information.[21] Another possibility is that we can trace the information in a biological system back to an abiotic source of exogenous information, but then cannot trace it back any further. Graham Cairns-Smith, for instance, has a clay-template theory for the origin of life in which self-replicating clays form templates for carbon-based life.[22] The Cairns-Smith theory is an abiotic infusion theory, with exogenous information represented in (abiotic) clays providing templates for carbon-based life. What the Cairns-Smith theory does not treat is how the exogenous information that was transmitted to carbon-based life from clay templates got into those clay templates in the first place. Needless to say, the Cairns-Smith theory is highly speculative. Still another possibility is that we can trace the information in a biological system all the way back to the initial conditions of the Big Bang.[23] But this approach remains scientifically sterile until a definite informational pathway can be traced back to the initial conditions of the universe (more on this in section 6.6).

In tracing back the informational pathways of life, evolutionary biology does well to avoid speculation and to follow only those informational pathways that can be rigorously traced. To take an analogy, I can rigorously trace the informational pathway that issued in my copy of *King Lear* through the various extant editions of the play spanning the last four centuries. On the other hand, I cannot even begin to trace the informational pathway that issued in some isolated first century papyrus fragment found in the Egyptian desert. Any story behind this fragment is lost and cannot be reconstructed. Likewise, evolutionary biology may trace an informational pathway back to an abiotic source of exogenous information. On the other hand, it may remain stuck at a given irreducibly complex biological structure, forever unable to trace it back to an abiotic source of exogenous information.

To sum up, a design-theoretic research program reconceptualizes evolutionary biology in information-theoretic terms. An evolutionary biology thoroughly cognizant of information theory is one whose chief task is to trace informational pathways. In tracing these pathways, evolutionary biology must place a premium on rigor. Detailed informational pathways need to be explicitly exhibited—just-so stories will not do. Moreover, informational pathways need to conform to biological reality, and not to the virtual reality residing in a computer.[24] Finally, empirical evidence—and not metaphysical prejudice or aesthetic preference—must decide whether an informational pathway exists at all. For instance, the Darwinian predisposition to cash out taxonomy in terms of genealogy must not be taken as evidence for common descent. To establish common descent requires showing that certain informational pathways connect all organisms.[25]

In this reconceptualization of evolutionary biology, many low-level facts of current evolutionary biology will remain unchanged. What's more, information theory is sufficiently flexible to accommodate the naturalistic mechanisms of evolutionary change proposed to date. Nonetheless, their adequacy will have to be evaluated in terms of the information-theoretic constraints to which they are subject. In particular, the claim that the Darwinian mechanism can account for the full diversity of living forms will have to be rejected inasmuch as this mechanism is unable to generate the specified complexity inherent in—to take the most popular example—irreducibly complex biochemical systems (see chapter 5). Many old questions will remain. Many new questions will arise. But some old questions will have to be discarded. In particular, all reductionist attempts to explain specified complexity in terms of something other than intelligence will have to go by the board.

## 6.3    The Incompleteness of Natural Laws

From the design theorist's perspective, intelligent design offers plenty of interesting scientific problems to work on and certainly enough to turn intelligent design into a fruitful and exciting program of scientific research. Even so, many scientists remain skeptical. I want next to address some of their worries. Let me begin with the concerns of Howard Van Till. Van Till is a physicist with strong theological interests. Though not as well known as some critics of intelligent design, he has engaged this position at length. What's more, as one who takes theology seriously, Van Till's criticisms cannot be ascribed to atheist antipathy that rejects intelligent design simply because it might foster belief in God. That said, Van Till adamantly opposes intelligent design, regarding it not only as bad science but also as bad theology. His criticism is therefore instructive on two counts.

Van Till and I have known each other since the mid-1990s, and have been corresponding about the coherence of intelligent design as an intellectual project for about the last three years. Van Till's unchanging refrain has been to ask for clarification about what design theorists mean by the term "design." The point at issue for him is this: Design is unproblematic when it refers to something being conceptualized by a mind to accomplish a purpose; but when one attempts to attribute design to natural objects that could not have been formed by an embodied intelligence, design must imply not just conceptualization but also extra-natural assembly. The very possibility that intelligent design might require extra-natural assembly is for Van Till especially problematic.[26]

Although design theorists are no fans of naturalism, Van Till claims to turn the tables on them, charging design theorists with being guilty of "punctuated naturalism"—the view that for the most part natural processes rule the day but then intermittently need to be "punctuated" by interventions from a designing intelligence.[27] Van Till likes to state his objection to intelligent design this way: Design can have two senses, a "mind-like" sense (referring merely to conceptualization) and a "hand-like" sense (referring also to the mode of assembly); is intelligent design using design strictly in the mind-like sense or also in the hand-like sense? And if the latter, are design theorists willing to come clean and admit that their position commits them to extra-natural assembly?[28]

Although Van Till purports to ask these questions simply as an aid to clarity, it is important to understand how Van Till's own theological and philosophical presuppositions condition his formulation of these questions. Indeed, these presuppositions must themselves be clarified. For instance, what is "extra-natural assembly" (the term is Van Till's)? It is not what is customarily meant by miracle or supernatural intervention. Miracles typically connote a violation or suspension or overriding of natural laws. To attribute a miracle is to say that a natural cause was all set to make X happen, but instead Y happened. As I have argued throughout my work, design does not require this sort of counterfactual substitution.[29] When humans, for instance, act as intelligent agents, there is no reason to think that any natural law is broken. Likewise, should an unembodied designer act to bring about a bacterial flagellum, there is no reason prima facie to suppose that this designer did not act consistently with natural laws. It is, for instance, a logical possibility that the design in the bacterial flagellum was front-loaded into the universe at the Big Bang and subsequently expressed itself in the course of natural history as a miniature outboard motor on the back of *E. coli.* Whether this is what actually happened is another question (see section 6.6), but it involves no contradiction of natural laws and gets around the usual charge of miracles.

Nonetheless, even though intelligent design requires no contradiction of natural laws, it does impose a limitation on natural laws, namely, it purports that they are incomplete. Think of it this way. There are lots and lots of things that happen in the world. For many of those things we can find causal antecedents that account for them in terms of natural laws. Specifically, the account can be given in the form of a set of natural laws (typically supplemented by some auxiliary hypotheses[30]) that relates causal antecedents to consequents (i.e., the things we are trying to explain). Now why should it be that everything that happens in the world should submit to this sort of causal analysis? It is certainly a logical possibility that we live in such a world. But it is hardly self-evident that we do. For instance, we have no evidence whatsoever that there is a set of natural laws, auxiliary hypotheses, and antecedent conditions that account for the writing of this book. If we did have such an account, we would be well on the way to reducing mind to body. But no such reduction is in the offing, and cognitive science is to this day treading water when it comes to the really big question of how brain enables mind.[31]

Intelligent design regards intelligence as an irreducible feature of reality. Consequently it regards any attempt to subsume intelligent agency under natural causes as fundamentally misguided and regards the natural laws that characterize natural causes as fundamentally incomplete. This is not to deny derived intentionality, in which artifacts, though functioning according to natural laws and operating by natural causes, nonetheless accomplish the aims of their designers and thus exhibit design. Yet whenever anything exhibits design in this way, the chain of natural causes leading up to it is incomplete and must presuppose the activity of a designing intelligence. Note that this is not to deny or endorse what philosophers call the *principle of sufficient reason*, the view that everything that exists has a sufficient reason for its existence (i.e., a reason that leaves nothing unaccounted for). It is simply to say that when anything exhibits design, its explanation remains fundamentally incomplete until one introduces an intelligence that is not reducible to natural causes or to the natural laws that describe them.

I will come back to what it means for a designing intelligence to act in the physical world, but for now I want to focus on the claim by design theorists that natural causes and the natural laws that characterize them are incomplete. It is precisely here that Van Till objects most strenuously to intelligent design and that his own theological and philosophical interests come to light. "Extra-natural assembly" for Howard Van Till does not mean a miracle in the customary sense but rather that natural causes were insufficient to account for the assembly in question. Van Till holds to what he calls a Robust Formational Economy Principle (abbreviated RFEP; "formational economy" refers to the capacities or causal powers embedded in nature). This is a metaphysical principle. According to this principle nature is endowed with all the (natural) causal powers it ever needs to accomplish all the things that happen in nature. Thus in Van Till's manner of speaking, it is within nature's formational economy for water to freeze when its temperature is lowered sufficiently. Natural causal powers are completely sufficient to account for liquid water turning to ice. What makes Van Till's formational economy *robust* is that everything that happens in nature is like this—even the origin and history of life. In other words, the formational economy is complete.[32]

But how does Van Till know that the formational economy is complete? Van Till holds this principle for theological reasons. According to him, for natural causes to lack the power to effect some aspect of nature would mean that God had not fully gifted the creation. Conversely, a creator or designer who must act in addition to natural causes to produce certain effects has denied the creation benefits it might otherwise possess. Van Till portrays his God as supremely generous whereas the God of the design theorists he portrays as a miser. Van Till even refers to intelligent design as a "celebration of gifts withheld."[33]

Though rhetorically shrewd, Van Till's criticism is hardly the only way to spin intelligent design theologically. Granted, if the universe is like a clockwork (cf. the design arguments of the British natural theologians), then it would be inappropriate for God, who presumably is a consummate designer, to intervene periodically to adjust the clock. Instead of periodically giving the universe the gift of "clock-winding and clock-setting," God should simply have created a universe that never needed winding or setting. But what if instead the universe is like a musical instrument? (Cf. the design arguments of the Church Fathers, like Gregory of Nazianzus, who compared the universe to a lute[34]—in this respect I much prefer the design arguments of antiquity to the design arguments of the British natural theologians.) Then it is entirely appropriate for God to interact with the universe by introducing design (or in this analogy, by skillfully playing a musical instrument). Change the metaphor from a clockwork to a musical instrument, and the charge of "withholding gifts" dissolves. So long as there are consummate pianists and composers, player-pianos will always remain inferior to real pianos. The incompleteness of the real piano taken by itself is therefore irrelevant here. Musical instruments require a musician to complete them. Thus, if the universe is more like a musical instrument than a clock, it is appropriate for a designer to interact with it in ways that affect its physical state. On this view, for the designer to refuse to interact with the world is to withhold gifts.

Van Till's Robust Formational Economy Principle is entirely consistent with the methodological naturalism embraced by most scientists (the view that the natural sciences must limit themselves to naturalistic explanations and must scrupulously avoid assigning any scientific meaning to intelligence, teleology, or actual design). Indeed, this principle provides a theological justification for science to stay committed to naturalism. It encourages science to continue business as usual by restricting itself solely to natural causes and the natural laws that describe them. But this raises the question why we should want science to continue business as usual. How do we know that the formational economy of the world is robust in Van Till's sense? How do we know that natural causes (whether instituted by God as Van Till holds or self-subsistent as the atheist holds) can account for everything that happens in nature? Clearly the only way to answer this question scientifically is to go to nature and see whether nature exhibits things that natural causes could not have produced.

## 6.4 Does Specified Complexity Have a Mechanism?

What are the candidates here for something in nature that is nonetheless beyond nature? In my view the most promising candidate is specified complexity. The term "specified complexity" has been in use for about thirty years. The first reference to it with which I am familiar is from Leslie Orgel's 1973 book *The Origins of Life*, where specified complexity is treated as a feature of biological systems distinct from inorganic systems.[35] Richard Dawkins also employs the notion in *The Blind Watchmaker*, though he does not use the actual term (he refers to "complicated things" that are "specifiable in advance").[36] In *The Fifth Miracle* Paul Davies claims that life is mysterious not because of its complexity per se but because of its "tightly specified complexity."[37] Stuart Kauffman in *Investigations* proposes a "fourth law" of thermodynamics to account for specified complexity.[38] Specified complexity is, as we have seen in the previous chapters, a form of information, though one richer than Shannon information. Shannon's theory focuses exclusively on the complexity of information without reference to its specification (see section 3.2). Consequently, Shannon's theory underwrites no design inference. By contrast, to identify specified complexity is logically equivalent to detecting and inferring design (significantly, this means that intelligent design can be conceived as a branch of information theory).

Most scientists familiar with specified complexity think that the Darwinian mechanism is adequate to account for it once one has differential reproduction and survival (in chapter 4 we saw that the Darwinian mechanism has no such power; even so, my argument here carries through without that result). But outside a context that includes replicators, no one has a clue how specified complexity occurs by purely natural means. This is not to say there has not been plenty of speculation (e.g., clay templates, hydrothermic vents, and hypercycles), but none of this speculation has come close to solving the problem. Unfortunately for naturalistic origin-of-life researchers, this problem seems not to be eliminable since simple replicators that do not exhibit specified complexity and that ideally could be bootstrapped to replicators that do exhibit specified complexity in fact never do anything biologically significant. As Brian Goodwin notes, for evolution to do anything biologically significant requires that a cellular context already be in place—and even the simplest cell requires vast amounts of specified complexity.[39] For this reason Paul Davies suggests that the explanation of specified complexity will require some fundamentally new kinds of natural laws.[40] But so far such laws remain completely unknown. Kauffman's reference to a "fourth law," unlike my proposal in chapter 3, merely cloaks the scientific community's ignorance about the naturalistic mechanisms supposedly responsible for the specified complexity in nature.

Van Till agrees that specified complexity is an open problem for science. At a symposium on intelligent design at the University of New Brunswick sponsored by the Center for Theology and the Natural Sciences (15–16

September 2000), Van Till and I took part in a panel discussion. When I asked him how he accounts for specified complexity in nature, he called it a mystery that he hopes further scientific inquiry will resolve. But resolve in what sense? On Van Till's Robust Formational Economy Principle, there must be some causal mechanism in nature that accounts for any instance of specified complexity. We may not know it and we may never know it, but surely it is there. For the design theorist to invoke an unembodied intelligence is therefore out of bounds.

But what happens once some causal mechanism is found that accounts for a given instance of specified complexity? Something that is specified and complex is highly improbable with respect to all causal mechanisms currently known. Consequently, for a causal mechanism to come along and explain something that previously was regarded as specified and complex means that the item in question is in fact no longer specified and complex with respect to the newly found causal mechanism. The task of causal mechanisms is to render probable what otherwise seems highly improbable. Consequently, the way naturalism explains specified complexity is by dissolving it. Intelligent design makes specified complexity a starting point for inquiry. Naturalism regards it as a problem to be eliminated. That is why, for instance, Richard Dawkins wrote *Climbing Mount Improbable*.[41] To climb Mount Improbable one needs to find a gradual route that breaks a horrendous improbability into a sequence of manageable probabilities each one of which is easily bridged by a natural mechanism.

Lord Kelvin once remarked, "If I can make a mechanical model, then I can understand; if I cannot make one, I do not understand."[42] Repeatedly, critics of design have asked design theorists to provide a causal mechanism whereby an unembodied designer inputs specified complexity into the world. This question presupposes a self-defeating conception of design and tries to force design onto a Procrustean bed sure to kill it. *Intelligent design is not a mechanistic theory!* Intelligent design regards Lord Kelvin's dictum about mechanical models not as a sound regulative principle for science but as a straitjacket that artificially constricts science. SETI researchers, for instance, are not invoking a mechanism when they explain a radio transmission from outer space as the result of an extraterrestrial intelligence.

To ask for a mechanism to explain the effect of an intelligence (leaving aside derived intentionality) is like Aristotelians asking Newton what keeps bodies in rectilinear motion at a constant velocity moving (for Aristotle the crucial distinction was between motion and rest; for Newton it was between accelerated and unaccelerated motion). This is simply not a question that arises within Newtonian physics. Newtonian physics proposes an entirely different problematic from Aristotelian physics. Similarly, intelligent design proposes an entirely different (and I would argue far richer) problematic from science committed to naturalism. Intelligent design is fully capable of accommodating mechanistic explanations. Intelligent design has no interest in dismissing mechanistic explanations. Such explanations are wonderful as far as they go. But they only go so far, and they are incapable of accounting for specified complexity.

In rejecting mechanical accounts of specified complexity, design theorists are not arguing from ignorance. Arguments from ignorance have the form "Not X, therefore Y." Design theorists are not saying that for a given natural object exhibiting specified complexity, all the natural causal mechanisms so far considered have failed to account for it and therefore it had to be designed. Rather they are saying that the specified complexity exhibited by a natural object can be such that there are compelling reasons to think that no natural causal mechanism is capable of accounting for it. Usually these "compelling reasons" take the form of an argument from contingency in which the object exhibiting specified complexity is compatible with but in no way determined by the natural laws relevant to its occurrence (see sections 1.3 and 2.6). For instance, for polynucleotides and polypeptides there are no physical laws that account for why one nucleotide base is next to another or one amino acid is next to another. The laws of chemistry allow any possible sequence of nucleotide bases (joined along a sugar-phosphate backbone) as well as any possible sequence of L-amino acids (joined by peptide bonds).

Design theorists are attempting to make the same sort of argument against mechanistic accounts of specified complexity that modern chemistry makes against alchemy. Alchemy sought to transform base into precious metals using very limited means like furnaces and potions (though not particle accelerators). We rightly do not regard the contemporary rejection of alchemy as an argument from ignorance. For instance, we do not charge the National Science Foundation with committing an argument from ignorance for refusing to fund alchemical research. Even so, it is evident that not every combination of furnaces and potions has been tried to transform lead into gold. But that is no reason to think that some combination of furnaces and potions might still constitute a promising avenue for effecting the desired transformation. We now know enough about atomic physics to preclude this transformation. So too, we are now at the place where transforming a biological system that does not exhibit an instance of specified complexity (say a bacterium without a flagellum) into one that does (say a bacterium with a flagellum) cannot be accomplished by purely natural means but also requires intelligence.

There are a lot of details to be filled in, and design theorists are working overtime to fill them in. What I am offering here is not the details but an overview of the design research program as it justifies the inability of natural mechanisms to account for specified complexity. This part of its program is properly viewed as belonging to science. Science is in the business of establishing not only the causal mechanisms capable of accounting for an object having certain characteristics but also the inability of causal mechanisms to account for such an object—or what Stephen Meyer calls "proscriptive generalizations." There are no causal mechanisms that can account for perpetual motion machines. This statement is a proscriptive generalization. Perpetual motion machines violate the second law of thermodynamics and can thus on theoretical grounds be eliminated. Design theorists are likewise offering in-principle theoretical objections for why the specified complexity in biological systems cannot be accounted for in terms of purely natural causal mechanisms. Such proscriptive generalizations are not arguments from ignorance.

Assuming such an in-principle argument can be made (and for the sequel I will assume that the previous chapters have established as much), the design theorist's inference to design can no longer be considered an argument from ignorance. With such an in-principle argument in hand, not only has the design theorist excluded all natural causal mechanisms that might account for the specified complexity of a natural object, but the design theorist has also excluded all explanations that might in turn exclude design. The design inference is therefore not purely an eliminative argument, as is so frequently charged. Rather, design inferences, by identifying specified complexity, exclude everything that might in turn exclude design (see section 2.10).

It follows that contrary to the frequently-leveled charge that design is untestable, design is in fact eminently testable. Indeed, specified complexity tests for design (see section 6.9). Specified complexity is a well-defined statistical notion. The only question is whether an object in the real world exhibits specified complexity. Does it correspond to an independently given pattern and is the event delimited by that pattern highly improbable (i.e., complex)? These questions admit a rigorous mathematical formulation and are readily applicable in practice.

Not only is design eminently testable, but to deny that design is testable commits the fallacy of *petitio principii*, that is, begging the question or arguing in a circle.[43] It may well be that the evidence to justify that a designer acted to bring about a given natural structure may be insufficient. But to claim that there could never be enough evidence to justify that a designer acted to bring about a given natural structure is insupportable. The only way to justify the latter claim is by imposing on science a methodological principle that deliberately excludes design from natural systems, to wit, methodological naturalism. But to say that design is not testable because we have defined it out of existence is hardly satisfying or legitimate. Darwin claimed to have tested for design in biology and found it wanting. Design theorists are now testing for design in biology afresh and finding that biology is chock-full of design.

Specified complexity is only a mystery so long as it must be explained mechanistically. But the fact is that we attribute specified complexity to intelligences (and therefore to entities that are not mechanisms) all the time. The reason that attributing specified complexity to intelligence for biological systems is regarded as problematic is because such an intelligence would in all likelihood have to be unembodied (though strictly speaking this is not required of intelligent design—the designer could in principle be an embodied intelligence, as with the panspermia theories[44]). But how does an unembodied intelligence interact with natural objects and get them to exhibit specified complexity? We are back to Van Till's problem of extra-natural assembly.

## 6.5   The Nature of Nature

The next question I want to take up is what nature must be like for an unembodied intelligence or designer to interact coherently with it. Before proceeding to this question, however, I need to say something about the distinction between embodied and unembodied intelligences. By an embodied intelligence (or designer) I mean an intelligence whose mode of operation is confined to some physical entity located within spacetime (note that on this definition the entire universe is not an embodied intelligence since it properly comprises spacetime and is therefore not located within spacetime). Human beings are embodied intelligences. Even suitably organized fields of force might constitute embodied intelligences. Essential to an embodied intelligence is a physical entity through which the intelligence is expressed.

By contrast, an unembodied intelligence retains no physical entity through which the intelligence is expressed. For instance, specified complexity due to an embodied intelligence can be traced back causally to the physical entity constituting its embodiment; but for an unembodied intelligence the causal trail does not terminate in a physical entity. Thus by an unembodied intelligence or designer I mean an intelligence whose mode of operation cannot be confined to a physical entity located within spacetime. The existence of such an intelligence is compatible with pantheism, panentheism, Stoicism, Neoplatonism, deism, and theism. It is incompatible with naturalism.

With the distinction between embodied and unembodied intelligences in place, let us now consider what the natural world must be like for an unembodied intelligence to interact coherently with it. Fortunately, we are not in the situation of Descartes seeking a point of contact between the material and the spiritual at the pineal gland. For Descartes the physical world consisted of extended bodies that interacted only via direct contact. Thus for a spiritual dimension to interact with the physical world could only mean that the spiritual caused the physical to move. In arguing for a substance dualism in which human beings consist of both spirit and matter, Descartes therefore had to argue for a point of contact between spirit and matter. He settled on the pineal gland because it was the one place in the brain where symmetry was broken and where everything seemed to converge (most parts of the brain have right and left counterparts).

Although Descartes's argument does not work, the problem it tries to solve is still with us and surfaces regularly in discussions about intelligent design. For instance, Paul Davies has expressed his doubts about intelligent design this way: "At some point God has to move the particles."[45] The physical world consists of physical stuff, and for a designer to influence the arrangement of physical stuff seems to require that the designer intervene in, meddle with, or in some way coerce this physical stuff. What is wrong with this picture of supernatural action by a designer? The problem is not a flat contradiction with the results of modern science. Take, for instance, the law of conservation of energy. Although the law is often stated in the form "energy can neither be created nor destroyed," in fact all we have empirical evidence for is the much weaker claim that "in an isolated system energy remains constant." Thus a supernatural action that moves particles or creates new ones is beyond the power of science to disprove because one can always claim that the system under consideration was not isolated.

There is no logical contradiction here. Nor is there necessarily a god-of-the-gaps problem here. It is certainly conceivable that a supernatural agent could act in the world by moving particles so that the resulting discontinuity in the chain of physical causality could never be removed by appealing to purely physical mechanisms (i.e., physical causes completely characterized by well-defined natural laws—these can be deterministic or indeterministic). The "gaps" in the god-of-the-gaps objection are meant to denote gaps of ignorance about underlying physical mechanisms. But there is no reason to think that all gaps must give way to ordinary physical explanations once we know enough about the underlying physical mechanisms. The mechanisms may simply not exist. Some gaps might constitute ontic discontinuities in the chain of physical causes and thus remain forever beyond the capacity of physical mechanisms.

Although an unembodied designer who "moves particles" is not logically incoherent, such a designer nonetheless remains problematic for science. The problem is that natural causes are fully capable of moving particles. Thus for an unembodied designer also to move particles can only seem like an arbitrary intrusion. The designer is merely doing something that nature is already doing, and even if the designer is doing it better, why did the designer not make nature better in the first place so that it can move the particles better? We are back to Van Till's Robust Formational Economy Principle.

But what if the designer is not in the business of moving particles but of imparting information? In that case nature moves its own particles, but an intelligence nonetheless guides the arrangement which those particles take. A designer in the business of moving particles accords with the following world picture: The world is a giant billiard table with balls in motion, and the designer arbitrarily alters the motion of those balls, or even creates new balls and then interposes them among the balls already present. On the other hand, a designer in the business of imparting information accords with a very different world picture: In that case the world becomes an information processing system that is responsive to novel information. Now the interesting thing about information is that it can lead to massive effects even though the energy needed to represent and impart the information can become infinitesimal. For instance, the energy requirements to store and transmit a launch code are minuscule, though getting the right code can make the difference between starting World War III and maintaining peace. Frank Tipler and Freeman Dyson have even argued that arbitrarily small amounts of energy are capable of sustaining information processing and in fact sustaining it indefinitely.[46]

When a system is responsive to information, the dynamics of that system will vary sharply with the information imparted and will largely be immune to purely physical factors (e.g., mass, charge, or kinetic energy). A medical doctor who utters the words "Your son is going to die" might trigger a nervous collapse in a troubled parent whereas uttering the words "Your son is going to live" might prevent it. Moreover, it is largely irrelevant how loudly the doctor utters one sentence or the other or what bodily gestures accompany the utterance. Consider another example. After killing the Minotaur on Crete and setting sail back for Athens, Theseus forgot to substitute a white flag for a black flag. Theseus and his father Aegeus had agreed that a black flag would signify that Theseus had been killed by the Minotaur whereas a white flag would signify his success in destroying it. Seeing the black flag hoisted on the ship at a distance, Aegeus committed suicide. Or consider yet another nautical example, in this case a steersman who guides a ship by controlling its rudder. The energy imparted to the rudder is minuscule compared to the energy inherent in the ship's motion, and yet the rudder guides its motion. It was this analogy that prompted Norbert Wiener to introduce the term "cybernetics," which is derived etymologically from the Greek and means steersman. It is no coincidence that in his text on cybernetics, Wiener writes about information as follows: "Information is information, not matter or energy. No materialism which does not admit this can survive at the present day."[47]

How much energy is required to impart information? We have sensors that can detect quantum events and amplify them to the macroscopic level. What's more, the energy in quantum events is proportional to frequency or inversely proportional to wavelength. And since there is no upper limit to the wavelength of, for instance, electromagnetic radiation, there is no lower limit to the energy required to impart information. In the limit, a designer could therefore impart information into the universe without inputting any energy at all.

Limits, however, are tricky things. To be sure, an embodied designer could impart information by employing arbitrarily small amounts of energy. But an arbitrarily small amount of energy is still a positive amount of energy, and any designer employing positive amounts of energy to impart information is still, in Paul Davies's phrase, "moving the particles." The question remains how can an unembodied designer influence the natural world without imparting any energy whatsoever. It is here that an indeterministic universe comes to the rescue. Although we can thank quantum mechanics for the widespread recognition that the universe is indeterministic, indeterminism has a long philosophical history, and appears in such diverse places as the atomism of Lucretius and the pragmatism of Charles Peirce and William James.

For now, however, quantum theory is probably the best place to locate indeterminism. True, there is a sense in which quantum mechanics is deterministic: The evolution of the state function by means of the Schroedinger equation is deterministic; that is, given the state function at a given time, the Schroedinger equation prescribes the exact state at some future time. Nonetheless, the state function itself characterizes a probability distribution, and all observation of quantum systems involves sampling from such probability distributions. An analogy may help here. Imagine an urn that always contains ten balls. On Monday there are two white balls and eight black balls in the urn, on Tuesday there are three white balls and seven black balls in the urn, . . . , and on Sunday there are eight white balls and two black balls in the urn. Day to day the number of balls in the urns is determined. But sampling from these urns any day of the week is probabilistic and therefore indeterministic.

I need here to add a word about quantum cosmology and the many-worlds interpretation of quantum mechanics. Many quantum cosmologists would cringe at my characterization of quantum mechanics. The emerging consensus among quantum cosmologists (and one now held by Murray Gell-Mann, Philip Anderson, Stephen Hawking, and Steven Weinberg) is that quantum mechanics is *completely deterministic*. Accordingly, the state function of quantum mechanics does not characterize a probability distribution—we only interpret it as a probability distribution from our limited vantage.[48] Instead, the state function describes an ensemble of universes. Thus, the emerging consensus among quantum cosmologists is a many-worlds view (see section 2.8).

Why has this view taken hold? In quantum cosmology, when trying to apply quantum mechanics to the universe as a whole, having the state function collapse, as it must within a probabilistic interpretation, leads to a break in the dynamics of the quantum equations. This is mathematically unappealing (and for many cosmologists also metaphysically unappealing since it gives up on full deterministic causality—this was Einstein's worry about quantum mechanics). Thus, instead of allowing for state-function collapse, quantum cosmologists have come to prefer an expanded ontology in which all possible histories or worlds consistent with quantum mechanics get lived out.

The totality of physical reality for quantum cosmologists is therefore vastly bigger than the world we think we inhabit. We think we live in a world where Hitler lost World War II—and we are right as far as that goes. Yet from a many-worlds point of view, it would be more precise to say that we live within a world that is but one among a multiplicity of worlds each of which is as real as ours and whose union properly constitutes the whole of reality. Moreover, the role of quantum theory is to coordinate all those worlds. Thus within our world, Hitler lost World War II. But presumably there are quantum events that could have changed the course of that war, on account of which Hitler would then have won World War II, and whose consequences are being fully worked out in that alternate world.[49] The totality of physical reality is thus no longer properly conceived as a universe in the traditional sense but as a multiverse consisting of multiple worlds, multiple histories, and multiple minds.[50]

There is an old joke: *There's speculation, wild speculation, and cosmology!* When Alan Guth first began proposing his inflationary cosmology, Lenny Susskind remarked, "You know, the most *amazing* thing is that they pay us for this."[51] Cicero likewise remarked, "There is nothing so ridiculous but that some philosopher has said it"—no doubt he included the natural philosophers and cosmologists of his day. I would update Cicero's dictum as follows: "No cosmological idea is so crazy but that it becomes plausible and even compelling once it is given an elegant mathematical formulation and shown to underwrite physics as the ultimate science." My aim with these remarks is not ridicule but a reality check. Cosmologists are notorious for beginning with physics and ending in metaphysics. The problem is that if you are willing to monkey with metaphysics, you can get any result you like.

Many-worlds purchase complete determinism, but at a huge metaphysical cost. Many-worlds vastly inflate our ontology. In fact, inflated ontologies have become a dominant theme of recent cosmological speculation. The point to realize is that there is no reason to give these inflated ontologies, especially in the case of quantum many-worlds, any allegiance except as speculative hypotheses that are of interest because of the insights they generate. Why? Not merely because data underdetermine theories but because, in the case of inflated ontologies, data could never even in principle adjudicate among such theories (see section 2.8). David Lindley made this point beautifully in *The End of Physics: The Myth of a Unified Theory.*[52] Lindley's choice of the word "myth" was well-considered. The bloated ontologies of contemporary cosmological speculation, like the myths of old, bring unity to our understanding but at the cost of severing us from the data of actual experience.

I could go on with this sociological commentary on quantum cosmology and many-worlds, but let us consider the many-worlds interpretation on its own terms. First off, the many-worlds interpretation of quantum mechanics is an interpretation of quantum mechanics and not quantum mechanics itself. My minimalist probabilistic interpretation is likewise an interpretation of quantum mechanics. Because it is minimalist, it is compatible with all interpretations of quantum mechanics that allow for a fundamental indeterminism in the world. It is incompatible only with interpretations that view quantum mechanics as completely deterministic, as with the many-worlds interpretation. It is absolutely crucial here to understand that interpretations of quantum mechanics are empirically indistinguishable. I already quoted Anthony Sudbery to that effect in chapter 2, but it is worth repeating the quote: "An interpretation of quantum mechanics is essentially an answer to the question 'What is the state vector?' Different interpretations cannot be distinguished on scientific grounds—they do not have different experimental consequences; if they did they would constitute different *theories*."[53]

How, then, do we decide between a minimalist probabilistic interpretation that stresses an indeterministic universe and a many-worlds interpretation that stresses a completely deterministic, albeit ontologically bloated, multiverse? Although empirics alone are not enough to distinguish the two, one consideration is, at least for me, decisive. That consideration centers on the priority of probabilities in quantum mechanics and on how we make sense of those probabilities. Historically, quantum mechanics did not begin with a many-worlds formulation and then derive probabilities. Historically, quantum mechanics began with trying to make sense of probabilistic phenomena. Then, because full deterministic causality had for centuries been elevated as a regulative ideal for physics, a way was found to interpret quantum mechanics nonprobabilistically via many-worlds. This historical priority of probabilities in the formulation of quantum mechanics suggests to me a conceptual and ontic priority: quantum mechanics is fundamentally a probabilistic theory describing an indeterministic world, and only with considerable finagling can it be interpreted as a completely deterministic theory. The minimalist probabilistic interpretation comes to terms with the probabilities arising out of quantum mechanics as such. The many-worlds interpretation begins with these same probabilities, must explain them away, but then must recover them for use in actual quantum mechanical experiments (it is the probabilities associated with measurements and not the many-worlds themselves that figure directly into quantum mechanical experiments).

Not only is the many-worlds interpretation parasitic on the minimal probabilistic interpretation; it is not even clear whether the many-worlds interpretation allows for a coherent recovery of probabilities. As Michael Dickson observes:

> Without a notion of identity across time of a world (or mind), it is unclear how probabilities can be made empirically manifest [within the many-worlds interpretation]; i.e., the connection between probabilities and relative frequencies (over time) is severed. Indeed, the very notion of performing an experiment (which inevitably takes time) is apparently unavailable without the prior notion of what constitutes *the same world* (or mind) over time.[54]

Within the many-worlds interpretation, worlds that constitute the multiverse are continually splitting in accord with the probabilities given by quantum theory. As a consequence, there is no experimental way to track those probabilities within a given world. To be sure, one can assign probabilities simply on the basis of quantum theory. But unless science is to become a purely rationalist enterprise, it is also necessary to ground those probabilities in experience. The many-worlds interpretation seems not to allow this. Indeed, experience can only take place within a world, not across worlds. There are other difficulties with the many-worlds interpretation. But my concern here is simply to address the growing sense that this interpretation is the only game in town. It is not. Yet even if it were, it would get around the problem of causal indeterminacy only to face a still deeper problem of contingency—why do we inhabit this world rather than another and why is our world chock-full of specified complexity while others are not?

Let us now return to the question of how an unembodied designer can impart information into the natural world without imparting any energy. Setting aside the many-worlds interpretation of quantum mechanics and the determinism it implies, let us grant that there is genuine indeterminism in the natural world. In that case, whether an unembodied designer works through quantum mechanical effects is not ultimately the issue. Certainly quantum mechanics is more hospitable to an information processing view of the universe than the older mechanical models. All that is needed, however, is a universe whose constitution and dynamics are not reducible to deterministic natural laws (indeed, given the imprecision inherent in our measurements, there is no way ever to establish determinism with finality). Such a universe will produce random events and thus have the possibility of producing events that exhibit specified complexity (i.e., events that stand out against the backdrop of randomness). As I have stressed throughout this book, specified complexity is a form of information, albeit a richer form than Shannon information, which trades purely in complexity (see chapter 3). What's more, I have argued that specified complexity is a reliable empirical marker of actual design. And indeed, our best empirical evidence confirms that we live in a nondeterministic universe that is open to novel information, that exhibits specified complexity, and that therefore offers convincing evidence of an unembodied designer who has imparted it with information.

Consider, for instance, a device that outputs 0s and 1s and for which our best science tells us that the bits are independent and identically distributed so that 0s and 1s each have probability 1/2. (The device is therefore an idealized coin tossing machine; note that quantum mechanics offers such a device in the form of photons shot at a polaroid filter whose angle of polarization is 45 degrees in relation to the polarization of the photons—half the photons will go through the filter, counting as a "1"; the others will not, counting as a "0.") Now, what happens if we control for all possible physical interference with this device, and nevertheless the bit string that this device outputs yields an English text-file in ASCII code that delineates the cure for cancer (and thus a clear instance of specified complexity)? We have therefore precluded that a designer imparted a positive amount of energy (however minuscule) to influence the output of the device. Nevertheless, there is no way to avoid the conclusion that a designer (presumably unembodied) influenced the output of the device despite imparting no energy to it. Note that there is no problem of counterfactual substitution here. It is not that the designer expended any energy and therefore did something physically discernible to the device in question. Any bit when viewed in isolation is the result of an irreducibly chance-driven process. And yet the arrangement of the bits in sequence cannot reasonably be attributed to chance and in fact points unmistakably to an intelligent designer.

It is at this point that critics of design typically throw up their hands in despair and charge that design theorists are merely evading the issue of how a designer introduces design into the world. Surely there must be some physical mechanism by which the information is imparted. Surely there are thermodynamic limitations governing the flow of information. Thermodynamic limitations do apply if we are dealing with embodied designers who need to output energy to transmit information. But unembodied designers who co-opt random processes and induce them to exhibit specified complexity are not required to expend any energy. For them the problem of "moving the particles" simply does not arise. Indeed, they are utterly free from the charge of counterfactual substitution, in which natural laws dictate that particles would have to move one way but ended up moving another because an unembodied designer intervened. Indeterminism means that an unembodied designer can substantively affect the structure of the physical world by imparting information without imparting energy.

It is worth noting that this design-theoretic use of indeterminism need not constitute a rejection of the principle of sufficient reason. Often indeterminism suggests acausality so that an event that is attributed to a random or indeterministic process is regarded as having no cause, or at best an incomplete cause (i.e., whatever we are calling a cause does not provide a complete account of the event in question). If one views chance as fundamental and specified complexity as an anomaly of chance, then this view follows. But one can also view design (i.e., the activity of intelligent agency) as fundamental and treat chance as an epiphenomenon of design. We considered this possibility in section 2.9, where letter distributions of English texts were seen to follow a well-defined probability distribution (i.e., the relative frequency of the letter *e* being approximately 13 percent, that of *t* approximately 9 percent, etc.) even though the texts themselves were the result of design. On such a view intelligent agency provides a sufficient reason for chance events.

Indeed, one can take this line of reasoning further and argue that chance and randomness do not even make sense apart from design.[55] Consider that for any chance process like tossing a coin, if the coin is tossed indefinitely, any finite sequence will not only appear once but also appear infinitely often (this follows from the Strong Law of Large Numbers).[56] Now consider further that we can have experience only of finite sequences of coin tosses. Suppose therefore that I come in at some arbitrary point in the tossing of a coin that is being tossed indefinitely. On what basis can I have confidence that the finite sequence I witness will in some way be "representative of chance"? For instance, on what basis should I expect approximately the same number of heads and tails? Since the coin is being tossed indefinitely, even if I witness a million coin tosses, there will be runs of a million heads in a row. What precludes me from witnessing such a sequence? To be sure, one can argue that it is highly unlikely that I witness a million heads in a row. But that merely restates the problem in terms of the problem of induction. And the problem of induction is itself unresolved.

But suppose chance and randomness is an epiphenomenon of design. Then the problem of chance inducing events that are unrepresentative of chance (e.g., a million heads in a row) can be effectively circumvented since design has a way of stabilizing and thereby justifying chance. To come back to the letter frequencies of English texts, although those frequencies follow a chance distribution, they are perfectly stable as a result of orthographic, syntactic, grammatical, and semantic constraints on English. For those letter frequencies to diverge sharply from the norm, as for instance with Ernest Vincent Wright's novel *Gadsby*, which contained no occurrence of the letter *e*, would therefore itself constitute an instance of design.[57] Though these remarks may appear speculative, in fact they provide the key to resolving certain longstanding paradoxes connected with the study of randomness.[58]

There is no logical inconsistency and no evading the hard problems of science in treating the world as a medium receptive to information from an unembodied intelligence. In requiring a mechanistic account of how an unembodied intelligence imparts information and thereby introduces design into the world, the critic of design exhibits a failure of imagination. Such a critic is like a physicist trained only in Newtonian mechanics and desperately looking for a classical account of how a single particle like an electron can go through two slits simultaneously to produce a diffraction pattern on a screen. (This is the famous double-slit experiment; technically, the diffraction pattern is a statistical distribution that results from multiple electrons being shot individually at the screen.) On a classical Newtonian view of physics, only a classical account in terms of sharply localized and individuated particles makes sense. And yet nature is unwilling to oblige any such account of the double-slit experiment (note that the Bohmian approach to quantum mechanics merely shifts what is problematic in the classical view to Bohm's quantum potential[59]). Richard Feynman was right when he remarked that no one understands quantum mechanics. The "mechanics" in "quantum mechanics" is nothing like the "mechanics" in "Newtonian mechanics." There are no analogies that carry over from the dynamics of macroscopic objects to the quantum level. In place of understanding we must content ourselves with knowledge. We do not *understand* how quantum mechanics works, but we *know* that it works. So too, we may not *understand* how an unembodied designer imparts specified complexity into the world, but we can *know* that such a designer imparts specified complexity into the world.

It follows that Howard Van Till's riddle to design theorists is ill-posed. Van Till asks whether the design that design theorists claim to find in natural systems is strictly mind-like (i.e., conceptualized by a mind to accomplish a purpose) or also hand-like (i.e., involving a coercive extra-natural mode of assembly that forcibly moves the particles). But Van Till has omitted a third option, namely, that design can also be word-like (i.e., imparting information to a receptive medium). In the liturgies of most Christian churches, the faithful pray that God keep them from sinning in "thought, word, and deed." Each element of this tripartite distinction is significant. Thoughts left to themselves are inert and never accomplish anything outside the mind of the individual who thinks them. Deeds, on the other hand, are coercive, forcing physical stuff to move now this way and now that way (it is no accident that the concept of *force* plays such a crucial role in the rise of modern science). But between thoughts and deeds are words. Words mediate between thoughts and deeds. Words give expression to thoughts and bring the self in contact with the other. On the other hand, words by themselves are never coercive; without deeds to back up words, words lose their power to threaten. Nonetheless, words have the power to engender deeds by finding a receptive medium that can then act on them. The power of words is in persuasion, not coercion.

## 6.6 Must All Design in Nature Be Front-Loaded?

But simply to allow that an unembodied designer has imparted information (and therewith design) into the natural world is not enough. There are many thinkers who are sympathetic to design but who prefer that all the design in the world be front-loaded.[60] The advantage of putting all the design in the world at, say, the initial moment of the Big Bang is that it minimizes the conflict between design and science as currently practiced. A designer who front-loads the design of the world imparts all the world's information before natural causes become operational and thus before natural causes have an opportunity to conflict with the introduction of novel information. This move restricts design to structuring the laws of nature and thereby precludes design from violating those laws and thus violating nature's causal structure. In effect, there is no need to think of the world as an informationally open system. Rather, we can still think of it mechanistically—like the outworking of a complicated differential equation, albeit with the initial and boundary conditions designed. The impulse to front-load design is deistic, and I expect any theories about front-loaded design to be just as successful as deism was historically, which always served as an unsatisfactory halfway house between theism (with its informationally open universe) and naturalism (which insists the universe remain informationally closed).[61]

There are no good reasons to require that the design of the universe must be front-loaded. Certainly maintaining peace with an outdated mechanistic view of science is not a good reason. Nor is the theological preference for a hands-off designer, even if it is couched as a Robust Formational Economy Principle. To be sure, front-loaded design is a logical possibility. But so is interactive design (i.e., the design that a designer introduces by imparting information over the course of natural history). The only legitimate reason to limit all design to front-loaded design is if there could be no empirical grounds for preferring interactive design to front-loaded design. Michael Murray attempts such an argument.[62] Accordingly, he argues that for an unembodied designer, front-loaded design and interactive design will be empirically equivalent. Murray's argument hinges on a toy example in which a deck of cards has been stacked by the card manufacturer before the deck gets wrapped in cellophane and distributed to cardplayers. Should a cardplayer now insist on using the deck just as it arrived from the manufacturer and should that player repeatedly win outstanding hands at poker, even if there were no evidence whatsoever of cheating, then the arrangement of the deck by the manufacturer would have to be attributed to design. Murray implies that all design attributed to unembodied designers is like this, requiring no novel design in the course of natural history but only at the very beginning when the deck was stacked.

But can all design not attributable to an embodied intelligence be dismissed in this way? Take the Cambrian explosion in biology, for instance.

Niles Eldredge, James Valentine, and even Stephen Jay Gould (when he is not fending off the charge of aiding creationists) admit that the basic metazoan body-plans all arose in a remarkably short span of geological time (5 to 10 million years) and for the most part without any evident precursors (there are some annelid tracks as well as evidence of sponges leading up to the Cambrian, but that is about it with regard to metazoans; single-celled organisms abound in the Precambrian).[63] Assuming that the animals fossilized in the Cambrian exhibit actual design, where did that design come from? To be committed to front-loaded design means that all these body-plans that first appeared in the Cambrian were in fact already built in at the Big Bang (or whenever that information was front-loaded), that the information for these body-plans was expressed in the subsequent history of the universe, and that if we could but uncover enough about the history of life, we would see how the information expressed in the Cambrian fossils merely exploits information that was already in the world prior to the Cambrian period. Now that may be, but there is no evidence for it. All we know is that information needed to build the animals of the Cambrian period was suddenly expressed at that time and with no evident informational precursors.[64]

To see what is at stake here, consider the textual transmission of a manuscript composed by an anonymous author, say the New Testament book of Hebrews. There is a manuscript tradition that allows us to trace this book (and specifically the information in it) back to at least the second century A.D. Some scholars think the book was written sometime in the first century by a colleague of the Apostle Paul. One way or another we cannot be certain of the author's identity. What's more, the manuscript trail goes dead in the first century A.D. Consequently, it makes no sense to talk about the information in this book being in some sense front-loaded at any time prior to the first century A.D. (much less at the Big Bang).

Now Murray would certainly agree (for instance, he cites the design of the pyramids as not being front-loaded). In the case of the transmission of biblical texts, we are dealing with human agents whose actions in history are reasonably well understood. But the distinction he would draw between this example, involving the transmission of texts, and the previous biological example, involving the origin of body-plans, cannot be sustained. Just because we do not have direct experience of how unembodied designers impart information into the world does not mean we cannot say where that information was initially imparted and where the information trail goes dead.

The key evidential question is not whether a certain type of designer (embodied or unembodied) produced the information in question, but how far that information can be traced back. With the Cambrian explosion the

information trail goes dead in the Cambrian. So too with the book of Hebrews it goes dead in the first century A.D. Now it might be that with the
Cambrian explosion, science may progress to the point where it can trace
the information back even further—say to the Precambrian or possibly even
to the Big Bang. But there is no evidence for it and there is no reason—other
than a commitment to methodological naturalism—to think that all naturally occurring information must be traceable back in this way. What's more,
as a general rule, information tends to appear discretely at particular times
and places. To require that the information in natural systems (and throughout this discussion the type of information I have in mind is specified
complexity) must in principle be traceable back to some repository of front-
loaded information is, in the absence of evidence, an entirely ad hoc restriction.

It is also important to see that there is more to theory choice in science
than empirical equivalence. The ancient Greeks knew all about the need for
a scientific theory to "save the phenomena" (Pierre Duhem even wrote a
delightful book about it with that title).[65] A scientific theory must save (or
be faithful to) the phenomena it is trying to characterize. That is certainly a
necessary condition for an empirically adequate scientific theory. What's
more, scientific theories that save the phenomena equally well are by definition empirically equivalent. But there are broader coherence issues that always arise in theory choice so that merely saving phenomena is not sufficient
for choosing one theory over another. Empirically equivalent to the theory
that the universe is 14 billion years old is the theory that it is only five
minutes old and that it was created with all the marks of being 14 billion
years old. Nonetheless, no one takes seriously a five-minute-old universe.
Also empirically equivalent to a 14 billion-year-old universe is a six thou-
sand-year-old universe in which the speed of light has been slowing down
and enough auxiliary assumptions are introduced to account for the data
from geology and archeology that are customarily interpreted as indicating a
much older earth. In fact, the scientific community takes young earth creationists to task precisely for making too many ad hoc assumptions that favor
a young earth. Provided that there are good reasons to think that novel
design was introduced into the world subsequent to its origin (as for instance
with the Cambrian explosion, where all information trails for metazoan
body-plans go dead in the Precambrian), it would be entirely artificial to
require that science nonetheless treat all design in the world as front-loaded
just because methodological naturalism requires it or because it remains a
bare possibility that the design was front-loaded after all.

Please note that I am not offering a theory about the frequency or intermittency with which an unembodied designer imparts information into the world. I would not be surprised if most of the information imparted by such a designer will elude us, not conforming to any patterns that might enable us to detect this designer (just as we might right now be living in a swirl of radio transmissions by extraterrestrial intelligences, though for lack of being able to interpret these transmissions we lack any evidence that embodied intelligences on other planets exist at this time). The proper question for science is not the schedule according to which an unembodied designer imparts information into the world, but the evidence for that information in the world, the times and locations where that information first becomes evident, and the informational pathways linking the origin of that information to the present. That is all empirical investigation can reveal to us. What's more, short of tracing the information back to the Big Bang (or wherever else we may want to locate the origin of the universe), we have no good reason to think that the information exhibited in some physical system was in fact front-loaded.

## 6.7   Embodied and Unembodied Designers

Even if we grant the possibility of an unembodied designer, why should we think that the design produced by such a designer could be accessible to scientific investigation in the same way as design produced by an embodied designer? This worry underlies the impulse to front-load all the design in nature. We all have experience with designers that are embodied in physical stuff, notably other human beings. But what experience do we have of unembodied designers? With respect to intelligent design in biology, for instance, Elliott Sober wants to know what sorts of biological systems should be expected from an unembodied designer. What's more, Sober claims that if the design theorist cannot answer this question (i.e., cannot predict the sorts of biological systems that might be expected on a design hypothesis), then intelligent design is untestable and therefore unfruitful for science.

As we saw in section 2.9, to place this demand on design hypotheses is ill-conceived. We infer design regularly and reliably without necessarily knowing the characteristics of the designer or being able to assess what the designer is likely to do. It is of no evidential significance whatsoever whether a designer is embodied or unembodied. The reason embodiment is irrelevant to design is because design is always an inference from empirical data and never a direct intuition of the designer's mental processes. We do not get into the mind of designers and thereby attribute design. Rather we look at effects in the physical world that exhibit clear marks of intelligence and from those marks infer to a designing intelligence. This is true even for those most uncontroversial of embodied designers, namely, fellow human beings. We recognize their intelligence not by performing a "Vulcan mind-meld" (as in the television series *Star Trek*) but by examining their actions and determining whether these actions display marks of intelligence. A human being who continuously mumbles the same nonsense syllable displays no intelligence and provides no warrant for attributing design.

The philosopher Thomas Reid made this same argument over 200 years ago:

> No man ever saw wisdom [read "design" or "intelligence"], and if he does not [infer wisdom] from the marks of it, he can form no conclusions respecting anything of his fellow creatures. . . . But says Hume, unless you know it by experience, you know nothing of it. If this is the case, I never could know it at all. Hence it appears that whoever maintains that there is no force in the [general rule that from marks of intelligence and wisdom in effects a wise and intelligent cause may be inferred], denies the existence of any intelligent being but himself.[66]

The aim of this book has been to elucidate, make precise, and justify the key empirical marker that reliably signals design, namely, specified complexity. Central to this task has been casting specified complexity into the idiom of modern information theory.

My critics remain unconvinced. Robert Pennock, for instance, maintains that design inferences must be "based upon known types of causal processes."[67] Moreover, he claims that design inferences become increasingly tenuous as the underlying causal processes depart from those that are best known. He writes:

> What about design explanations? Here there are fewer constraints, but in certain contexts, if we stick to our ordinary, natural notion of intentional design, we can still make some headway; when archeologists pick out something as an artifact or suggest possible purposes for some unfamiliar object they have excavated they can do so because they already have some knowledge of the causal processes involved and have some sense of the range of purposes that could be relevant. It gets more difficult to work with the concept when speaking of extraterrestrial intelligence, and harder still when considering the possibility of animal or machine intelligence. But once one tries to move from natural to supernatural agents and powers as creationists desire, "design" loses any connection to reality as we know it or can know it scientifically.[68]

While it is not my aim to argue for creationism, it is important to see how Pennock's argument against creationism here falters. Certainly animal learning theorists, SETI researchers, and proponents of strong artificial intelligence would reject his claim that intelligence as it arises in their respective disciplines is more tenuously inferred than with humans (in the case of strong artificial intelligence, for instance, machine intelligence is regarded as the genus and human intelligence as the species). The crucial divide here, as always, is between embodied and unembodied intelligences (or as Pennock calls them, "supernatural agents"). Pennock's attempt to confine design to "known types of causal processes" is merely another instance of defining intelligent design out of existence by presupposing naturalism. Intelligent agency is a known type of causal process. Moreover, if an unembodied intelligent agent is indeed responsible for the specified complexity that we observe in biological systems, then we have plenty of experience of that agent's actions (our very bodies providing a case in point).

Our inability to reduce the actions of such agents to purely natural causes is therefore no argument at all against their detectability or against the validity of inferring design for instances of specified complexity that could only be plausibly attributed to an unembodied designer. When Pennock requires that design inferences be "based upon known types of causal processes," he really means that the underlying causal processes must be fully reducible to natural causes before a design inference may legitimately be drawn. But that is precisely the point at issue, namely, whether intelligent agency reduces to or transcends natural causes. Specified complexity as a criterion for detecting design allows that question to be assessed without prejudice. Pennock, on the other hand, by presupposing naturalism has stacked the deck so that only one answer is possible.

Larry Arnhart likewise remains unconvinced that a design inference can validly infer to an unembodied intelligence. Arnhart maintains that our knowledge of design arises in the first instance not from any inference but from introspection of our own human intelligence. As a consequence, he concludes that we have no empirical basis for inferring design whose source is unembodied.[69] Though at first blush plausible, this argument quickly collapses when probed. Jean Piaget, for instance, would have rejected it on developmental grounds: Babies do not make sense of intelligence by introspecting their own intelligence but by coming to terms with the effects of intelligence in their external environment. For example, they see the ball in front of them and then taken away, and learn that Daddy is moving the ball—thus reasoning from effect to intelligence. Introspection (always a questionable psychological category) plays at best a secondary role in how initially we make sense of intelligence and design.[70]

I would argue, however, that even later in life, when we have attained full self-consciousness and when introspection can be performed with varying degrees of reliability, design is inferred. Indeed, introspection must always remain inadequate for assessing intelligence and attributing design. By definition intelligence presupposes the power or facility to choose between options—this coincides with the Latin etymology of "intelligence," namely, "to choose between." Introspection is entirely the wrong instrument for assessing this aspect of intelligence. For instance, I cannot by introspection assess my intelligence at proving theorems in differential geometry. How do I know that I can *choose* the right sequence of steps in, say, the proof of the Nash embedding theorem? It has been over a decade since I have proven any theorems in differential geometry. I need to get out paper and pencil and actually try to prove some theorems in that field. How I do—and not my introspective memory of how well I did in the past—will determine whether and to what degree intelligence can be attributed to my theorem proving. The only way to assess intelligence is to test it and see what it does. And the primary thing that intelligences do is generate specified complexity.[71]

I therefore continue to maintain that intelligence is always inferred, that we infer it through well-established methods, and that there is no principled way to distinguish design due to embodied and unembodied designers so that one is empirically accessible and the other is empirically inaccessible. This is the rub. And this is why intelligent design is such an intriguing intellectual possibility—it threatens to make the ultimate questions real. Convinced Darwinists like Pennock and Arnhart therefore need to block the design inference whenever it threatens to implicate an unembodied designer. Embodied designers are okay. That is why Francis Crick can get away with his directed panspermia theory in which intelligent aliens (who are embodied designers) seed the world with life from outer space.[72] So long as the designer is embodied, Darwinists can claim that the designer is an evolved intelligence that arose via the Darwinian mechanism. Unembodied designers, however, are strictly outside the bounds of the Darwinian mechanism and therefore strictly proscribed.

Not only is there no evidential significance to whether a designer is embodied or unembodied, but refusing to countenance the possibility of unembodied designers impedes scientific inquiry. To see this, consider the following variant of the Explanatory Filter (see section 1.3). I call this variant the Naturalized Explanatory Filter (see figure 6.1). Since the most typical candidate for an unembodied designer is God, I have formulated the Naturalized Explanatory Filter with explicit reference to God. Nevertheless, in place of God one is free to substitute any unembodied entity that is unacceptable on naturalistic grounds and yet implicated by a design inference. The

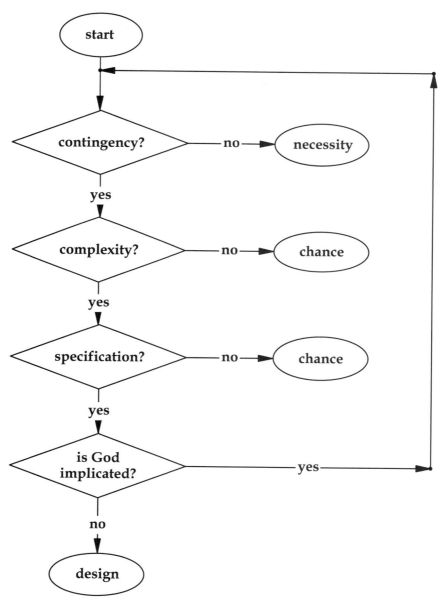

**Figure 6.1.  The Naturalized Explanatory Filter.**

Naturalized Explanatory Filter accurately captures how scientific naturalism tries to account for the design of natural systems for which no embodied designer can plausibly be invoked as an explanation.

Like the Explanatory Filter in section 1.3, the Naturalized Explanatory Filter assesses contingency, complexity, and specification. Moreover, so long as no naturalistically unacceptable entity (like God) is implicated, it properly sorts through the three primary modes of explanation—necessity, chance, and design. Nonetheless, as soon as a naturalistically unacceptable entity is implicated, instead of coming to terms with it and squarely acknowledging that there is now a design problem here, the Naturalistic Explanatory Filter conveniently adds a fourth decision node that cycles the explanatory analysis back to the beginning of the flowchart and thereby ensures that only necessity and chance will receive further consideration. Thus, even if an unembodied intelligence is responsible for the design displayed in some object, a science committed to the Naturalized Explanatory Filter will be sure never to discover it. A science that on a priori grounds refuses to consider the possibility of unembodied designers therefore artificially limits what it can discover. Instead of rejecting design as the conclusion of a sound scientific argument, this approach stipulates it out of existence. Essential to science is a spirit of free and open inquiry. Scientific naturalism, and the Naturalized Explanatory Filter that it endorses, betrays that spirit.

The Naturalized Explanatory Filter makes clear how naturalism has been used historically to derail the design inference whenever design has come too close to challenging naturalism. Because the Naturalized Explanatory Filter is a subterfuge, once exposed it requires further rationalization to keep it afloat. Wesley Elsberry performs the needed damage control by proposing still another variant of the Explanatory Filter:

> My explanatory filter has one more alternative classification than Dembski's, that of unknown causation. This alternative recognizes that the set of knowledge used to make a classification can alter the classification. By allowing an event to be classified as due to unknown causation, I simultaneously reduce the number of false classifications that will later be overturned due to the availability of additional information and also identify those events whose circumstances require further study in order to resolve a causative factor. The use of unknown causation as a category is common in those day-to-day operations of humans looking for design in events, such as forensics. Forcing final classification of events under limited knowledge ensures that mistakes in classification will be made in Dembski's explanatory filter.[73]

Elsberry casts himself and his modified filter as defending scientific rigor and caution. But in fact he is offering nothing more than the Naturalized

Explanatory Filter. The category of "unknown causation" is a provisional holding category introduced to avoid the conclusion of design so long as designers unacceptable to naturalism are implicated. As soon as some explanation acceptable to naturalism becomes available, that category is immediately discarded.

## 6.8   Who Designed the Designer?

According to Richard Dawkins, to explain by means of an unembodied designer is "to explain precisely nothing, for it leaves unexplained the origin of the Designer. You have to say something like 'God was always there', and if you allow yourself that kind of lazy way out, you might as well just say 'DNA was always there', or 'Life was always there', and be done with it."[74] Dawkins takes this line because he, like many scientists and philosophers, is convinced that proper scientific explanations must be reductive, moving from the complex to the simple. Thus Dawkins writes, "The one thing that makes evolution such a neat theory is that it explains how organized complexity can arise out of primeval simplicity."[75] Dawkins explicitly equates proper scientific explanation with what he calls "hierarchical reductionism," according to which "a complex entity at any particular level in the hierarchy of organization" must properly be explained "in terms of entities only one level down the hierarchy."[76] Thus embodied designers are okay because ultimately they promise to submit to reductive explanations (Dawkins is a universal Darwinist, so any designers anywhere in the universe must result via Darwinian evolution—embodied designers are for Dawkins evolved designers). On the other hand, unembodied designers are not okay because they never can submit to reductive explanations.

In responding to the who-designed-the-designer question, it is therefore best first to dispense with Dawkins's reductionist view of science. This is easily done. While no one will deny that reductive explanation is extremely effective within science, it is hardly the only type of explanation available to science. The divide-and-conquer mode of analysis behind reductive explanation has strictly limited applicability within science. Complex systems theory has long since rejected a reductive bottom-up approach to complex systems.[77] To properly understand a complex system requires a top-down approach that focuses on global relationships between parts as opposed to analysis into individual parts. The Santa Fe Institute, which thus far is no fan of intelligent design, was founded to study systems of "simple interacting elements that [produce] through their aggregate behavior a global emergent order unpredictable simply through analysis of low-level interactions."[78] Likewise, a reductive mode of analysis is incapable of making headway with specified complexity (cf. CSI holism as described in section 3.9). Intelligent design is an integrative, top-down theory of complex structures. Integration is as much a part of science as reduction and analysis.

The who-designed-the-designer question invites a regress that is readily declined. The reason this regress can be declined is because such a regress arises whenever scientists introduce a novel theoretical entity. For instance, when Ludwig Boltzmann introduced his kinetic theory of heat back in the late 1800s and invoked the motion of unobservable particles (what we now call atoms and molecules) to explain heat, one might just as well have argued that such unobservable particles do not explain anything because they themselves need to be explained.[79]

It is always possible to ask for further explanation. Nevertheless, at some point scientists stop and content themselves with the progress they have made. Boltzmann's kinetic theory explained things that the old phenomenological approaches to heat failed to explain—for instance, why shaking a container filled with a gas caused the temperature of the gas to increase. Whereas the old phenomenological approach provided no answer, Boltzmann's kinetic theory did: shaking the container caused the unobservable particles making up the gas to move more quickly and thus caused the temperature to rise.

So too with design, the question is not whether design theorists have resolved all lingering questions about the designing intelligence responsible for specified complexity in nature. Such questions will always remain. Rather, the question is whether design does useful conceptual work, a question that Dawkins's criticism leaves unanswered. Design theorists argue that intelligent design is a fruitful scientific theory for understanding systems like Michael Behe's irreducibly complex biochemical machines. Such an argument has to be taken on its own merits. Moreover, it is a scientific argument.

Intelligent design as a scientific research program looks for empirical markers of intelligence in nature. Irreducible complexity is one such marker in biology, and it reliably points to an intelligence. Now the designer who is behind that design is as far as we can tell not part of nature (at least as nature is now understood by the scientific community). Consequently there is no "marker" attached to this designer indicating that this designer is in turn designed. The theory of intelligent design therefore avoids the "design regress" in which we must—to stay consistent with our own principles—answer whether the designer is designed. The designer is not an event, object, or structure about which a scientific theory of design requires us to consider whether the designer is in turn designed. This question does, however, arise for embodied designers as in Francis Crick's directed panspermia theory.

It is important to understand that design-theoretic explanations are proximal or local explanations rather than ultimate explanations. Design-theoretic explanations are concerned with determining whether some particular event, object, or structure exhibits clear marks of intelligence and can thus be legitimately ascribed to design. Consequently, design-theoretic reasoning does not require the who-designed-the-designer question to be answered for a design inference to be valid. As Jay Richards remarks, "If a detective explains a death as the result of a murder by, say, Jeffrey Dahmer, no one says, 'OK, then who made Jeffrey Dahmer?' If someone explains some buried earthenware as the result of artisans from the second century B.C., no one complains, 'Yeah, but who made the artisan?'"[80] Likewise in biology design inferences are not invalidated for failing to answer Dawkins's who-designed-the-designer question.

The who-designed-the-designer question can also be interpreted as a metaphysical rather than scientific question. As such it is a call for ultimate rather than proximal explanation. Proximal explanations are contextual and local, focusing on particular features of the world at particular times and places. Ultimate explanations, on the other hand, are global and encompassing, focusing on the entire world across time. Now, as Jay Richards notes, "Every ultimate explanation posits some final resting place of explanation, beyond which one cannot go."[81] The naturalist is likely to posit Nature (writ large) or the Universe (also writ large) or mass-energy or superstrings or some such entity as the final resting place for explanation. Likewise, the design theorist is likely to posit a generic designer or specified complexity or immanent teleology or God as the final resting place of explanation. This does not mean that all ultimate explanations are equivalent. But judging their merits goes beyond the remit of science. That is not to say that science is irrelevant to deciding among ultimate explanations. Darwinism conduces toward naturalism whereas intelligent design, at least in contemporary western culture, conduces toward theism. The crucial point for this discussion, however, is that design-theoretic explanations can be coherent without being tied to ultimate explanations.

## 6.9   Testability

Eugenie Scott is a physical anthropologist who, as director of the National Center for Science Education, travels the United States warning audiences about the threat of intelligent design to science. Scott's key criticism against intelligent design since the early 1990s has been that intelligent design is untestable. For instance, in an exchange with Stephen Meyer in 1994 in *Insight* magazine, Scott remarked that until design theorists develop a "theo-meter" (the neologism is hers) to test for design, they are not doing science.[82] More recently she has charged that intelligent design does not propose any "testable model."[83]

The testability objection to intelligent design can be interpreted in two ways. One is to claim that intelligent design is in principle untestable. This seems to have been Scott's line in the early 1990s. Certainly it is a hallmark of science that any of its claims be subject to revision or refutation on the basis of new evidence or further theoretical insight. If this is what one means by testability, then design is certainly testable. Indeed, it was in this sense that Darwin tested William Paley's account of design and found it wanting. It simply will not wash to say that design is not testable and then in the same breath say that Darwin tested design and refuted it.

The other way to interpret the testability objection is to claim that intelligent design may in principle be testable, but that no tests have been proposed to date.[84] This seems to be Scott's line currently. Indeed, if the testability objection is to bear any weight, its force must reside in the absence of concrete proposals for testing intelligent design. Are such proposals indeed lacking? Rather than looking solely at the testability of intelligent design, I want simultaneously to consider the testability of Darwinism. By comparing the testability of the two theories, it will become evident that even the more charitable interpretation of Scott's testability objection does not hold up.

In relation to science, testability is a very broad notion. It certainly includes Karl Popper's notion of falsifiability, but it is hardly coextensive with it and can apply even if falsifiability does not obtain.[85] Testability as well covers confirmation, predicability, and explanatory power. At the heart of testability is the idea that our scientific theories must make contact with and be sensitive to what is happening in nature. What is happening in nature must be able to affect our scientific theories not only in form and content but also in the degree of credence we attach to or withhold from them. For a theory to be immune to evidence from nature is a sure sign that we are not dealing with a scientific theory.

What then are we to make of the testability of both intelligent design and Darwinism taken not in a vague generic sense but concretely? What are the specific tests for intelligent design? What are the specific tests for Darwinism? And how do the two theories compare in terms of testability? To answer these questions, let us run through several aspects of testability, beginning with falsifiability.

Is intelligent design falsifiable? Is Darwinism falsifiable? Yes to the first question, no to the second. Intelligent design is eminently falsifiable. Specified complexity in general and irreducible complexity in biology are within the theory of intelligent design the key markers of intelligent agency. If it could be shown that biological systems like the bacterial flagellum that are wonderfully complex, elegant, and integrated could have been formed by a gradual Darwinian process (which by definition is nontelic), then intelligent design would be falsified on the general grounds that one does not invoke intelligent causes when purely natural causes will do. In that case Occam's razor finishes off intelligent design quite nicely.

I am being a bit fast and loose in my use of falsifiability here. Strictly speaking, if complex biological systems like the bacterial flagellum could be shown to result from purely natural causes (like the Darwinian mechanism), it would not follow as a logical entailment that the intelligent design of life is false in the sense of being necessarily untrue or decisively refuted. I am using the term "falsifiable" not just in the strict sense where claims get eliminated because they are demonstrated to be false but also in the looser sense where claims get eliminated because they lack warrant or are superfluous. The main point of Popper's criterion of falsifiability was not so much that scientific claims must have the possibility of being demonstrated false as that they must have the possibility of being eliminated as the result of new evidence. If Behe's irreducibly complex biochemical machines suddenly submit to purely naturalistic explanations, design would become superfluous and drop out of scientific discussion. It is in this sense that I am using the term "falsifiable."[86]

Even with this broadened definition of falsifiability, falsifying Darwinism seems effectively impossible. The problem is that Darwinists tend to raise the standard for falsification too high. It is certainly possible with the use of invariants to show that no Darwinian pathway could reasonably be expected to lead to a given biological structure (see section 5.8). But Darwinists typically want something much stronger, namely, to show that no conceivable Darwinian pathway could have led to a given biological structure. Such a demonstration requires an exhaustive search that is effectively impossible to carry out. What's more, Darwinists are apt to retreat into the murk of historical contingency to shore up their theory. For instance, Allen Orr in his critique of Behe's work on irreducibly complex biochemical systems remarked, "We have no guarantee that we can reconstruct the history of a biochemical pathway."[87] What Orr conceded with one hand, however, he was quick to retract with the other. He added, "But even if we can't, its irreducible complexity cannot count against its gradual evolution."[88]

The fact is that for complex systems like the bacterial flagellum no biologist has or is anywhere close to reconstructing its history in Darwinian terms (not just its actual history but any conceivable detailed Darwinian history). Is Darwinian theory therefore falsified? Hardly. I have yet to witness one committed Darwinist concede that any feature of nature might even in principle provide countervailing evidence to Darwinism. This is not merely to say that Darwinists have not found anything in nature that they regard as providing counterevidence to Darwinism. Rather, it is to say that they cannot even imagine anything in nature that might provide such counterevidence. Where logically one expects a concession that Darwinism might not be the whole story, one is instead treated to an admission of ignorance. Thus it is not that Darwinism has been falsified or disconfirmed, but that we simply do not know enough about a biological system and its history to determine how the Darwinian mechanism might have produced it.

For instance, to neutralize the challenge that the irreducible complexity of the bacterial flagellum raises against Darwinism, Kenneth Miller points to our ignorance of how the flagellar motor works: "Before [Darwinian] evolution is excoriated for failing to explain the evolution of the flagellum, I'd request that the scientific community at least be allowed to figure out how its various parts work."[89] Even so, we know enough about the bacterial flagellum to know that it is irreducibly complex. Miller's appeal to ignorance obscures just how much we know about the flagellum, how compelling the case is for its design, and how unfalsifiable Darwinism is when Darwinists proclaim that the Darwinian selection mechanism can account for it despite the absence of any identifiable biochemical pathway.

What about positive evidence for intelligent design and Darwinism? From the design theorist's perspective, the positive evidence for Darwinism is confined to small-scale evolutionary changes like insects developing insecticide resistance. This is not to deny large-scale evolutionary changes, but it is to deny that the Darwinian mechanism can account for them. Evidence like that for insecticide resistance confirms the Darwinian selection mechanism for small-scale changes but hardly warrants the grand extrapolation that Darwinists want. It is a huge leap going from insects developing insecticide resistance via the Darwinian mechanism of natural selection and random variation to the very emergence of insects in the first place by that same mechanism.

Darwinists invariably try to minimize the extrapolation from small-scale to large-scale evolution, arguing that it is a failure of imagination on the part of critics to appreciate the wonder-working power of the Darwinian mechanism. From the design theorist's perspective, however, this is not a case of failed imagination but of the emperor's new clothes. Yes, there is positive evidence for Darwinism, but the strength and relevance of that evidence on behalf of large-scale evolution is very much under dispute, if not within the Darwinian community then certainly outside of it.

What about the positive evidence for intelligent design? It seems that here we may be getting to the heart of Eugenie Scott's concerns. I submit that there is indeed positive evidence for intelligent design. To see this, let us recall the example in section 1.3 from the movie *Contact*. In the movie, radio astronomers determine that they have established contact with an extraterrestrial intelligence after they receive a long sequence of prime numbers, represented as a sequence of bits. Although in the actual SETI program, radio astronomers look not for something as flamboyant as prime numbers but something much more plebeian, namely, narrow bandwidth transmissions (as occur with human radio transmissions), the point nonetheless remains that SETI researchers would legitimately count a sequence of prime numbers (and less flamboyantly though just as assuredly a narrow bandwidth transmission) as positive evidence of extraterrestrial intelligence. No such conclusive signal has yet been observed, but if it were observed, Eugenie Scott would not be protesting that SETI has not proposed any "testable models." Instead she would rejoice that the model had been tested and decisively confirmed.

Now what is significant about a sequence of prime numbers from outer space is that they exhibit specified complexity—there has to be a long sequence (hence complexity) and it needs to display an independently given pattern (hence specificity). But what if specified complexity is also exhibited in actual biological systems? In fact it is—notably in the bacterial flagellum (see chapter 5). Even so, it appears that Eugenie Scott would not be entirely happy admitting that intelligent design is positively confirmed once some clear-cut instances of specified complexity are identified in biological systems. Why not? According to her, design theorists "never tell you what happened."[90] Yet neither do SETI researchers. If a SETI researcher discovers a radio transmission of prime numbers from outer space, the inference to an extraterrestrial intelligence is clear, but the researcher does not know "what happened" in the sense of knowing any details about the radio transmitter or for that matter any details about the extraterrestrial that built the radio transmitter and sent the radio transmission.

Ah, but we have experience with radio transmitters. At least with extraterrestrial intelligences we can guess what might have happened. But we do not have any experience with unembodied designers, and that is clearly what we are dealing with when it comes to design in biology. Actually, if an unembodied designer is responsible for biological complexity, then we do have quite a bit of experience with such a designer through the designed objects in nature (not least ourselves) that confront us all the time. On the other hand, it is true that we possess very little insight at this time into how such a designer acted to bring about the complex biological systems that have emerged over the course of natural history.

Darwinists take this present lack of insight into the workings of an unembodied designer not as remediable ignorance on our part and not as evidence that the designer's capacities far outstrip ours, but as proof that there is no unembodied designer—period. By the same token, if an extraterrestrial intelligence communicated via radio signals with earth and solved computational problems that exceeded anything an ordinary or quantum computer could ever solve, we would have to conclude that we were not really dealing with an intelligence because we have no experience of super-mathematicians who can solve such problems. My own view is that with respect to biological design humans are in the same position as William James's dog who would sit at his master's feet and study James while James was studying a book. Our incomprehension over biological design is the incomprehension of a dog trying to understand its master's actions. Significantly, Darwinists regularly sing the praises of natural selection and the wonders it has wrought while admitting that they have no comprehension of how those wonders were wrought. Natural selection, we are assured, is cleverer than we are or can ever hope to be. Darwinists have merely swapped one form of awe for another. They have not eliminated it.

It is no objection at all that we do not at this time comprehend how an unembodied designer produced biological systems exhibiting specified complexity. We know that specified complexity is reliably correlated with the effects of intelligence. The only reason to insist on looking for nontelic explanations to explain the complex specified structures in biology is because of a prior commitment to naturalism that perforce excludes unembodied designers. It is illegitimate, scientifically and rationally, to claim on a priori grounds that such entities do not exist, or if they do exist that they can have no conceivable relevance to what happens in the world. Do such entities exist? Can they have empirical consequences? Are they relevant to what happens in the world? Such questions cannot be prejudged except on metaphysical grounds. To prejudge these questions the way Eugenie Scott does is therefore to make certain metaphysical commitments about what there is and what has the capacity to influence events in the world. Such commitments are utterly gratuitous to the practice of science. Specified complexity confirms design regardless whether the designer responsible for it is embodied or unembodied.

Another aspect of testability is predictability. A good scientific theory, we are told, is one that predicts things. If it predicts things that do not happen, then it is tested and found wanting. If it predicts things that do happen, then it is tested and regarded as successful. If it does not predict anything, however, what then? Often with theories that try to account for features of natural history, prediction gets generalized to include retrodiction, in which a theory also specifies what the past should look like. Darwinism is said to apply retrodictively to the fossil record and predictively in experiments that place organisms under selection pressures and attempt to induce adaptive changes.

But in fact Darwinism does not retrodict the fossil record. Natural selection and random variation applied to single-celled organisms offers no insight at all into whether we can expect multicelled organisms, much less whether evolution will produce the various body-plans of which natural history has left us a record. At best one can say that there is consilience, that the broad sweep of evolutionary history as displayed in the fossil record is consistent with Darwinian evolution. Design theorists strongly dispute this as well (pointing especially to the Cambrian explosion). But detailed retrodiction and detailed prediction are not virtues of Darwin's theory. Organisms placed under selection pressures either adapt or go extinct. Except in the simplest cases where there is, say, some point mutation that reliably confers antibiotic resistance on a bacterium, Darwin's theory has no way of predicting just what sorts of adaptive changes will occur. "Adapt or go extinct" is not a prediction of Darwin's theory but an axiom that can be reasoned out independently of the theory.

Challenging me in *American Outlook*, physical anthropologist Alex Duncan remarked: "A scientific theory makes predictions about the world around us, and enables us to ask and answer meaningful questions. If we ask why do polar bears have fur while penguins have feathers, given the similar nature of their environments and lifestyles, evolution provides an answer to this question. The only answer creationism (or intelligent design) provides is 'Because God made them that way.'"[91] Actually, evolution, whether Darwinian or otherwise, makes no predictions about there being bears or birds at all or for that matter bears having fur and birds having feathers. Once bears or birds are on the scene, they need to adapt to their environment or die. Intelligent design can accommodate plenty of evolutionary change (including common descent) and allows for natural selection to act as a conservative force to keep organisms adapted to their environments. Contrary to Duncan's remark, intelligent design does not push off all explanation to the inscrutable will of God. On the other hand, intelligent design utterly rejects natural selection as a creative force capable of bringing about the specified complexity we see in organisms.

Darwin's theory has virtually no predictive power. Insofar as it offers predictions, they are extremely general, concerning the broad sweep of natural history and in that respect quite questionable. (Why else would Stephen Jay Gould and Niles Eldredge need to introduce punctuated equilibria if the fossil record were such an overwhelming vindication of Darwinism?)[92] Or else, when the predictions are not extremely general, they are extremely specific and picayune, dealing with small-scale adaptive changes. Newton was able to predict the precise pathways that planets trace out in cosmic history. Darwinists can neither predict nor retrodict the precise pathways that organisms trace out in the course of natural history.

But what about the predictive power of intelligent design? Intelligent design offers one obvious prediction, namely, that nature should be chock-full of specified complexity and therefore should contain numerous pointers to design (see the very end of section 5.10). This prediction is increasingly being confirmed. What's more, once designed systems are in place, operational, and interacting (as with an economy or ecosystem), intelligent design predicts certain patterns of technological evolution, notable among these being sudden emergence, convergence to ideality, and extinction.[93] Although research in this area is only now beginning, preliminary indications are that biology confirms these patterns of technological evolution. Significantly, these patterns are non-Darwinian.[94]

Nonetheless, there is a sense in which to require prediction of intelligent design fundamentally misconstrues intelligent agency and design. To require of intelligent design that it predict specific instances of design in nature is to put design in the same boat as natural laws, locating their explanatory power in an extrapolation from past experience. This is to commit a category mistake. To be sure, designers, like natural laws, can behave predictably (designers can institute policies that end up being rigidly obeyed and often follow routinized procedures in problem solving). Yet unlike natural laws, which are universal and uniform, designers are also innovators. Innovation, the emergence to true novelty, eschews predictability. Designers are inventors. We cannot predict what an inventor would do short of becoming that inventor. Intelligent design presents a radically different problematic from a mechanistic science wedded solely to undirected natural causes. It offers predictability concerning the presence of design and the evolution of already existing designs, but it offers no predictability about fundamentally novel designs.

According to Darwin the great advantage of his theory over William Paley's theory of design was that Darwin's theory managed to account for a wide diversity of biological facts that Paley's theory could not. Darwin's theory was thus thought to have greater explanatory power than Paley's, and this relative advantage could be viewed as a test of the two theories. Underlying explanatory power is a view of explanation known as inference to the best explanation, in which a "best explanation" always presupposes at least two competing explanations. Consequently, a "best explanation" is one that comes out on top in a competition with other explanations (see section 2.9). Design theorists see advances in the biological and information sciences as putting design back in the saddle and enabling it to outperform Darwinism, thus making design rather than natural selection currently the best explanation of biological complexity. Darwinists of course see the matter differently.

What I want to focus on here, however, is not the testing of Darwinism and intelligent design against the broad body of biological data, but the related question of which theoretical framework can accommodate the greater range of biological possibilities. Darwinism and intelligent design are not just theories that make claims about the world (claims that can be either true or false, assertible or unassertible), but also theoretical frameworks offering certain explanatory tools and strategies. Darwinism's framework is thoroughly naturalistic and limits itself to a certain set of theoretical options. Intelligent design's framework is nonnaturalistic and offers a different set of theoretical options. Are there things that might occur in biology for which a design-theoretic framework could give a better, more accurate account than a purely Darwinian and therefore nonteleological framework? The answer is yes.

First off, let us be clear that intelligent design, conceived now not as a theory but as a theoretical framework, can accommodate all the results of Darwinism. To be sure, as scientific theories, Darwinism and intelligent design contradict each other since intelligent design claims biology exhibits actual design whereas Darwinism claims biology exhibits only apparent design. But as a theoretical framework, intelligent design incorporates all the tools of Darwinism. Intelligent design assigns a very high place to natural causes and mechanisms. Insofar as these operate in nature, intelligent design wants to understand them and give them their due. But intelligent design also regards natural causes as incomplete and wants to leave the door open to intelligent causes. Intelligent design therefore does not repudiate the Darwinian mechanism. It merely assigns it a lower status than Darwinism does.

The Darwinian mechanism does operate in nature and insofar as it does, intelligent design can live with its deliverances. Even if the Darwinian mechanism could be shown to do all the design work for which design theorists want to invoke intelligent causation (say for the bacterial flagellum and systems like it), a design-theoretic framework would not destroy any valid findings of science. To be sure, design would then become a largely superfluous component of this framework (though according to chapter 4 there would still be an ineliminable aspect of design in the "well-wrought fitness functions" that enable the Darwinian mechanism to produce increasing biological complexity). But a design-theoretic framework would not on this account become self-contradictory or incoherent.

The worst that can happen to a design-theoretic framework is that design ends up being superfluous. The worst that can happen to a Darwinian framework is that it blinds itself to facts staring it in the face and fundamentally misconstrues reality. The dangers that confront science by adopting a Darwinian framework and the naturalism it presupposes therefore far outweigh the dangers that confront science by adopting a design-theoretic framework. To see this, suppose that I were a supergenius molecular biologist and that I invented some hitherto unknown molecular machine, far more complicated and marvelous than the bacterial flagellum. Suppose further I inserted this machine into a bacterium, set this genetically modified organism free, allowed it to reproduce in the wild, and destroyed all evidence of my having created the molecular machine. Suppose, for instance, the machine is a stinger that injects other bacteria and explodes them by rapidly pumping them up with some gas (I am not familiar with any such molecular machine in the wild), thereby allowing bacteria endowed with my invention to consume their unfortunate prey.

Now let us ask the question, If a Darwinist came upon this bacterium with the novel molecular machine in the wild, would that machine be attributed to design or to natural selection? When I presented this example to a noted Darwinist at a conference some time back, he shrugged it off and remarked that natural selection created us and so by extension also created my novel molecular machine. But of course this argument will not wash since the issue is whether natural selection could indeed create us. What's more, if Darwinists came upon my bacterial stinger in the wild, they would not look to design but would reflexively turn to natural selection. But, if we go with the story, I designed the bacterial stinger and natural selection had nothing to do with it. Moreover, intelligent design, by focusing on the stinger's specified complexity, would confirm the stinger's design whereas Darwinism never could. It follows that a design-theoretic framework could account for biological facts that would forever remain invisible within a Darwinian framework. It seems to me that this possibility constitutes a joint test of Darwinism and intelligent design that strongly supports intelligent design, if not as the truth then certainly as a live theoretical option that must not be precluded for a priori philosophical reasons like naturalism.

To sum up, there is no merit to Eugenie Scott's claim that intelligent design is untestable or has not put forward any "testable models." Intelligent design's claims about specified and irreducible complexity are in close contact with the data of biology and open to refutation as well as confirmation. What's more, as a framework for doing science intelligent design is more robust and sensitive to the possibilities that nature might actually throw our way than Darwinism, which must view everything through the lens of chance and necessity and take a reductive approach to all signs of teleology in nature.

But is not intelligent design just a stone's throw from all sorts of religious craziness? Even if a theory of intelligent design should ultimately prove successful and supersede Darwinism, it would not follow that the designer posited by this theory would have to be a transcendent deity or for that matter be real in some ontological sense. One can be an antirealist about science and simply regard the designer as a regulative principle—a conceptually useful device for making sense out of certain facts of biology—without assigning the designer any weight in reality. Wittgenstein, for instance, regarded the theories of Copernicus and Darwin not as true but as "fertile new points of view."[95]

Ultimately, the main question that confronts scientists working on a theory of intelligent design is whether design provides powerful new insights and fruitful avenues of research. The metaphysics underlying such a theory, and in particular the ontological status of the designer, can then be taken up by philosophy and theology. Indeed, one's metaphysics ought to be a matter of indifference to one's scientific theorizing about design. The fact that it is not for Eugenie Scott says more about her own biases than about the biases of design theorists, whose primary task is to explore the fruitfulness of design for science. Yes, we have our work cut out for us. But instead of facilitating that work, Scott and her National Center for Science Education are far more interested in relegating that work to oblivion. Design is too powerful an idea and too perennial an issue to suffer that fate.

## 6.10   Magic, Mechanism, and Design

In concluding this book I want to address one last criticism of intelligent design, namely, that it substitutes magic for mechanism, or alternatively that it invokes a supernatural cause where an ordinary natural cause will do. To understand his criticism we need briefly to review how the intelligent design community conceives its task. Proponents of intelligent design regard it as a scientific research program that investigates the effects of intelligent causes. Note that intelligent design studies the *effects* of intelligent causes and not intelligent causes *per se*. Intelligent design does not try to get into the head of a designing intelligence; rather, it looks at what a designing intelligence does and therewith draws inferences.

Intelligent design is at once old and new. It is old because many special sciences already fall under it. Forensic science, intellectual property law, cryptography, random number generation, and SETI all look at certain features of the world and try to infer an intelligent cause responsible for those features. Where intelligent design gets controversial is when one takes its methods for detecting design in human contexts and shifts them to the natural sciences where no embodied, reified, or evolved intelligence could have been present. What if, for instance, the methods of intelligent design are applied to biology and show that biological systems are in fact designed? The application of intelligent design to the natural sciences is both novel and threatening, and has prompted full-scale rebuttals like those by Robert Pennock and Kenneth Miller.[96]

Why is intelligent design so threatening to the scientific community? Ever since Darwin, science has acted as though no divine architect was needed to start creation on its course. Consequently, any designing agents, including ourselves, must result from a long evolutionary process that itself was not designed. Designing agents like ourselves therefore occur at the end of (for all we know) an undesigned natural process and cannot be prior to it. But if there is design in biology and cosmology, then that design could not result from an evolved intelligence. Rather, it must be an unembodied intelligence. Enter "the big G." If there is a designer behind biology and cosmology, the options for who that designer is are quite limited, with God being the preferred option for most people. But for God to play a substantive role in science is unacceptable to many scientists, who have come to regard science as a purely naturalistic enterprise focused exclusively on the operation of natural causes.

Hence the increasing attacks against intelligent design, like those by Pennock and Miller. What underlies these critiques is one main worry: To permit an unembodied designer into science will destroy science, reintroducing all sorts of magical, superstitious, and occult entities that modern science long ago banished from our understanding of the world. Pennock gives particularly apt expression to this worry in his criticism of Phillip Johnson (a law professor at the University of California at Berkeley who is also an outspoken critic of Darwinism and advocate of intelligent design). According to Pennock, Johnson's position on intelligent design raises a particularly worrisome legal consequence. As Pennock sees it, Johnson advocates "that science admit the reality of supernatural influences in the daily workings of the world."[97] But what if these same supernatural influences were admitted into Johnson's own area of specialization—the law? Here's the concern as Pennock lays it out in *Tower of Babel*:

> For the law to take [Johnson's view] seriously as well, it would have to be open to both suits and defenses based on a range of possible divine and occult interventions. Imagine the problems that would result if the courts had to accept legal theories of this sort. How would the court rule on whether to commit a purportedly insane person to a mental hospital for self-mutilation who claims that the Lord told her to pluck out her eye because it offended her? How would a judge deal with a defendant, Abe, accused of attempted murder of his son, Ike, who claims that he was only following God's command that he kill Ike to prove his faith?[98]

Implicit in this passage and throughout Pennock's book is a forced choice between mechanism and magic: Either the world works by mechanisms that obey inviolable natural laws and that admit no break in the chain of natural causation, or pandemonium breaks loose and the world admits supernatural interventions that ruin science and our understanding of the world generally (and legal studies in particular). Pennock is offering his readers mechanism. Johnson is offering them magic. Any reasonable person knows which option to choose.

But as with most forced choices, there is a *tertium quid* that Pennock has conveniently ignored, and that when properly understood shows that the real magician here is in fact Pennock and not Johnson. The *tertium quid* is design, which is entirely separable from magic. Pennock, as a trained philosopher, knows that design requires neither magic nor miracles nor a creator—the ancient Stoics, for instance, had design without supernatural interventions or a transcendent deity.[99] Design is detectable; we do in fact detect it; we have reliable methods for detecting it; and its detection involves no recourse to the supernatural. As I have argued throughout this book, design is common, rational, and objectifiable.

The real magician in Pennock's *Tower of Babel* is not Phillip Johnson and his fellow design theorists, but Pennock himself and his fellow evolutionary naturalists. Pennock, like most Darwinists, subscribes to a "free-lunch" form of magic in which it is possible to get something for nothing. To be sure, the "nothing" here need not be an absolute nothing. What's more, the transformation of nothing into something may involve minor expenditures of effort. For instance, the magician may need to utter "abracadabra" or "hocus-pocus." Likewise, the Darwinian just-so stories that attempt to account for complex, information-rich biological structures are incantations that give the illusion of solving a problem but in fact merely cloak ignorance (see section 1.10).

Darwinists, for instance, explain the human eye as having evolved from a light sensitive spot that successively became more complicated as increasing visual acuity conferred increased reproductive capacity on an organism.[100] In such a just-so story, all the historical and biological details in the eye's construction are lost. How did a spot become innervated and thereby light-sensitive? How did a lens form within a pinhole camera? What changes in embryological development are required to go from a light-sensitive sheet to a light-sensitive cup? None of these questions receives an answer in purely Darwinian terms. Darwinian just-so stories have no more scientific content than Rudyard Kipling's original just-so stories about how the elephant got its trunk or the giraffe its neck.[101] To be sure, such stories are entertaining, but they hardly engender profound insight.

The great appeal behind the "free-lunch" form of magic is the offer of a bargain—indeed an incredible bargain for which no amount of creative accounting can ever square the books. The idea of getting something for nothing has come to pervade science. In cosmology, Alan Guth, Lee Smolin, and Peter Atkins all claim that this marvelous universe could originate from quite unmarvelous beginnings (a teaspoon of ordinary dust for Guth, black-hole formation for Smolin, and set-theoretic operations on the empty set for Atkins).[102] In biology, Jacques Monod, Richard Dawkins, and Stuart Kauffman claim that the panoply of life can be explained in terms of quite simple mechanisms (chance and necessity for Monod, cumulative selection for Dawkins, and autocatalysis for Kauffman).[103]

We have become so accustomed to this something-for-nothing way of thinking that we no longer appreciate just how deeply magical it is. Consider, for instance, the following evolutionary account of neuroanatomy by Melvin Konner, an anthropologist and neurologist at Emory University: "Neuroanatomy in many species—but especially in a brain-ridden one like ours—is the product of sloppy, opportunistic half-billion year [evolution] that has pasted together, and only partly integrated, disparate organs that evolved in different animals, in different eras, and for very different purposes."[104] And since human consciousness and intelligence are said to derive from human neuroanatomy, it follows that these are themselves the product of a sloppy evolutionary process.

But think what this means. How do we make sense of "sloppy," "pasted together," and "partly integrated," except with reference to "careful," "finely adapted," and "well integrated"? To speak of hodge-podge structures presupposes that we have some concept of carefully designed structures. And of course we do. Humans have designed all sorts of engineering marvels, everything from Cray supercomputers to Gothic cathedrals. But that means, if we are to believe Melvin Konner, that a blind evolutionary process (what Richard Dawkins calls the "blind watchmaker") cobbled together human neuroanatomy, which in turn gave rise to human consciousness, which in turn produces artifacts like supercomputers, which in turn are not cobbled together at all but instead are carefully designed. Out pop purpose, intelligence, and design from a process that started with no purpose, intelligence, or design. This is magic.

Of course, to say this is magic is not to say it is false. It is after all a logical possibility that purpose, intelligence, and design emerge by purely mechanical means out of a physical universe initially devoid of these. Intelligence, for instance, may just be a survival tool given to us by an evolutionary process that places a premium on survival and that is itself not intelligently guided. The basic creative forces in nature might be devoid of intelligence. But if that is so, how can we know it? And if it is not so, how can we know that? It does no good simply to presuppose that purpose, intelligence, and design are emergent properties of a universe that otherwise is devoid of these.

The debate whether nature has been invested with purpose, intelligence, and design is not new. Certainly the ancient Epicureans and Stoics engaged in this debate. The Stoics argued for a design-first universe: the universe starts with design and any subsequent design results from the outworkings of that immanent design (they resisted subsequent novel infusions of design). The Epicureans, on the other hand, argued for a design-last universe: the universe starts with no design and any subsequent design results from the interplay of chance and necessity.[105] What is new, at least since the Enlightenment, is that it has become intellectually respectable to cast the design-first position as disreputable, superstitious, and irrational; and the design-last position as measured, parsimonious, and alone supremely rational. Indeed, the charge of magic is nowadays typically made against the design-first position and not against the design-last position, as I have done here.

But why should the design-first position elicit the charge of magic? Historically in the West, design has principally been connected with Judeo-Christian theism. The God of Judaism and Christianity is said to introduce design into the world by intervening in its causal structure. But such interventions cannot be anything but miraculous. And miracles are the stuff of magic. So goes the argument. The argument is flawed because there is no necessary connection between God introducing design into the world and God intervening in the world in the sense of violating its causal structure. One way around this problem is to conceive of God as front-loading all the design in nature (see section 6.6). Another way is to conceive of nature as not totally under the sway of natural laws but rather as a medium receptive to novel information (see section 6.5).

Paradoxically, the very clockwork universe of the British natural theologians, which they used to buttress the design-first position, was probably more responsible than anything for in the end undermining that position and promoting the design-last position. The early British natural theologians, and especially Robert Boyle, embraced the mechanical philosophy, the view that the world is an assemblage of material entities interacting by purely mechanical means. Boyle advocated the mechanical philosophy because he saw it as refuting the immanent teleology of Aristotle and the Stoics for which design arose as a natural outworking of natural forces. For Boyle this was idolatry, identifying the source of creation not with God but with nature. The mechanical philosophy offered a world operating by mechanical principles and processes that could not be confused with God's creative activity and yet allowed such a world to be structured in ways that clearly indicated the divine handiwork and therefore design. What's more, the British natural theologians always retained miracles as a mode of divine interaction that could bypass mechanical processes. Over the subsequent centuries, however, what remained was the mechanical philosophy and what dropped out was the need to invoke miracles or God as designer. Henceforth, purely mechanical processes could themselves do all the design work for which the Stoics had required an immanent natural teleology (cf. their "world soul") and for which Boyle and the British natural theologians required God.[106]

Fortunately, as is by now evident, design can be formulated without presupposing the mechanical philosophy and without falling prey to the charge of magic. By contrast, the a priori exclusion of design has a much harder time resisting the charge of magic. Indeed, the design-last position is inherently magical. Consider the following remark by Harvard biologist Richard Lewontin in *The New York Review of Books*:

> We take the side of science *in spite of* the patent absurdity of some of its constructs, *in spite of* its failure to fulfill many of its extravagant promises of health and life, *in spite of* the tolerance of the scientific community for unsubstantiated just-so stories,

because we have a prior commitment, a commitment to materialism [i.e., natural-ism]. It is not that the methods and institutions of science somehow compel us to accept a material explanation of the phenomenal world, but, on the contrary, that we are forced by our a priori adherence to material causes to create an apparatus of investigation and a set of concepts that produce material explanations, no matter how counterintuitive, no matter how mystifying to the uninitiated.[107]

If this is not magic, what is?

Even so, the scientific community continues to be skeptical of design. The worry is that design will give up on science. In place of a magic that derives something from nothing, design substitutes a designer who explains every-thing. Magic gets you something for nothing and thus offers a bargain. De-sign gets you something by presupposing something unimaginably bigger and thus asks you to sell your scientific soul. At least so the story goes. But design can be explanatory without giving away the store. Certainly this is the case for human artifacts, which are properly explained by reference to design. Nor does design explain everything: There is no reason to invoke design to explain a random inkblot; but a Dürer woodcut is something else altogether. As a research program, intelligent design extends design from the realm of human artifacts to the natural sciences. The program may ultimately fail, but it is only now being tried, and it is certainly worth a try. Moreover, this program has a rigorous information-theoretic underpinning.

Bargains are all fine and well, and if you can get something for nothing, go for it. But there are situations that admit no free lunch, where you get what you pay for, and in which at the end of the day there has to be an accounting of the books. The big question confronting science is whether design can be gotten on the cheap or must be paid for in kind. In this book I have argued that design admits no bargains. Specified complexity, that key indicator of design, has but one known source, namely, intelligence. Speci-fied complexity cannot be purchased without it. Indeed, all attempts to pur-chase specified complexity without intelligence end in a sterile reductionism that tries to make natural causes do the work of intelligent causes. It is time to come clean about what natural causes can and cannot accomplish. They cannot substitute for intelligent causes. They are not a free lunch.

## Notes

1. See Charles Darwin, On the Origin of Species, facsimile 1st ed. (1859; reprinted Cambridge, Mass.: Harvard University Press, 1964), ch. 1, titled "Variation under Domes-tication."

2. Interestingly, my most severe critics have been philosophers (for instance, Elliott

Sober and Robin Collins—see chapter 2). Mathematicians and statisticians have been far more receptive to my codification of design inferences. Take, for instance, the positive notice of *The Design Inference* in the May 1999 issue of the *American Mathematical Monthly* as well as mathematician Keith Devlin's appreciative remarks about my work in his July/August 2000 article for *The Sciences* titled "Snake Eyes in the Garden of Eden": "Dembski's theory has made an important contribution to the understanding of randomness—if only by highlighting how hard it can be to differentiate the fingerprints of design from the whorls of chance." See http://www.nyas.org/books/sci/sci_0700_devl. html (last accessed 11 June 2001).

3. In this section I address concerns raised by Eugenie Scott and the National Center for Science Education about what intelligent design means for the teaching of evolution. See Scott's piece titled "The Big Tent and the Camel's Nose," *Metaviews* 008 (12 February 2001): http://www.metanexus.net (last accessed 11 June 2001).

4. Note that neither designer nor creator need be personal or transcendent. To be sure, within traditional theism the creator is a personal God who transcends the physical world. But design and creation make sense apart from traditional theism. Design is fundamentally concerned with arrangements of preexisting stuff that signify intelligence. Creation is fundamentally concerned with the source of being of the world.

5. See, for instance, Peter Douglas Ward, *On Methuselah's Trail: Living Fossils and the Great Extinctions* (New York: W. H. Freeman, 1992), 29. Ward writes, "The seemingly sudden appearance of skeletonized life has been one of the most perplexing puzzles of the fossil record. How is it that animals as complex as trilobites and brachiopods could spring forth so suddenly, completely formed, without a trace of their ancestors in the underlying strata? If ever there was evidence suggesting Divine Creation, surely the Precambrian and Cambrian transition, known from numerous localities across the face of the earth, is it." Note that Ward is a well-known expert on ammonite fossils and does not favor a creation-based view.

6. David Berlinski, "Denying Darwin: David Berlinski and Critics," *Commentary* (September 1996): 24.

7. Thomas H. Clark and Colin W. Stearn, *The Geological Evolution of North America* (New York: Ronald Press, 1960), 43.

8. See Robert F. DeHaan and John L. Wiester, "The Cambrian Explosion: The Fossil Record and Intelligent Design," 145–156 in *Signs of Intelligence: Understanding Intelligent Design*, eds. W. A. Dembski and J. M. Kushiner (Grand Rapids, Mich.: Brazos Press, 2001), 155.

9. For the countervailing evidence see for instance Michael Denton, *Evolution: A Theory in Crisis* (Bethesda, Md.: Adler & Adler, 1985) and Jonathan Wells, *Icons of Evolution* (Washington, D.C.: Regnery, 2000).

10. See William L. Craig, "Naturalism and Cosmology," 215–252 in *Naturalism: A Critical Analysis*, eds. W. L. Craig and J. P. Moreland (London: Routledge, 2000) and John Leslie, *Universes* (London: Routledge, 1989).

11. Hubert Yockey, for instance, treats the specified complexity in living systems as "axiomatic" and leaves it at that. Similarly, Francis Crick refers to the specified complex-

ity in the genetic code as a "frozen accident." See Hubert Yockey, *Information Theory and Molecular Biology* (Cambridge: Cambridge University Press, 1992), 335 and Francis Crick, "The Origin of the Genetic Code," *Journal of Molecular Biology* 38 (1968): 367–379.

12. Carlos F. Amábile-Cuevas, Maura Cárdenas-García, and Mauricio Ludgar, "Antibiotic Resistance," *American Scientist* 83 (1995): 324.

13. See W. F. Doolittle, "Lateral Gene Transfer, Genome Surveys, and the Phylogeny of Prokaryotes," *Science* 286 (1999): 1443; and C. R. Woese, "Interpreting the Universal Phylogenetic Tree," *Proceedings of the National Academy of Sciences* 97 (2000): 6854–6859.

14. Lynn Margulis, *Symbiosis in Cell Evolution: Microbial Communities in the Archean and Proterozoic Eons*, 2nd ed. (New York: Freeman, 1993).

15. Franklin Harold, "From Morphogenes to Morphogenesis," *Microbiology* 141 (1995): 2765.

16. Scott Gilbert, *Developmental Biology*, 3rd ed. (Sunderland, Mass.: Sinauer, 1991), pt. I.

17. See Jacques Monod, *Chance and Necessity* (New York: Vintage, 1972), 12.

18. For a more thorough account of what I am calling "modification," see John A. Endler and Tracy McLellan, "The Processes of Evolution: Toward a Newer Synthesis," *Annual Review of Ecology and Systematics* 19 (1988): 396, table 1; and James D. Watson, Nancy H. Hopkins, Jeffrey W. Roberts, Joan A. Steitz, and Alan M. Weiner, *Molecular Biology of the Gene*, 4th ed. (Menlo Park, Calif.: Benjamin/Cummings, 1987), chs. 10–12.

19. Amábile-Cuevas et al., "Antibiotic Resistance," 324.

20. Dawkins's metaphor of gradually wending one's way up a mountain is therefore inappropriate when it comes to Michael Behe's irreducibly complex biochemical systems. Dawkins's Mount Improbable cannot be climbed for such systems; in that case its face is sheer and the Darwinian mechanism is incapable of scaling it. See Richard Dawkins, *Climbing Mount Improbable* (New York: Norton, 1996).

21. This is by far the most common case in biology—see Michael Behe, *Darwin's Black Box* (New York: Free Press, 1996), ch. 8.

22. See Alexander G. Cairns-Smith, *Seven Clues to the Origin of Life* (Cambridge: Cambridge University Press, 1985); and Alexander G. Cairns-Smith and H. Hartman, eds., *Clay Minerals and the Origin of Life* (Cambridge: Cambridge University Press, 1986).

23. This is Michael Corey's preferred option. See Michael Corey, *Back to Darwin: The Scientific Case for Deistic Evolution* (Lanham, Md.: University Press of America, 1994).

24. See Stuart Kauffman, *The Origins of Order: Self-Organization and Selection in Evolution* (Oxford: Oxford University Press, 1993); Stuart Kauffman, *At Home in the Universe* (Oxford: Oxford University Press, 1995); and Christopher Langton, ed., *Artificial Life III: Proceedings of the Workshop on Artificial Life held June, 1992 in Santa Fe, New Mexico*, in *Santa Fe Institute Studies in the Sciences of Complexity*, vol. 17 (Redwood City, Calif.: Addison-Wesley, 1994).

25. See Paul Nelson, *On Common Descent*, University of Chicago Evolutionary Monographs (2001): in press.

26. See Howard J. Van Till, "Does 'Intelligent Design' Have a Chance? An Essay Review," *Zygon* 34(4) (1999): 667–675 as well as Van Till's response to Stephen Meyer's

article in Richard F. Carlson, ed., *Science and Christianity: Four Views* (Downers Grove, Ill.: InterVarsity, 2000), 188–194.

27. "Is ID 'Punctuated Naturalism'?" *Newsletter of the American Scientific Affiliation and Canadian Scientific & Christian Affiliation* 42(5) (September/October 2000): 5.

28. See Van Till in Carlson, *Science and Christianity*, 190 as well as Howard J. Van Till, "Basil and Augustine Revisited: The Survival of Functional Integrity," *Origins & Design* 19(1) (1998): 34–35.

29. See William A. Dembski, *Intelligent Design: The Bridge between Science and Theology* (Downers Grove, Ill.: InterVarsity, 1999), chs. 2 and 3.

30. Auxiliary hypotheses typically regulate the applicability of the natural laws in relation to antecedent conditions.

31. See Gazzaniga's preface in Michael S. Gazzaniga, ed., *The Cognitive Neurosciences* (Cambridge, Mass.: MIT Press, 1995), xiii.

32. For Van Till's Robust Formational Economy Principle see his chapter in Carlson, *Science and Christianity*, 216–220.

33. Thus Van Till will charge that intelligent design "focuses on formational gifts withheld and welcomes empirical evidence for the absence of such gifts." Quoted in Van Till, "Basil and Augustine Revisited," 35.

34. Gregory of Nazianzus writes, "For every one who sees a beautifully made lute, and considers the skill with which it has been fitted together and arranged, or who hears its melody, would think of none but the lutemaker, or the luteplayer, and would recur to him in mind, though he might not know him by sight. And thus to us also is manifested That which made and moves and preserves all created things, even though He be not comprehended by the mind. And very wanting in sense is he who will not willingly go thus far in following natural proofs." Gregory of Nazianzus, *The Second Theological Oration*, 288–301 in *The Nicene and Post-Nicene Fathers of the Christian Church*, 2nd series, vol. 7, eds. P. Schaff and H. Wace (Grand Rapids, Mich.: Eerdmans, 1989), 290.

35. Leslie Orgel, *The Origins of Life* (New York: Wiley, 1973), 189.

36. Richard Dawkins, *The Blind Watchmaker* (New York: Norton, 1987), 9.

37. Paul Davies, *The Fifth Miracle* (New York: Simon & Schuster, 1999), 112.

38. Stuart Kauffman, *Investigations* (New York: Oxford University Press, 2000), 160. He actually proposes four variants. See also section 3.10.

39. See Brian Goodwin, *How the Leopard Changed Its Spots: The Evolution of Complexity* (New York: Scribner's, 1994), 35–36.

40. Davies, *The Fifth Miracle*, 17.

41. Richard Dawkins, *Climbing Mount Improbable* (New York: Norton, 1996).

42. This is how the quote has come down popularly. The exact quote reads: "I never satisfy myself until I can make a mechanical model of a thing. If I can make a mechanical model I can understand it. As long as I cannot make a mechanical model all the way through I cannot understand." Lord Kelvin, *Baltimore Lectures* (Baltimore: Publication Agency of Johns Hopkins University, 1904), 270.

43. Robert Larmer developed this criticism effectively at the New Brunswick symposium adverted to earlier in this section.

44. See, for instance, Francis Crick and Leslie E. Orgel, "Directed Panspermia," *Icarus* 19 (1973): 341–346.

45. Remark made by Paul Davies at a symposium titled "Complexity, Information, and Design: A Critical Appraisal," sponsored by the Templeton Foundation and occurring in Santa Fe, New Mexico, 15–16 October 1999. Davies organized this symposium.

46. Tipler refers to this possibility as the "Eternal Life Postulate." See Frank Tipler, *The Physics of Immortality* (New York: Random House, 1994), 108, 116–119. See also Freeman Dyson, "Time Without End: Physics and Biology in an Open Universe," *Reviews of Modern Physics* 51 (1979): 447–460 and Freeman Dyson, *Infinite in All Directions* (New York: Harper & Row, 1988).

47. Norbert Wiener, *Cybernetics*, 2nd ed. (Cambridge, Mass.: MIT Press, 1961), 132.

48. I am grateful to Frank Tipler for forcing this clarification. Personal communication, 11 May 2001.

49. For a less dramatic example of this same idea, see David Deutsch, *The Fabric of Reality: The Science of Parallel Universes—and Its Implications* (New York: Penguin, 1997), 52–53.

50. Ibid., 45–46.

51. Alan Guth, *The Inflationary Universe: The Quest for a New Theory of Cosmic Origins* (Reading, Mass.: Addison-Wesley, 1997), 179.

52. David Lindley, *The End of Physics: The Myth of a Unified Theory* (New York: Basic Books, 1993).

53. Sudbery, *Quantum Mechanics and the Particles of Nature* (Cambridge: Cambridge University Press, 1984), 212.

54. W. Michael Dickson, *Quantum Chance and Non-Locality: Probability and Non-Locality in the Interpretations of Quantum Mechanics* (Cambridge: Cambridge University Press, 1998), 62.

55. See my article "Randomness by Design," *Nous* 25(1) (1991): 75–106.

56. Heinz Bauer, *Probability Theory and Elements of Measure Theory*, trans. R. B. Burckel, 2nd English ed. (New York: Academic Press, 1981), 172.

57. Ernest Vincent Wright, *Gadsby* (Los Angeles: Wetzel, 1939).

58. Dembski, "Randomness by Design."

59. See Peter Holland, *The Quantum Theory of Motion: An Account of the de Broglie-Bohm Causal Interpretation of Quantum Mechanics* (Cambridge: Cambridge University Press, 1993), 72–74.

60. E.g., Richard Swinburne and Paul Davies. See Richard Swinburne, *The Existence of God* (Oxford: Clarendon, 1979), ch. 8 entitled "Teleological Arguments" and Paul Davies, *The Mind of God* (New York: Touchstone, 1992), ch. 8 entitled "Designer Universe."

61. See Dembski, *Intelligent Design*, ch. 3.

62. Michael Murray, "Natural Providence," unpublished typescript. This paper was presented at the Wheaton Philosophy Conference, 28 October 2000. For the conference schedule, though not Murray's paper, see http://www.wheaton.edu/philosophy/conference.html (last accessed 11 June 2001).

63. Gould writes, "Nonetheless, these exciting finds in Precambrian paleontology do not remove the problem of the Cambrian explosion, for they include only the simple bacteria and blue-green algae, and some higher plants such as green algae. The evolution of complex Metazoa seems as sudden as ever. (A single Precambrian fauna has been found at Ediacara in Australia. It includes some relatives of modern fan corals, jellyfish, worm-like creatures, arthropods, and two cryptic forms unlike anything alive today. Yet the Ediacara rocks lie just below the base of the Cambrian and qualify as Precambrian only by the slimmest margin. A few more isolated finds from other areas around the world are likewise just barely Precambrian.) If anything, the problem is increased because exhaustive study of more and more Precambrian rocks destroys the old and popular argument that complex Metazoa are really there, but we just haven't found them yet." Quoted from Stephen Jay Gould, *Ever Since Darwin: Reflections in Natural History* (New York: W. W. Norton 1977), 121. Compare the following remark by Niles Eldredge: "Most families, orders, classes, and phyla appear rather suddenly in the fossil record, often without anatomically intermediate forms smoothly interlinking evolutionarily derived descendant taxa with their presumed ancestors." Quoted from Niles Eldredge, *Macro-Evolutionary Dynamics: Species, Niches, and Adaptive Peaks* (New York: McGraw-Hill 1989), 22. Finally, consider the following remark by Valentine et al.: "Taxa recognized as orders during the (Precambrian-Cambrian) transition chiefly appear without connection to an ancestral clade via a fossil intermediate. This situation is in fact true of most invertebrate orders during the remaining Phanerozoic as well. There are no chains of taxa leading gradually from an ancestral condition to the new ordinal body type. Orders thus appear as rather distinctive subdivisions of classes rather than as being segments in some sort of morphological continuum." Quoted from J. W. Valentine, S. M. Awramik, P. W. Signor, and P. M. Sadler, "The Biological Explosion at the Precambrian-Cambrian Boundary," *Evolutionary Biology* 25 (1991): 284.

64. I do not want to be dogmatic here. Simon Conway Morris, for instance, in a paper titled "Nipping the Cambrian 'Explosion' in the Bud," concludes that recent work by Budd and Jensen constitutes "a landmark in attempting to restore some degree of biological credibility to this fast moving, complex and fascinating field." According to Conway Morris, Budd and Jensen argue plausibly that "much of what we see in the Cambrian is telling us that the given body plans were assembled by familiar processes in a believable biological fashion on a credible geological time scale." See Simon Conway Morris, "Nipping the Cambrian 'Explosion' in the Bud," *BioEssays* 22 (December 2000): 1053–56. For the article Conway Morris cites, see G. E. Budd and S. A. Jensen, "A Critical Reappraisal of the Fossil Record off the Bilaterarian Phyla," *Biological Reviews* 75 (2000): 253–295. But compare Conway Morris's earlier writings: "The 'Cambrian explosion' is a real evolutionary event, but its origins are obscure. At least 20 hypotheses have been proposed, and although arguments linking diversification to oxygen levels, predation, faunal provinciality and ocean chemistry all attract support, it is the case that 'The emergence of Metazoa remains the salient mystery in the history of life.'" Quoted from Simon Conway Morris, "The Fossil Record and the Early Evolution of the Metazoa," *Nature* 361 (21 January 1993): 222. Is there any reason to expect that the Budd and Jensen proposal

will be any more successful at accounting for the Cambrian Explosion than the twenty hypotheses Conway Morris cites?

65. Pierre Duhem, *To Save the Phenomena: An Essay on the Idea of Physical Theory from Plato to Galileo*, trans. E. Dolan and C. Maschler (1908; reprinted Chicago: University of Chicago Press, 1969).

66. Thomas Reid, *Lectures on Natural Theology*, eds. E. Duncan and W. R. Eakin (1780; reprinted Washington, D.C.: University Press of America, 1981), 56.

67. Robert Pennock, "The Wizards of ID," *Metaviews* 089 (12 October 2000): http://www.metanexus.net (last accessed 11 June 2001).

68. Ibid.

69. Larry Arnhart, Michael J. Behe, and William A. Dembski, "Conservatives, Darwin & Design: An Exchange," *First Things* (November 2000): 23–31.

70. For a nice summary of Piaget's model for the development of human intelligence, see James E. Loder and W. Jim Neidhardt, *The Knight's Move* (Colorado Springs: Helmers & Howard, 1992), 148–151. See also Barbara Inhelder and Jean Piaget, *The Growth of Logical Thinking from Childhood to Adolescence*, trans. A. Parsons and S. Melgram (New York: Basic Books, 1958).

71. See Douglas S. Robertson, "Algorithmic Information Theory, Free Will, and the Turing Test," *Complexity* 4(3) (1999): 25–34. According to Robertson, the defining feature of intelligence is the "creation of new information," by which is properly understood the creation of specified complexity.

72. Crick and Orgel, "Directed Panspermia."

73. Quoted from http://inia.cls.org/~welsberr/zgists/wre/papers/dembski7.html (last accessed 11 June 2001). See also John S. Wilkins and Wesley R. Elsberry, "The Advantages of Theft over Toil: The Design Inference and Arguing from Ignorance," *Biology and Philosophy*, forthcoming. Elsberry's modified filter has been widely cited as effectively rebutting my original filter. For instance, the National Center for Science Education officially endorses Elsberry's modified filter—see http://www.ncseweb.org/resources/rncse_content/vol19/5927_ithe_design_inferencei_by_12_30_1899.asp (last accessed 11 June 2001). See as well Taner Edis, "Darwin in Mind: 'Intelligent Design' Meets Artificial Intelligence," *Skeptical Inquirer* 25(2) (March/April 2001): 36. Besides adding another terminal node to my filter (i.e., "unknown causation"), Elsberry's modified filter also concedes to natural selection events that end up at the design node. In fact, Elsberry's modified filter miscarries in both places where it challenges my original filter: As we saw in chapters 4 and 5, natural selection cannot account for complex specified events that end up at the design node. What's more, Elsberry's category of unknown causation amounts to a disingenuous display of ignorance—intelligence and only intelligence is known to generate specified complexity, and to pretend otherwise by cloaking knowledge in ignorance is tendentious (in this case serving to promote naturalism, which is a metaphysical position, in the name of science).

74. Dawkins, *The Blind Watchmaker*, 141.

75. Ibid., 316.

76. Ibid., 13.

77. See David Berlinski, *On Systems Analysis: An Essay Concerning the Limitations of Some Mathematical Methods in the Social, Political, and Biological Sciences* (Cambridge, Mass.: MIT Press, 1976).

78. Stefan Helmreich, *Silicon Second Nature: Culturing Artificial Life in a Digital World* (Berkeley, Calif.: University of California Press, 1998), 44.

79. Ernst Mach actually did make such an argument. See Lawrence Sklar, *Physics and Chance: Philosophical Issues in the Foundations of Statistical Mechanics* (Cambridge: Cambridge University Press, 1993), 32–34, 131–132.

80. Personal communication, 30 January 2001.

81. Ibid.

82. Eugenie Scott, "Keep Science Free from Creationism," *Insight* (21 February 1994): 30. Strictly speaking, Scott was here criticizing scientific creationism rather than intelligent design. Nonetheless, she applies the same criticism to intelligent design.

83. Lecture presented 18 January 2001 at the University of California at Berkeley and sponsored by the department of integrative biology—see http://ib.berkeley.edu/seminars/index.html (last accessed 11 June 2001). Scott's talk was titled "Icons of Creationism: The New Anti-Evolutionism and Science."

84. Elliott Sober and Philip Kitcher have both offered such a qualified concession in regard to creationism. Sober writes, "Perhaps one day, creationism will be formulated in such a way that the auxiliary assumptions it adopts are independently supported. My claim is that no creationist has succeeded in doing this yet." Likewise Kitcher writes, "Even postulating an unobserved Creator need be no more unscientific than postulating unobservable particles. What matters is the character of the proposals and the ways in which they are articulated and defended." See respectively Elliott Sober, *Philosophy of Biology* (Boulder, Colo.: Westview, 1993), 52 and Philip Kitcher, *Abusing Science: The Case Against Creationism* (Cambridge, Mass.: MIT Press, 1982), 125.

85. For a nice summary of Popper's mature views on falsification, see Karl Popper, "The Problem of Demarcation" (1974), in *Popper Selections*, ed. D. Miller (Princeton: Princeton University Press, 1985), 118–130.

86. I am indebted to Todd Moody for this clarification.

87. H. Allen Orr, "Darwin v. Intelligent Design (Again)," *Boston Review* (December/January 1996–1997): 30.

88. Ibid.

89. Kenneth Miller, *Finding Darwin's God* (New York: HarperCollins, 1999), 148.

90. That is how she put it in her U. C. Berkeley lecture—see note 73.

91. Alex Duncan, "Creative Ignorance," *American Outlook* (March/April 2001): 20.

92. Niles Eldredge and Stephen Jay Gould, "Punctuated Equilibria: An Alternative to Phyletic Gradualism," 82–115 in *Models in Paleobiology*, ed. T. J. M. Schopf (San Francisco: Freeman, 1973).

93. The work of Genrich Altshuller is seminal in this regard. Altshuller, a Russian engineer and scientist, analyzed over 400,000 patents from different fields of engineering. He was especially interested in tracking the evolution of technological systems. From the trends he observed, he found that the evolution of engineering systems is not random but

obeys certain laws. He codified these laws under the acronym TRIZ. TRIZ corresponds to a Russian phrase that in English means "Theory of Inventive Problem Solving" and is sometimes given the acronym TIPS. Even though TRIZ is widely employed in industry, its applications to biology are only now becoming evident. See http://www.triz.org and http://www.triz-journal.com (both sites last accessed 11 June 2001). I am indebted to Terry Rickard for drawing my attention to the connection between intelligent design and TRIZ.

94. William A. Dembski, "ID as a Theory of Technological Evolution," *Interpreting Evolution* (advanced summer workshop sponsored by the American Association for the Advancement of Science, the Center for Theology and the Natural Sciences, and the Philadelphia Center for Religion and Science), paper delivered at Haverford College, 17 June 2001 (paper available upon request from author: William_Dembski@baylor.edu). See also Genrich Altshuller, *The Innovation Algorithm: TRIZ, Systematic Innovation and Technical Creativity* (Worcester, Mass.: Technical Innovation Center, 1999); Semyon Savransky, *Engineering of Creativity: Introduction to TRIZ Methodology of Inventive Problem Solving* (Boca Raton, Fla.: CRC Press, 2000); and Ellen Domb's introduction to TRIZ at http://www.triz-journal.com/whatistriz/index.htm (last accessed 11 June 2001).

95. Ludwig Wittgenstein, *Culture and Value*, ed. G. H. von Wright, trans. P. Winch (Chicago: University of Chicago Press, 1980), 18e.

96. Robert Pennock, *Tower of Babel* (Cambridge, Mass.: MIT Press, 1999); Miller, *Finding Darwin's God*.

97. Pennock, *Tower of Babel*, 295.

98. Ibid.

99. This is made abundantly clear in F. H. Sandbach, *The Stoics*, 2nd ed. (Indianapolis: Hackett, 1989), especially ch. 4.

100. Dawkins, *The Blind Watchmaker*, 85–86.

101. Rudyard Kipling, *Just So Stories* (Garden City, N.Y.: Doubleday, 1912).

102. See respectively Alan Guth, *The Inflationary Universe* (Reading, Mass.: 1997); Lee Smolin, *The Life of the Cosmos* (New York: Oxford University Press, 1997); Peter Atkins, *Creation Revisited* (Harmondsworth: Penguin, 1994).

103. See respectively Jacques Monod, *Chance and Necessity* (New York: Vintage, 1972); Dawkins, *The Blind Watchmaker*; Stuart Kauffman, *At Home in the Universe* (New York: Oxford University Press, 1995).

104. Quoted in Moshe Sipper and Edmund Ronald, "A New Species of Hardware," *IEEE Spectrum* 37(4) (April 2000): 59.

105. See Sandbach, *The Stoics*, 14–15. Also, be looking for Ben Wiker's forthcoming *Moral Darwinism* with InterVarsity.

106. I am grateful to Ted Davis for pointing out to me Boyle's theological motivation for the mechanical philosophy. See Robert Boyle, *The Works of Robert Boyle*, vol. 14, eds. M. Hunter and E. B. Davis (London: Pickering & Chatto, 2000), 147–155 as well as Robert Boyle, *A Free Enquiry into the Vulgarly Received Notion of Nature*, eds. M. Hunter and E. B. Davis (Cambridge: Cambridge University Press, 1996).

107. Richard Lewontin, "Billions and Billions of Demons," review of *The Demon-Haunted World: Science as a Candle in the Dark* by Carl Sagan, *The New York Review of Books* (9 January 1997): 31. Emphasis in the original.

~

# Index

*Numbers in italics refer to figures.*

~

# About the Author

A mathematician and a philosopher, William A. Dembski is associate research professor in the conceptual foundations of science at Baylor University and senior fellow with Discovery Institute's Center for the Renewal of Science and Culture in Seattle. Dr. Dembski previously taught at Northwestern University, the University of Notre Dame, and the University of Dallas. He has done postdoctoral work in mathematics at MIT, in physics at the University of Chicago, and in computer science at Princeton University. A graduate of the University of Illinois at Chicago where he earned a B.A. in psychology, an M.S. in statistics, and a Ph.D. in philosophy, he also received a doctorate in mathematics from the University of Chicago in 1988 and a master of divinity degree from Princeton Theological Seminary in 1996. He has held National Science Foundation graduate and postdoctoral fellowships. Dr. Dembski has published articles in mathematics, philosophy, and theology journals and is the author of seven books. In *The Design Inference: Eliminating Chance Through Small Probabilities* (Cambridge University Press, 1998), he examines the design argument in a post-Darwinian context and analyzes the connections linking chance, probability, and intelligent causation. The present volume, which is the sequel to *The Design Inference*, critiques Darwinian and other naturalistic accounts of evolution.